Lecture Notes in Computer Science

Edited by G. Goos and J. Hartmanis

339

M. Rafanelli J.C. Klensin
P. Svensson (Eds.)

Statistical and Scientific Database Management

Fourth International Working Conference SSDBM
Rome, Italy, June 21–23, 1988
Proceedings

Springer-Verlag
Berlin Heidelberg New York London Paris Tokyo

Editors

Maurizio Rafanelli
I.A.S.I. – CNR
Viale Manzoni, 30, I-00185 Roma, Italy

John C. Klensin
Massachusetts Institute of Technology,
Room 20A–226
77 Massachusetts Avenue, Cambridge, MA 02139, USA

Per Svensson
Swedish Defence Research Establishment
S-10254 Stockholm, Sweden

CR Subject Classification (1987): H, G.3, I.3.5, J.2–3

ISBN 3-540-50575-X Springer-Verlag Berlin Heidelberg New York
ISBN 0-387-50575-X Springer-Verlag New York Berlin Heidelberg

Printing and binding: Druckhaus Beltz, Hemsbach/Bergstr.
2145/3140-543210

FOREWORD

The Fourth International Working Conference on Statistical and Scientific Data Base Management (IV SSDBM) held on June 21 - 23, 1988 in Rome, Italy, continued the series of conferences initiated in California in December 1981 by the Lawrence Berkeley Laboratory (LBL) of the University of California. The purpose of this conference was to bring together database researchers, users and system builders, working in this specific field, to discuss the particular points of interest, to propose new solutions to the problems of the domain and to expand the topics of the previous conferences, both from the theoretical and from the applicational point of view.

Much to our pleasure we succeeded in transforming the workshop into a working conference in order to provide an environment in which those who are interested in this field may freely exchange their ideas. At the same time we intend to select the contributions very carefully in order to obtain a better quality and scientific respectability.

Why "working conference" ?

"Working" : If on the one hand the transformation of the "working groups" into "open panels" requested a greater effort of preparation, on the other hand more qualified and *participating* people joined the open panels. In each of the three scheduled panels a chair discussed the problems regarding a selected topic (presenting a paper included in the proceedings) and gave proposals and suggestions about the possible solutions. Subsequently, he opened a discussion with some panelists previously chosen from the experts of the area and also with the audience in the conference hall.

"Conference" : A rigorous selection has been made of the submitted papers. This fact requested a stronger engagement from all: each paper has been referred by three to six members (we have sent papers of European authors to three overseas members and viceversa, and, to obtain a more uniform and impartial evaluation, a subset of the papers were sent both to three European and to three overseas members).

The structure of this conference, based on the experience of the previous events, included four scientific sessions in which eleven selected scientific papers were presented and discussed by the authors. These papers dealt with the following topics: knowledge base and expert system, data model, natural language processing, query language, time performance, user interface, heterogeneous data classification, storage constraints, automatic drawing, ranges and trackers, and arithmetic coding.

Also two special sessions had been scheduled, in which six other scientific papers, containing important points of originality, were presented and discussed as work in progress. In particular, the topics of these papers are geographical data modeling, spatial database queries, user interface in an Object Oriented SDB, interpretation of queries, graphical query languages, and knowledge browsing - front ends.

The following paper by Maurizio Rafanelli provides both an introduction on the general concepts helpful to people outside the field and a survey of all the papers in these proceedings.

The conference also had three invited papers regarding topics of particular interest such as "Temporal Data", "Statistical Data Management Requirements" and "Knowledge Based Decision Support Systems", included in this volume.

Furthermore, as already said, in three open panels relevant topics have been discussed regarding statistical relational model, geographical database systems and statistical information, and database management systems for statistical and scientific applications. Papers by the chairmen, contributions by the panelists and a summary of the respective discussions which followed the presentations of the chairmen are included in this volume, too.

As General Chairman and Program Chairmen of the conference, we wish to thank the Istituto di Analisi dei Sistemi ed Informatica for the financial and organizational support given to the conference. Our thanks also go to Arie Shoshani for his valuable suggestions, as well as to Roger Cubitt for his useful advice and for the immediate (and fundamental) finantial support which proved to be essential for the start of the conference. At last, we are also grateful to Andrew Westlake for the circulation of the final program in the Statistical Software Newsletters journal. Nearly one thousand people in more than seventy countries of the worrld were informed.

We would like to extend our thanks to all the authors who submitted papers for consideration; we are particularly pleased by the fact that among them there are researchers from countries which had not been represented in previous workshops.

We repeat our thanks to all the Program Committee Members (R.A. Becker, F.Y. Chin, R.E. Cubitt, H. Hinterberger, N.C. Lauro, H. Lutz, F.M. Malvestuto, M. McLeish, Z. Michalewicz, R.J. Muller, S.B. Navathe, S. Nordbotten, F. Olken, G. Ozsoyoglu, F.L. Ricci, H. Sato, A. Shoshani, B. Sundgren, Y. Vassiliou and K.M. Wittkowski) and the other referees (T. Carey, K.W. Chen, R. Dawson, D. Defays, G. Di Battista, P. Di Felice, G. Gambosi, S. Knudsen, R. Marti, J.C. Nordbotten, H. Pilat, D. Rotem, T. Saydam, A. Segev, M. Theodoridou, Y. Yeorgavoudakis, R. Zahir) for their effective and efficient work.

At last, we would like to thank all the sponsors of the conference, that is the Universities of Rome "La Sapienza" and of Naples, ISTAT (Central Institute of Statistics), ENEA (National Committee for Research and Development of Nuclear Energy and Alternative Sources), IASC (International Association for Statistical Computing), the Swedish Defense Research Establishment, EUROSTAT (Commission of the European Communities Statistics Office) and IBM.

Our thanks also go to ACM (association for Computing Machinery) and to IEEE (the Institute of Electrical and Electronics Engineers - Middle and South Italian Section) for their cooperation and support, and the Italian Society of Statistics and the Ministry of Scientific and Technological Research for the high patronage given to the conference.

On a more personal level we would like to thank Nella Ricci for the most valuable cooperation given both in the conference organization and to us personally.

John C. Klensin	Maurizio Rafanelli	Per Svensson
Co-chairman	General Chairman	Co-chairman

CONTENTS

WORK IN PROGRESS PAPERS

Research Topics in Statistical and scientific database management : the IV SSDBM

Maurizio Rafanelli
IASI - viale Manzoni 30, 00185 Roma, Italy

I. INTRODUCTION

Statistical and scientific databases (SSDB) are generally defined as databases that support statistical and scientific applications. In recent years there has been a growing interest in this area, due, in part, to the inadequacy of commercial database management systems to support statistical and scientific applications, and in part to the important properties which characterize the statistical and scientific data and their representation and manipulation with respect to the other types of data [Shoshani 82], [Brown 83], [Shoshani 85/b], [Rafanelli 88].

In the literature the term "statistical databases" (SDB) refers to databases that represent statistical or summary information and are used for statistical analysis [Shoshani 82], [Wong 82], [Ozsoyoglu 85]. Many application areas include examples of SDB: for instance, socio-economic databases (such as "population or income counts", "energy production and consumption", "health and medical statistics"), business databases (such as "financial summary reports" and "sales forecasting",), and so on.

SDB are quite different from conventional databases. The main reasons are:

a) The "data structure": For example, most of the existing models for conventional databases support merely "simple data structures", generally represented by "relations", whereas SDB need to support "complex data structures", generally expressed in the form of tables (time series, data matrices, contingency tables) [Brown 83], [Shoshani 85/b]. Moreover the characteristics of such data require more complete description of quantitative information (numerical data inside the tables) in the database (for instance, the inclusion of appropriate metadata) [McCarthy 82], [Bishop 83], [Mark 86], [Ghosh 88]. Finally, the tables are often characterized by large amounts of missing data or by sparse structure [Shoshani 82].

b) The "data manipulation": Boolean operations between data are not of prime importance for the statistical use and the traditional operations of data updating and deletion are rare or forbidden [Ghosh 86]. Data are stable, because they refer to events already happened. New information can only be added to the previous data to enlarge the database [Shoshani 82], [Rafanelli 85/a], [Rafanelli 85/b]. More exactly, data are rarely updated, except for corrections, and the updates do not need to be performed in a transaction mode. Instead the most common manipulation operations are connected to the encoding or summarization of data, or to the reclassification of the variables that describe the summary data, or manipulation of two or more tables to obtain another resulting table. Such manipulation has to consider both the above mentioned complex data structure, and the data type (for instance, "average" or "rate") of the summary data [Rafanelli 83], [Fortunato 86]. Furthermore, statistical analysis requires different types of operations over the data, and the multidimentional data structures need: i) at physical level, special compression techniques and access methods; ii) at logical level, special model to represent multidimentional aspects of the data; iii) at data management level, new operators (and, then, new query languages) able to consider operations particularly important for SSDB, but not traditionally considered in the conventional database theory [Klug 81], [Ozsoyoglu 84], [Rafanelli 85/b], [Shoshani 85/b], [Ozsoyoglu 87], [Rafanelli 88].

c) The "data processing": The statistician uses the techniques of "missing data imputation and "outlier elimination or adjustment" for modifying the data. In general, data are not updated except to correct original coding errors. Statistical packages or programs are required for data analysis, and are rarely manolithically integrated with SSDB systems in an attempt to provide all capabilities for all users, and there is some controversy over whether such integration would be desirable. Furthermore, when a statistical evaluation is to be done on a file that contains sensitive information, the question of the "data privacy" arises [Chin 78], [Denning 80], [Denning 83], [Ghosh 86], [Denning 87], [Michalewicz 87/a], [Palley 87].

Generally there are two broad classes of Statistical Databases (SDB): *micro* and *macro* SDBs [Wong 84]. The former refers to SDB containing disaggregate data (or microdata), that is records of individual entities (such as patients' medical record in hospitals, or various census data regarding the single citizen, etc.) . It is used primarily for statistical analysis. The latter refers to SDB containing aggregated data (or macrodata), typically the result of the earlier application of simple functions such as sums or counts on the microdata. These functions are followed by other, often more complex, procedures to obtain, for instance, mortality tables by sex, type of disease and year, or tables (with data type = "percentage"), regarding the "consumption of energy" by type of energy supply, year and country, or tables in which the "average" of the salary by type of industry, year and employee qualification, etc. appears.

The term "scientific databases" refers to databases that include scientific data, most of which result from *experiments* and *simulations* ; we call both of them "experimental data". In addition there exist data in *support* of the experiments and data that are *generated* from the experiment data; we refer to them collectively as "associated data" [Shoshani 84].

The experiment data can be classified according to two important characteristics: *regularity* (which refers to the pattern of the points or coordinates for which values are measured or computed) and *density* (which indicates whether all the potential data points have actual values associated with them). Sparsity implies a large number of null values which may be compressed out; furthermore, the quantity of data contained in scientific databases can increase rapidly (often several billion for an entire experiment) [Shoshani 85/a].

The data in *support* of the experiments fall into two types, called *configuration data* (which describe the initial structure of an experiment or simulation) and *instrumentation data* (which are descriptions of different instruments used in an experiment and their changes over time). A more detailed description of characteristics and properties of scientific data is reported in [Shoshani 84], [Shoshani 85/a].

Statistical and Scientific Database Management Systems (SSDBMS) may, instead, be defined as database management systems able to model, store and manipulate data in a manner suitable for the needs of the statisticians and to apply statistical data analysis techniques to the stored data. Most of the commercial systems available today were designed primarily to support transactions for business applications (such as banking or marketing), while statistical and scientific applications have different data structures, characteristics and properties [Shoshani 85/b]. They also require different operations over the data and involve different processing requirements.

2. BACKGROUND ON THE PREVIOUS CONFERENCES

At the beginning of this decade some researchers began to look at the difficulty of representing (by a suitable model) and of manipulating (by adequate operators) data which, for their use, could be called "statistical data", and also at the problem of inference control (for the compromise of confidential information) or "data privacy" [Kam 77], [Denning 80], [Schlorer 80], [Johnsson 81], [Chan 81], [Sato 81], [Klug 82], [Shoshani 82], [Denning 83], [Su 83]. The topics which appeared in the first papers were "data model", "data manipulation", "security" and, in general, *lists* of characteristics and of typical problems of the area.

The first Workshop on SDB [Proc. 1th SDBM], followed by a second Workshop after two years [Proc. 2th SDBM] widened the area to "special data types", "operators", "metadata management", "security", "user interface" and "physical storage and implementation"; this fact allowed the first definition of a set of topics characterizing this area [Shoshani 82], [Bates 82], [Denning 84].

The success of these meetings encouraged the researchers in the area to organize the third Workshop three years later [Proc. 3th SSDBM]. New topics were added, including "Expert Systems" and "Object Oriented (Systems)". More importantly, "Scientific Data" were also considered, because of the emerging similarities in the DBMS problems and requirements [Cubitt 87]. The fourth event in the series was decided to be a Conference and not a Workshop anymore. The validity of this choice has been confirmed by the quality of the accepted papers, which show a real maturity in the field that was not there in the previous meetings. This fact appeared to all the participants particularly exciting. To strengthen this impression contributed also the qualified and active participation to the discussions happened at the end of each Open Panel. They (see [Ghosh], [Westlake] and [Svensson]) have regarded the following topics: "Statistical relational model", "Geographical database systems and statistical information" and

"Database management systems for statistical and scientific applications", and in this volume a summary of the above mentioned discussions are reported too. At the conference three invited speakers presented their papers. In particular, the hypothesis that operations historically carried out in a specifically-scientific or -statistical database management context can reasonably be accommodated by the relational model is discussed in [Klensin]. The topic regarding knowledge based support for the management of statistical databases is discussed in [Wittkowski]. In this paper the attention is concentrated on the intermediate "semantics" and "strategy" layers, respect to the six layers (with different levels of abstraction) in which the relevant knowledge is classified. Finally, the representation of a temporal data model in relational environment is proposed in [Segev] (and presented at the conference by A. Shoshani). In this paper the authors show that the concept of a temporal relation needs to be introduced, and define a temporal normal form for it.

In this paper the topics of research will be summarized and brief comments on the papers presented will be made, in the conviction that in this meeting the most significant and recent works in this area (some of these represent research papers that describe ongoing work).

3. THE IV SSDBM: THE SCIENTIFIC PAPERS

The collection of papers presented reflects the diversity of approaches used to solve the problems considered: from the data classification to the use of methodologies of artificial intelligence, from data model to statistical query language, from security to coding, from handling time to automatic drawing, etc., the proposals made characterize the topics in which, at present, the researchers are involved the most.

One of the topics investigated regards the use of A.I. techniques. In particular, the paper of [McLeish] is described the conversion of three traditional databases to one integrated knowledge base, implemented as a fuzzy relational database. The research refers to the application of "fuzzy relations" in (veterinary) medical diagnosis for the implementation of an Expert System. It emphasizes how such implementation can be achieved by giving relations in the database, the formulae for the fuzzy relation system and the EQUEL code. The procedure used, based on fuzzy set theory, should be compared with more advanced techniques of handling uncertainty, but the achieved knowledge base/data base is suitable for scientific and statistical computations.

In [Sato] a model is proposed for describing statistical data and classifications and knowledge about it is described in a frame system; furthermore, an experimental natural query language is discussed. The proposed object oriented data design consists of several levels of abstraction (data model level, conceptual level, DB level) and distinguishes between data conceptually obtainable in the object world (conceptual file) and data actually stored in a database (database file). Moreover it specifies relationships among classifications and categories. Then the statistical (natural) query language is discussed; it uses a semantic grammar based on the statistical data model and has the capability of presenting neighboring data even if right-fit data does not exist in the DB. Furthermore it uses three kinds of knowledge bases: grammatical rule base, inference rule base and three dictionaries. This is an important task of representing and using knowledge about statistical data and classifications and, also if the final solution has by no means yet been found, the paper represents a step in the right direction.

Another topic investigated was "query processing". Srivastava and Rotem discuss in [Srivastava] the problem that arises when, for statistical analysis or operational purposes, the system may be required to compute the answer. A solution is to maintain a set of "precomputed" information (as aggregated data). The answer provided is often approximate, but acceptable under several (defined) circumstances, and the trade-off is that the processing of queries is made faster at the expense of the precision. The work makes a significant contribution to the design of statistical abstracts for dynamic databases.

Also in [Barcaroli] the authors examine the time performance optimization of the statistical database system. A convenient solution could be to store the derived tables in a secondary memory, but a trade-off between storage resources and time performance arises (especially when the storage space is constrained). The authors present a mathematical model for solving this problem (even though the definition of savings complicates the formulation, rather than simply minimizing the cost of table derivations). An heuristic algorithm to compute (in a polynomial time) a good solution to the problem is shown too. The discussed approach constitutes a basic step towards a methodology for logical design of statistical databases.

Bassiouni, in [Bassiouni/a] discuss an approach to handle time in a relational query language. Also "temporal database" is discussed, the semantics of temporal data is examined, and an extended logic for comparison and temporal boolean operators is proposed. The paper contains original ideas, even though it is more useful for managing microdata rather than for statistical data, expecially of aggregate type. It is a good basis for discussion of an extension, for instance, to a situation with missing data during some intervals (at present the results are not applicable in such a situation).

The problem of combining summary data sets, coming from different data sources and with different classifications, is discussed in [Malvestuto]. In this paper the authors present a way to infer summary data by refinement when the underlying database consists of two distributions $x=f_p(p)$ and $x=f_q(q)$. Using the proposed procedure, it is possible in some cases to answer the query of a user, even though the required data are not explicitly contained in the database. After having dealt with the algebraic structure of all possible partitions of a given value domain (by the algebra of partitions), the problem of the measurability of an arbitrary partition is discussed and a way to refine a database when it is not perfectly refinable is presented. As the authors themselves observe, this paper is a good basis for the next step, concerned with the cases of multidimentional databases, as well as with the definiteness constraints.

An algorithm to solve a conjunctive query search problem is presented in [Andersson]. The authors combine the compressed fully transposed ordered file (CFTOF) storage structure used in the data analysis system Cantor and a modified interpolation search algorithm to solve the abovementioned problem. The performance of the algorithm is discussed for an important subclass, called "orthogonal range search" and results from the evaluation of a probabilistic performance model are shown. The basic idea contained in this paper is novel and useful. Even though the idea is simple, the performance analysis seems to be positive and promising, and the results appear valid for orthogonal range queries over the storage structures examined.

In [Chang] the authors attempt to extend Aho and Ulmann's multi-attribute file system design for partial match queries to orthogonal range queries [Aho 79]; an algorithm which can be applied to produce optimal integer solutions is presented.

The paper contains a neat result, which shows that files which are optimal for partial match queries are not necessarily optimal for range queries.

An efficient layout algorithm which allows the automatic drawing of statistical database diagram is presented in [Di Battista]. In this paper a direct, acyclic, connected graph G(V,E) is used to represent a statistical schema (similar to other papers [Chan 81], [Rafanelli 83], [Rafanelli 85/a]) and three steps of the proposed algorithm (vertical assignment, planarization and horizontal assignment) are examined. The paper does make a reasonable contribution to the development of a statistical DBMS interface.

"The goal of statistical database is to provide statistics about groups of individuals, while protecting their privacy"; this sentence in [Michalewicz] introduces the inference problem, that is, the problem of being able to infer individual, confidential information by correlating enough statistics. The paper proposes an improved protection mechanism (an inference control) for statistical queries; it appears to be a useful extension of earlier work by the same authors [Michalewicz 87/b] on query-set size inference control. In it a further investigation regarding the consequences of non-uniform distribution of ranges, for which queries are unanswerable, is made and a list of other open interesting problems is reported.

A data compression technique based on a modification of arithmetic coding is presented in [Bassiouni/b], and a simple model of improving arithmetic coding for certain types of data is proposed. A discussion for the VLSI design of the proposed algorithm is also included, with the aim of making a contribution to reducing the cost of data trasmission and data access within statistical database systems. The results are applicable and useful for general purposes databases, but even more so for SSDB.

4. THE WORK IN PROGRESS

Two special sections present and discuss other scientific papers which, rather than reporting on finished work or final results, report various stages of original work in progress. Some of them are on topics, at the margin of the SSDBM area, but involve related issues and characteristics. These include papers about geographically-organized data, a relatively new topic at these conferences.

In particular, a conceptual model for representing statistical-scientific data in the field of geographical information systems is presented in [Gambosi]. It contains original ideas and opens some disputable issues in the topic. The approach discussed allows the modeling of both the descriptive and the spatial-geometric aspects of information. The more interesting results seem to be an easier formulation of statistical queries (containing references to geographical attributes of data) and a way for an easy linking between statistical data and their geographical reference.

Another topic regards the spatial database queries. In [Laurini] the authors present some considerations on the resolution of some types of spatial queries in a relational framework when one also needs to use computational geometry algorithms. A number of praxis-oriented problems in the area of town planning, CAD-CAM up to computer vision are treated, but similar problems arise in other fields, such as geography, robotics, and so on. The authors discuss a wireframe model, the Peano relations model and a mixed model, and show that only the description of spatial objects based on Peano Relations allows the simple using of (Peano tuple) algebra for solving an important subclass of spatial queries.

The problem of helping a user to obtain a statistical table from a relational database by describing only elements that compose the table is discussed in [D'Atri]. The authors consider other aspects, such as logical indipendence and universal relation interfaces. The latter need, for statistical database, to work on a richer data model dealing with functions and unnormalized relations. The system proposed perform a weak interaction between a traditional universal relation interface and a separate knowledge-based module that handle statistical queries according to a logical independence approach. This last point regards the very complex process of interpretation of a statistical query based on dialogue, logically independent and friendly.

Another aspect of the user-interface problem is described in [Malmborg]. In this paper the author attempts to integrate ideas from several systems to come up with its own design (not a novel approach, but interesting). The paper represents an excellent use of plans for state of the art software to provide scientists with a more useful set of tools. In particular, in the first part background material on Object-Oriented Statistical Databases and on the interaction between statistical query languages and metadata is given. In the second part the author describes the design of a software system (TBE-2) user interface (graphical metadata browser and table design language). The concepts described are interesting, but, obviously, they can be evaluated only when implemented.

The problem addressed in [Stephenson] regards the use of knowledge engineering techniques to improve access to statistical data when the potential user lacks detailed knowledge of the available data. The goal of this work is to provide a front-end to a statistical database (even though the system proposed is not complete and the solution presented, although interesting and relevant to statistical databases, is partial) and a basis for evaluating the techniques proposed. The work reported on is part of a large project and the database is a collection of social statistics coming, mainly, from two publications of the Commission of the European Communities. An outline hypertext database, which sets out the methodology used and suggests the future steps, has been produced in the first phase.

A graphical query language for manipulating a semantic database that covers primary logic formations is described in [Du]. The paper deals with a large number of derivable features of query languages and their graphical representation. It also uses a comprehensive model as a basis which includes aggregation, generalization, and relationships over other relationships. Although the model is described in terms of conventional databases, it seems to be suggestive for SSDB interface. There is some question, however, whether the approach is too complicated for either statistical or commercial users. Finally, it shows how to graphically express simple aggregate functions (e.g., "sum" and "count") (see also [Ghosh] on this topic).

5. CONCLUSIONS

The papers presented at the IV SSDBM describe the different results reached in the context of Statistical and Scientific Database Management. In addition to the traditional topics (e.g., data model, physical organization and data retrieval, user interface and security) new topics emerged which appear to be growing in importance. Those include temporal data, statistical knowledge based decision support systems, geographical data and object oriented user interface. The interest aroused by and the success of the conference, confirmed by the constant, unfailing presence of all participants, is now an established event, held every two years,

thus becoming a fixed reference-point for researchers and users in this field.
The purpose of this paper is to bring up relevant issues arised in the above mentioned conference. I hope that, remembering what Shoshani and Wong wrote in [Shoshani 85], these proceedings will bring to the attention of researchers both the gap that still exists between available commercial data management software and the tools needed to support SSDB applications, and the problems (both theoric and applicative), at present still existing in this area.

Acknowledgement

I thank John Klensin for his helpful comments and suggestions on a draft of this paper.

BIBLIOGRAPHY

[Andersson] M.Andersson, P.Svensson *"A Study of Modified Interpolation Search in Compressed, Fully Transposed, Ordered Files"* These Proceed.

[Barcaroli] G.Barcaroli, G.Di Battista, E.Fortunato, C.Leporelli *"Design of Statistical Information Media: Time performances and Storage Constraints"* These Proceed.

[Bassiouni/a] M.A.Bassiouni, M.Llewellyn *"Handling Time in Query Languages"* These Proceed.

[Bassiouni/b] M.A.Bassiouni, N.Ranganathan, A.Mukherjee *"Software and Hardware Enhancement of Arithmetic Coding"* These Proceed.

[Bates 82] D.Bates, H.Boral, D.J.DeWitt *"A Framework for Research in Database Management for Statistical Analysis or A Primer on Statistical Database Management Problems for Computer Scientists"* Proc. of ACM-SIGMOD, Int. Conf. on Manag. of Data, Orlando, FL, June 1982

[Bishop 83] Y.M.Bishop, S.R.Freedman *"Classification of Metadata"* Proceed. of the II Intern. Work. on SDBM, Sept.1983

[Brown 83] V.A.Brown, S.B.Navathe, S.Y.W.Su *"Complex Data Types and Operators for Statistical Data and Metadata"* Proceed. of the II Intern. Work. on SDBM, Sept.1983

[Chan 81] Chan P., Shoshani A. *"SUBJECT: a Directory Driven System for Organizing and Accessing Large Statistical Databases"* Proc. VII Int. Conf. on Very Large Data Bases, Cannes, France, Sept. 1981

[Chang] C.C.Chang, C.Y.Chen *Orthogonal Range Retrieval using Bucket Address Hashing"* These Proceed.

[Chin 78] F.Y.Chin *"Security in Statistical Databases for Queries with Small Counts"* ACM Transactions on Database Systems, Vol.3, No.1, March 78

[Cubitt 87] R.Cubitt, A.Westlake *"Report on the Third Intern. Work. on SSDBM"* Statistical Software Newsletter, Vol.13, No.1, April 1987

[D'Atri] A.D'Atri, F.L.Ricci *"Interpretation of Statistical Queries to Relational Databases"* These Proceed.

[Denning 80] D.E.Denning *"Secure Statistical Databases under Random Sampling Queries"* ACM Trans. on Database Systems, Vol.5, N.8, Sept. 1980

[Denning 83] D.E.Denning, J.Schlorer *"Inference Controls for Statistical Databases"* IEEE Computer, Vol.16, No.17, July 1983

[Denning 84] D.Denning, W.Nicholson, G.Sande, A.Shoshani *"Research Topics in Statistical Database Management"* A Quaterly Bulletin of the IEEE Computer Society Tech. Comm. on Database Engineering, Vol.7 No.1, March 1984

[Denning 87] D.E.Denning and oth. *"Views for Multilevel Database Security"* IEEE Transactions on Software Engineering, Vol. SE-13, No.2, Febr. 1987

[Di Battista] G.Di Battista *"Automatic Drawing of Statistical Diagrams"* These Proceed.

[Du] H.Du, M.Azmoodeh *"GQL, a Graphical Query Language for Semantic Databases"* These Proceed.

[Fortunato 86] E.Fortunato, M.Rafanelli, F.L.Ricci, A.Sebastio *"A Logical Model and an Algebra for Statistical Databases"* Proc. in Computational Statistics, sh.com., Roma, 1986

[Gambosi] G.Gambosi, E.Nardelli, M.Talamo *"A Conceptual Model for the Representation of Statistical Data in Geogrephical Information Systems"* These Proceed.

[Ghosh] S.P.Ghosh *"Statistical Relational Model"* These Proceed.

[Ghosh 86] S.P.Ghosh *"Statistical Relational Tables for Statistical Database Management"* IEEE Transactions on Software Engineering, Vol. SE-12, No.12, Dec.1986

[Ghosh 88] S.Ghosh *"Statistics Metadata"* in Kotz-Johnson Encyclopedia of Statistical Science, Vol.8, John Wiley & Sons Inc. Publ., 1988

[Johsson 81] R.R.Johnson *"Modelling Summary Data"* Proceed. of the ACM-SIGMOD, Int. Conference on Manag. of Data, Ann Arbor, MI, April 1981

[Kam 77] J.B.Kam, J.D.Ullman *"A Model of Statistical Databases and their Security"* ACM Trans. on Database Systems, Vol.2, N.1, March 1977

[Klensin] J.C.Klensin *"Statistical Data Management Requirements and the SQL Standards--An Evolving Comparison"* These Proceed.

[Klug 81] A.Klug *"ABE - A Query Language for Costructing Aggregates-by-exaple"* Proceed. of the 1th LBL Workshop on Statistical Database Management, Menlo Park, CA, Dec. 1981

[Klug 82] A.Klug *"Access Paths in the ABE Statistical Query Facility"* Proc. of ACM-SIGMOD, Int. Conf. on Manag. of Data, Orlando, FL, June 1982

[Laurini] L.Laurini, F.Milleret *"Spatial Data Base Queries: Relational Algebra versus Computational Geometry"* These Proceed.

[Malmborg] E.Malmborg *"Design of the User-Interface for an Object-orienteo Statistical Database"* These Proceed.

[Malvestuto] F.M.Malvestuto, C.Zuffada *"The Classification Problem with Semantically Heterogeneous Data"* These Proceed.

[Mark 86] L.Mark and N.Roussopoulos *"Metadata Management"* IEEE Computer, dec.1986

[McCarthy 82] J.McCarthy *"Metadata Management for Large Statistical Databases"* Proceed. of the VIII Intern. Conf. on VLDB, Mexico, Sept. 1982

[McLeish] M.McLeish, M.Cecile, A.Lopez-Suarez *"Database Issues for a Veterinary Medical Expert System"* These Proceed.

[Michalewicz] Z.Michalewicz, K.W.Chen *"Ranges and Trackers in Statistical Databases"* These Proceed.

[Michalewicz 87/a] Z.Michalewicz *"Functional Dependencies and their Connection with Security of Statistical Databases"* Inform. Systems, Vol.12, No.1, 1987

[Michalewicz 87/b] Z.Michalewicz, A.Yeo *"Multiranges and Multitrackers in Statistical Databases"* Fundamenta Informaticae, Vol.X, No.4, Dec. 1987

[Ozsoyoglu 84] Z.M..Ozsoyoglu, G.Ozsoyoglu *"STBE - A Database Query Language for Manipulating Summary Data"* Proceed. IEEE COMPDEC Conference, Cicago, Nov. 1984

[Ozsoyoglu 85] G.Ozsoyoglu, Z.M.Ozsoyoglu *"Statistical Database Query Languages"* IEEE Transactions on Software Engineering, Vol. SE-11, No.10, Oct.1985

[Ozsoyoglu 87] G.Ozsoyoglu, Z.M.Ozsoyoglu, V.Matos *"Extending Relational Algebra and Relational Calculus with Set-Valued Attributes and Aggregate Functions"* ACM Trans. on Database Systems, Vol.12, No.4, Dec. 1987

[Palley 87] M.A.Palley, J.S.Simonoff *"The Use of Regression Methodology for the Compromise of Confidential Information in Statistical Databases"* ACM Trans. on Database Systems, Vol.12, N.4, Dec. 1987

[Proc. 1st SDBM] Proceed. of the 1th LBL Workshop on Statistical Database Management, Menlo Park, CA, Dec. 1981

[Proc. 2nd SDBM] Proceed. of the 2th LBL Workshop on Statistical Database Management, Los Altos, CA, Sept. 1983

[Proc. 3rd SSDBM] Proceed. of the 3th Intern. Workshop on Statistical and Scientific Database Management, Luxembourg, July 1986

[Rafanelli 83] M.Rafanelli, F.L.Ricci *"Proposal of a Logical Model for Statistical Data Base"* Proc. 2th Int. Work. on Statistical Database Management, Los Altos, CA, Sept. 1983

[Rafanelli 85/a] M.Rafanelli *"A Management System for Statistical Databases"* in "The Role of Data in Scientific Progress" P.S.Glaeser (ed.), Elvisier Science Publ. B.V. (North Holland), 1985

[Rafanelli 85/b] M.Rafanelli, F.L.Ricci *"STAQUEL: a Query Language for Statistical Database Management Systems"* Proceed. CIL '85, Barcelona, April 1985

[Rafanelli 88] M.Rafanelli *"Statistical and Scientific Database Management Systems"* Encyclopedia of Computer Science and Technology, A.Kent & J.G.Williams Ed.s, M.Dekker Inc. Pub., New York 1988 (in press)

[Sato] H.Sato *"A Data Model, Knowledge Base, and Natural Language Processing for Sharing a Large Statistical Database"* These Proceed.

[Sato 81] H.Sato *"Handling Summary Information in a Database: Derivability"* Proc. ACM-SIGMOD Int. Conf. on Management of Data, 1981

[Schlorer 80] J.Schlorer *"Disclosure from Statistical Databases: Quantitative Aspect of Trackers* ACM Trans. on Database Systems, Vol.5, N.4, Dec. 1980

[Segev] A.Segev, A.Shoshani *"The Representation of a Temporal Data Model in the Relational Environment"* These Proceed.

[Shoshani 82] A.Shoshani *"Statistical Databases: Characteristics, Problems and some Solutions"* Proceed. of the 8th Intern. Conf. on Very Large Data Bases, Mexico, Sept. 1982

[Shoshani 84] A.Shoshani, F.Olken, H.K.T.Wong: *"Characteristics of Scientific Databases"* Proceed. of the Tenth Intern. Conf. on Very Large Data Bases, Singapore, August 1984

[Shoshani 85/a] A.Shoshani, F.Olken, H.K.T.Wong *"Data Management Perspective of Scientific Data"* in "The Role of Data in Scientific Progress" P.S.Glaeser (ed.), Elvisier Science Publ. B.V. (North Holland), 1985

[Shoshani 85/b] A.Shoshani, H.K.T.Wong: *"Statistical and Scientific Database Issues"* IEEE Transactions on Software Engineering, vol.SE-11, N°10, Oct. 1985

[Stephenson] G.Stephenson *"Knowledge Browsing-Front End to Statistical Databases"* These Proceed.

[Srivastava] J.Srivastava, D.Rotem *"Precision-Time Tradeoff: A Paradigm for Processing Statistical Query Language on Databases"* These Proceed.

[Su 83] S.Y.W.Su *"SAM* : A Semantic Association Model for Corporate and Scientific-Statistical Databases"* Information Sciences Vol.29, No. 2-3, 1983

[Svensson] P.Svensson *"Database Management Systems for Statistical ana Scientific Applications: Are Commercially Available DBMS Good Enough ?"* These Proceed.

[Westlake] A.Westlake *"Geographical Database Systems and Statistical Information"* These Proceed.

[Wittkowski] K.M.Wittkowski *"Knowledge Based Support for the Management of Statistical Databases"* These Proceed.

[Wong 82] H.K.T. Wong: *"Statistical Database Management"* ACM-SIGMOD, Conf. on Data Management, 1982

[Wong 84] H.K.T. Wong: *"Micro and Macro Statistical/Scientific Database Management"* 1° Intern. Conf. on Data Engineering, Los Angeles, March 1984

STATISTICAL DATA MANAGEMENT REQUIREMENTS AND THE SQL STANDARDS--AN EVOLVING COMPARISON

John C. Klensin and Roselyn M. Romberg
Room 20A-226, Massachusetts Institute of Technology
Cambridge, MA 02139 USA

INTRODUCTION

This paper evaluates the hypothesis that operations historically carried out in a specifically-scientific or -statistical database management context can reasonably be accommodated by the relational model. Such a critique is particularly relevant at this time. With the adoption of SQL as the international standard for relational database management [ISO 87], with ongoing development of an extended version of SQL called SQL2[1] [Melt 88], and with the general pressures of the marketplace, we seem to be moving rapidly toward a period in which there will be few available database management systems that are not relational and SQL-based. The field has gone through a phase in which the developers of even classical statistical packages felt a need to claim "relational form" or "relational-like" (e.g., COMP 80), to a series of announcements of "SQL interfaces" from vendors of distinctly non-relational systems [Asht 88; Lotu 88]. In the interim, many papers at these workshops have shifted from "how to design a system to accomplish this end" to "how to force this problem into the relational model". If the SQL-based relational model is not optimal for statistical and scientific databases, its strengths and limitations should at least be understood and documented.

To some extent, the present pattern may parallel a trend in the early days of machine-assisted statistical computing: at the time a few "packages" became dominant, a shift occurred between the early practice of constructing and using software that was carefully matched to the problem at hand and the now-current practice of figuring out how to make the problem fit the available software packages [Ynte 70; Ande 72]. In the process, some data analysis procedures became more difficult, or fell into disuse, precisely because they did not match well with the more formal statistical procedures supported well in the model supplied by the packages [Tuke 65; Klen 81]. Some package developers went so far as to suggest that only those studies be conducted that would permit analysis with the package of choice [Nie 75] which is, a priori, not a good idea scientifically.

In this paper, we examine, and respond to, the following three questions. The intent is to expose the areas where SQL and, in some cases, relational systems generally, are strong and where they are weak as supporting models for fairly complex statistical and scientific data and to suggest some of the implications of the weaknesses.

---

[1] The technical working group that develops SQL standards in the USA, and that has generated the base documents for the ISO standardization effort, is known as X3H2. "X3H2" documents referenced below are working documents produced by, or submitted to, that working group. They have no status as standards, only as proposals.

(i) What characteristics, discussed above, map directly between a statistical/scientific system and a SQL one? What other characteristics, not directly supported, can be replaced by other approaches, with adequate functionality, that can readily be expressed in SQL?

(ii) What features can be represented in the SQL-based system, albeit potentially with great trouble and discomfort?

(iii) What features cannot be accommodated at all because they contradict the relational model or SQL restrictions in one way or another?

We begin with a general review of the differences between commercially-oriented and statistically-oriented databases and database management systems. While we focus on the relational model and on SQL-based relational systems in particular, many of the contrasts between systems oriented in different ways stem from the more general commercial data base management and processing paradigms, of which most of the relational literature are clearly components. We focus specifically on the patterns of use, significant data types and structures, restrictions, and analytic operations required by statistical databases, in separate sections. Each of those sections offers a critical evaluation of SQL and the relational model as applied to those issues. Statistical database management systems and commercial ones also have much in common; this paper does not, in general, list the similarities.

Several fairly complete critiques of SQL relative to the relational model have appeared (see, for example, the comments by Date [Date 87] and Martin [Mart 86]); we do not address those issues except insofar as they have specific statistical database implications. Instead, we rely heavily on the social and behavioral science literature for discussion of data types and have deliberately chosen early papers to emphasize that the requirements and problems are not newly-identified. We also rely heavily on our experiences with a variety of analyses and databases. The examples are drawn from specifically designed SSDBM systems [Klen 81a; LAP 75] and from databases in the Oracle system [Orac 86][2]. Because statistical data dominate our experience, we draw more from statistical databases and software to manage them than from scientific ones.

DATABASE SYSTEM USAGE PATTERNS

The differences between commerical, especially relational, systems and statistical and scientific ones can be thought of as falling into two categories: those that are a consequence of differences between how the *systems* are used, and those that result from differing requirements for *data types, representa-*

---

[2] The paper is not a critical review of Oracle, but of SQL and the relational model: Oracle is, for the purpose of the analysis presented here, merely a largely standard-conforming test bed and source of convenient examples. Examples and conclusions based on Oracle should apply to a considerable degree to most SQL systems.

tions, and operators in the statistical and scientific system. The distinctions in usage patterns have been thoroughly reported [Teit 82; McIn 70; Shos 83; Klen 83; Ozso 83; Olke 86]. This section builds upon this existing database management literature, with a different perspective. The subsequent three sections draw upon the methodological literature to define requirements for statistical database management systems consequent of the nature of the data themselves. Throughout the paper, a "statistical database management system" is not only one in which "statistical data"--data representing statistics about some population--can be stored, retrieved, and manipulated, but also includes systems that provide database management support for statistical (or similar types of) analysis.

Transactions and Data Cleaning

Commercially-oriented database systems most often operate in a multiple-access, concurrent-update, transaction environment. Deleting and updating are frequent, and a commercially-oriented system must address potential anomalies which might result from frequent database modifications. Verification of data usually occurs at the time of data entry. Queries tend to consist of reports on many attributes of a small number of records. Data analysis, if performed at all, generally includes summary statistics and rarely anything else.

By contrast, the typical statistical database is constructed once, "cleaned" of coding or representational errors, and subsequently used only as a base from which data are inspected, extracted, analyzed, or otherwise reported upon. Statistical databases are characterized by a high ratio of queries to modifications; there are typically no modifications made to a completed statistical or scientific database. When statistical databases are modified (to correct further errors--"data editing"--or add new data), the usual procedures are akin to classical master file updating: the modifications are prepared and carefully audited, then the database is made unavailable for other use, updated, and audited again [Rian 86]. Queries tend to consist of complex statistical or mathematical analysis of a few attributes for many records or cases (or for an entire dataset).

The relational model can handle the data editing operations reasonably well. However, SQL provides only an UPDATE operation, which makes it quite difficult to interactively and selectively edit data in the style that is usually desired. For example, a system designed with this purpose in mind might support a dialogue similar to the following (text typed by the user in italics):

change_attribute age, grade in class prompt student for grade=missing;

William: *13, passed*
David: *14, passed*
Denise: *11, failed*

In this example, the first line is the command line entered by the user, instructing the system to search for instances where grade is missing and then prompt (with the student's name) for values for age and grade. Data editing is performed interactively [LAP 75]. The same changes in a SQL-based system would look like:

```
UPDATE CLASS SET AGE=13, GRADE='PASSED'
    WHERE STUDENT='WILLIAM' AND GRADE IS NULL;
UPDATE CLASS SET AGE=14, GRADE='PASSED'
    WHERE STUDENT='DAVID' AND GRADE IS NULL;
UPDATE CLASS SET AGE=11, GRADE='FAILED'
    WHERE STUDENT='DENISE' AND GRADE IS NULL;
```

One might need to perform a SELECT on GRADE IS NULL to examine AGE prior to performing the editing of the data as well, depending on the size or complexity of the database and the number of missing values.

Storage Management, Auditing, and Data Structure Stability

The management of a commercially-oriented database may require special attention to such issues as storage organization or auditing of changes or access. The structure of those databases, however, tends to change little over time; once a database for a commercially-oriented application has been designed and implemented, it may remain in use for many years. Commercial database management systems rarely include techniques which are out-of-the-ordinary for the proper and efficient handling of the commercially-oriented data, and the SQL standard is no exception.

By contrast, scientific and statistical databases tend to have characteristics which require special management techniques. For example, many databases contain large numbers of missing values or fixed-length text attributes with extraneous blanks. Special compression techniques may be necessary or appropriate. Statistical databases range in size from the moderate to the very large, and frequently contain widely varying value lengths or types, requiring compression or special storage and handling procedures as well. Many databases also contain large amounts of descriptive information along with the actual data values [Chan 83; DeNS 83; Ozso 83; LAP 75]. Although compression and storage algorithms are not discussed further in this paper, any one of these characteristics can require special procedures for efficient storage and handling of the data. These are, of course, also implementation issues, rather than SQL issues, unless the language makes efficient implementation excessively difficult.

Users of both types of systems require facilities to record the operations performed with them and on the data they contain. With commercial systems, "auditing" facilities are used to keep records of access and modifications to the database. For scientific, and especially statistical, systems, the requirements are different: as data are computed and derived from other data, it is quite important to track the origins of the derived values. One should be able to ask the question "where did this set of values come from" and receive an answer from the system. While it is well beyond today's state of the art for other than trivial cases, the software of an ideal system would recognize that the process of data analysis and deriving intermediate values inherently contains considerable procedural inefficiency and backtracking: the most useful history information would be canonicalized so that data derivations can be efficiently compared and reproduced.

Security, Privacy, and Confidentiality

Both commercial and scientific systems may require measures to protect data, and its privacy or confidentiality, but often with a slightly different orientation. This protection in commercial systems is typically provided by user authorization files, password protection at various levels, and protected views of the data, and SQL provides versions of all of these. Protection against imputations from data that are themselves to be made public is also required for some statistical databases [Ande 72; Denn 83], but automatic operations to make and verify the needed tests are rare, if reliable ones exist at all.

SIGNIFICANT DATA TYPES AND STRUCTURES

Social science and statistical requirements for special data types are frequently more extensive than are encountered in commercial, and some scientific, applications. Because of the broad mathematics implied by the definition of a domain in the relational papers, the term "data type" refers here to both the conventional concept of a type (e.g., text, dates, numbers) and the common usage of "structured", "aggregate", or "composite" data (e.g., arrays and sets of values). At present, it is at best difficult and at worst impossible to define several data types in common use in social science and statistical applications in SQL.

Data Types and Characteristics

Commercially-oriented databases typically have need of only a few data types. Text, integer, real, and date types usually meet the needs of most common business applications. SQL supports these types with CHARACTER of specified length, four closely-related exact numeric types (NUMERIC, DECIMAL, INTEGER, and SMALLINT), and three approximate numeric types (FLOAT, REAL, and DOUBLE PRECISION). Some SQL implementations, including Oracle, extend this list with additional forms including DATE, LONG (for extended character data), RAW and LONG RAW (for raw binary data). As discussed below, a few of these have been proposed for, and are likely to be included in, the SQL2 standard.

The SQL2 development process has included considerable discussion of extended types. For example, a proposal was made to, and discussed by, X3H2 to permit SQL data types to include a variable length character string type, a national character string type, an array type, a datetime type, an interval type, and an enumerated type (see discussion below). The prospects for these features are not bright: at the December 1987 meeting of X3H2, there was no clear majority to defer or to include national character string (national character set support), unstructured, and enumerated data types. A clear majority preferred deferring variable length or national character strings and array data types to some future revision of the standard beyond the draft proposed SQL2 [Melt 87]. More complex data forms, such as nominal, multiple-valued nominal, and ordinal data types, are as yet unrepresented in any proposals. Standards development activities often defer action on features on which agreement cannot be reached: these deferrals are not commitments to do anything but to review the topic in the undefined future; many, perhaps most, deferred proposals are never incorporated into standards; even some accepted ones are later rejected (for examples, see [Date 87]).

By contrast, experience with statistical databases indicates that they require a larger number of data types than the typical commercially-oriented database. Even the proposed SQL2 extension will not satisfy the requirements. For example, complex data types such as text with significantly varying lengths, ordered and unordered sets, vectors or arrays, and matrix data types are frequently needed to accommodate social or physical science data [Stam 73; Brow 83; Klen 83]. Stone and his colleagues discuss even more extreme textual cases [Ston 66], which a statistical database management system should still be able to accommodate. Several of these types are discussed in more detail below. Finally, time-dependent databases (as distinct from merely databases that contain dates) appear frequently in the econometric, and some sociological and political, literature [Brod 77; PSD2 83; Pell 80].

Extended Types for Measured Data--An Overview

Special types are also often desirable to distinguish among data at different levels and types of measurement. While the typical traditional database management system only distinguishes between "character" or "text" data and "numerical" data (sometimes expanded to "exact value" and "inexact value", as in SQL), additional distinctions are needed for data that will be treated with advanced statistical or data analysis techniques. While these distinctions can be made either by treating the values as of different types, or as descriptive metadata, they must be visible in the external schema. They also affect the validity of operators of the data manipulation language (e.g., comparisons have different values in SQL depending on whether the values being compared are exact-numeric or imprecise-numeric).

Stevens [Stev 46] suggests that four such types are important, types that he designated as "nominal" (i.e., categorical values with no intrinsic ordering), "ordinal" (i.e., categorical values with intrinsic ordering), and two numeric types, called "interval" and "ratio", differing depending on whether or not there is an intrinsic and meaningful zero value. From an operator standpoint, sorting is not defined for nominal values, and the only permitted comparisons are equality or inequality. A list of colors of objects would be a typical nominal variable. Ordinal values can be sorted, and greater than and less than comparisons can be performed, but they cannot be added or subtracted. Categorical education levels (e.g., no education, 1-6 years of school, 7-12 years of school, started university, completed university, etc.) constitute a common example of ordinal values. Ordinary textual values in database systems are typically treated as ordinal, not nominal, since sorting and inequality comparison operations are defined, based on alphabetical or other collation order. Similarly, the differences between interval and ratio level information govern whether the values of the numbers can be interpreted, or only the distances between them. Most economic and social data are, at best, interval; ratio-level data are not often seen in social phenomena.

In any event, if a statistical database system is going to maintain the statistical validity of its data then these different data types must be treated as if they are true types, with carefully-worked-out constraints on conversion and comparison among them.

Other researchers, e.g., Coombs, in his monumental work [Coom 67], have pointed out that even the Stevens categorization is inadequate, that additional subtle distinctions are needed. For example, there is considerable sociological and market research data expressed in terms of "preference scales": the answers to questions such as "on a scale of one to ten, how do you feel about...". But some research and experience indicate that close values may not be distinguishable, e.g., that there is a clear preference expressed when one item receives a score of eight and another receives a score of two, but there may not be a clear preference when one receives a two and another a three (see the discussion of a closely-related problem, arising from Guttman's work in [Shye 78]). These values take on not only some of the properties of ordinal coding (it is not possible to add or subtract them), but also some of the properties of numbers with imprecise comparisons. Other distinctions are made when "distances" between ordinal points can be estimated as equal (or at least ordinal ranks) and when they cannot. Sonquist [Sonq 77] provides additional applications-oriented discussion in these issues.

Enhanced Domains and Extended Data Types

Support for some additional complex data types could be provided in a relational context, if the concept of domain definitions were [re-]expanded to permit the "nonsimple" domains discussed in the original relational papers [Codd 70]. With extended domains, so-called "nominal" values are fairly easy to support: they map onto ideas of domains as enumerated sets (see below). "Ordinal" data values are more difficult, since both the value (a member of an enumerated set) and its inherent sequence must be maintained. Extended integrity constraints must be enforced for both: nominal is much like text, but there is a strong theoretical argument that neither comparison operations other than "equal/not-equal" nor sorting, nor even string operations, should be permitted. Ordinal is even more difficult, since sorting is sensible and "less/greater" comparisons can be made, but these must be enforced on the sequence, not on the names. That sequence may not participate in arithmetic, nor may the names participate in string operations as if they were ordinary text. As mentioned above, none of this is provided for, or clearly planned for, in SQL or its successors. The subsections that follow discuss a few representative statistical data types, some of which would be problematic even with an expanded domain concept.

"Enumerated Sets" versus Nominal and Ordinal Data

An enumerated type has been introduced into several systems, often as a restricted alternative to a "text" form. Another variation, defined as "a list of distinct identifiers that represents an ordered set of values", was proposed for SQL2. In the proposed SQL2 version, all values for a given enumerated type are comparable and values of two different enumerated types are not comparable. However, as for other complex data types, consideration of the enumerated type, even in its simplest form, has been deferred to a version of the standard after SQL2 has been completed [Melt 87]. In any event, that proposal solves only a small fraction of the problem: in particular, it does not make distinctions between unordered and ordered values, nor does it make "fuzzy equality" selections on adjacent ranges of ordered values convenient. More important, for statistical purposes, it is quite important to define an enumerated set domain so as to clearly distinguish between a "text" use and a "category name" use. If, for example, the members of an enumerated set are seen as "words"--character strings--then it might be sensible to permit operations that extract parts of those words, or search for substrings within them. By contrast, if the set members are strictly identifiers in the sense of category names, then they should be treated as atomic and not decomposable by string operations.

Array Aggregate Data Types

SQL is designed to operate on scalar values only, and the present draft SQL2 proposal does not address the issue of non-scalar types. Some scientific and statistical databases have used arrays and matrices for describing and storing data, frequently treating vectors or matrices as single entities (primary data elements) during analysis. Information about data type, dimensions, and bounds must be stored with the array along with the actual data values which comprise its elements. One proposal for array support in the third version of SQL or later suggests a number of possibilities, including the following structure and language [Hirs 87].

Example:

> Create a table of results for 20 questions, each ranked on a scale of -2 to +2 with room for the average, median, maximum, minimum, and count of responses for each question.

```
CREATE TABLE SURVEYRESULTS
   (QUESTION INTEGER,
    RESPONSECOUNT(-2:2) INTEGER,
    AVERAGE NUMERIC(6,2),
    MEDIAN NUMERIC(6,2),
    RANGE(0:1) INTEGER);
```

This proposed array scheme still does not address the issue of identifying, with labels, the rows, columns, planes, etc. of the arrays, often an important issue.

Additional complications may arise if arrays of some more complex data types, such as those above, are considered, and when the operation space is defined. For example, can one add, cellwise, a pair of four-dimensional arrays? A scalar to an array? A plane to a four-dimensional array? Is matrix multiplication to be permitted within the database system? Are array composition operations--joining two arrays of compatible dimensions--to be permitted? and so forth. Each of these operations has important uses in statistical work; each would complicate a database management system in significant ways.

Other aspects of the array type manifest themselves when one considers analytic procedures, particularly the tabulation of field values by categories. When the tabulation involves only a pair of category fields, the result is easily stored as a "flat" table, although that may not be an optimal solution. However, when three or more category fields are involved, the natural result is a multidimensional array or tightly-bound hierarchical structure [Ande 77]. Forcing that array into two dimensions not only is less suggestive, but may interfere with, or vastly complicate algorithms for, certain types of subsequent analysis, such as multivariate categorical analysis procedures that require reweighting or other dynamic adjustment procedures (see, e.g., Parta et al. [Part 82] or Mosteller's earlier work [Most 68] for a discussion of such a procedure). This places a strong premium on a database system's being able to retain multidimensional data.

Another Aggregate Type: Nominal Multiple-response

Data type problems arise when a field must contain the answer to a multiple-choice, multiple-answer question, e.g., "please indicate all of the following that apply". In some cases it is possible to handle this type of problem with an extended vector domain; in other cases, that approach and others do so much violence to both the user's model of the data and the requirements for data analysis to be both cumbersome and inappropriate.

A fairly lengthy, but still trivial, example will illustrate this point. Consider a simple table containing two columns: person names and colour choices, where any person may have multiple colour choices and any colour may be associated with multiple people. The first few rows of such a table might look like:

```
John   blue
Ann    green
Sally  blue
John   orange
```

A system with explicit maps and set-valued attributes might answer the query "display a list of all people who chose more than one colour" in the following way:

```
create TempTable := unique(Colour) in ColourChoices;
create_attribute NameCount :=
   tally_thru_relation(Name in ColourChoices);
create_attribute NameList := infer(Name in ColourChoices);
display Colour,NameCount,NameList in TempTable
   for (NameCount > 1);
```

One creates a temporary table using the unique values of the attribute "Colour", then creates an attribute for a count of the names per colour, and an attribute for a list of the names associated with each colour, then displays that list for all cases where the number of choices exceeds 1. Such a display might appear as:

```
Colour     NameCount       NameList
 =====     ==========       =========
 blue          2            John, Sally
 orange        3            Mark, John, Karen
 yellow        4            Lisa, John, Mark, Debra
```

In SQL, one is faced with the limits of the relational model in any attempt to structure the data in non-normal form. Since the values of expressions cannot be stored back into the database, the column "NameCount" would have to be created by hand in a two-step process, e.g., (i) display the values on the terminal using a SELECT COUNT(COLOUR), then (ii) use multiple INSERTs into the database on a colour-by-colour basis, rekeying the previously-displayed values. Likewise, since SQL does not support a set-valued "list" data type the attribute "NameList" could not be created as above. One could move beyond SQL and produce a display which would look similar to the one above with a relatively powerful report writer. This would not leave the values available for subsequent analysis, however. One could also produce a sorted display, e.g., "SELECT NAME FROM COLOURCHOICE GROUP BY NAME HAVING COUNT(COLOUR) > 1", which would have the same effect on analysis.

Less trivial variations on the example above are moderately common in data analysis contexts. In certain cases, the sets of values or entities must be bound to a particular ordering (a separate order-defining field in the case of entities), and that ordering preserved across selected operations. This was fairly easy in relational systems such as the early MacAIMS [Gold 70] and its successors and nearly-relational variants (see "entity numbers", "entity sets", and [ordered] "list attributes" in Janus [LAP 75]), but requires (at least) complex tricks in SQL (tricks that may violate normalization constraints).

DOMAINS AND RESTRICTIONS ON RELATIONSHIPS AMONG VALUES

Domains imply restrictions on the kinds of values that can be stored in particular fields (even if the domains themselves are implicit in the data types, as in SQL). By contrast, restrictions on the relationships among values associated with different attributes are generally implemented at the time of data entry, rather than being enforced dynamically. The approach of relying on checks during data entry is especially prevalent in commercial systems, where derived data without functional relationships to original data are rarely or never introduced into databases. In statistical databases, such derived data are the rule, rather than the exception. Since SQL does not support any form of assertion mechanism, integrity checks involving multiple columns, and even comparative range tests, are frequently imposed by an external forms manager or other data entry software at the time of (usually human) data entry.

The lack of more automatic constraint checking in SQL-based systems is apparently due to a similar assumption: that integrity checks on the data themselves (as distinct from their storage or values resulting from manipulation) are performed separately, external to the database system itself. Constraint-checking assertions have been proposed for SQL2, but postponed. They would permit any selection that could be specified in a WHERE clause to be enforced.

Domain-definition Restrictions: Enforcing Data Types

As discussed above, SQL does provide adequate support for data type enforcement with true, single-column, restrictions on domains. For example, a typical SQL CREATE statement defines each column specifically and independently, with all of the restrictions on the data types (and their sizes):

```
CREATE TABLE COMPANYDATABASE
(EMPLOYEE CHAR(20),
 JOB CHAR(10),
 DEPARTMENTNUMBER INTEGER,
 SALARY DECIMAL (9,2),
 HIREDATE DATE);
```

Verifying Relationships Among Values

Other restrictions or qualifications on values would be implemented at display or report-writing time, typically by applying conditional clauses to a SQL SELECT statement:

```
SELECT EMPLOYEE FROM COMPANYDATABASE
    WHERE JOB = 'SALES' AND SALARY < 1200.50;
```

The value-checking occurs only when the user inspects the results, here seeking a sales employee with a salary less than 1200.50.

Statistical Data, Metadata, and Meaning

For statistical or scientific data, restrictions on data values often form an integral component of the meaning of the data, and so must be understandable and obtainable at the time of data analysis or data exchange. Much more effective and precise restrictions are required to permit adequate constraints on data values and operators. Constraints on domains, such as complex range specifications and enumeration and references to other domains or attributes, are critical for many statistical database applications [Brow 82; McIn 70; Mull 83]. In addition to wanting to be able to restrict possible values in certain columns based on values in other columns, for example, one would also want to be able to look at the specific restrictions themselves. Obviously, restrictions imposed in an external data entry package are lost to the database itself unless included as some type of useable metadata in the database.

Using the example above, for instance, no metadata is incorporated into either the CREATE or the SELECT statements, and no precise domain restrictions apply. Because SQL does not provide for the binding of metadata to particular columns, our queries are limited to simple displays, such as the one above. So, for example, we could not incorporate either of the two types of potentially significant information with that column specification: (i) that "SALARY reflects the existing pay scale at the time of the questionnaire, but the organization was undergoing a review and reorganization on a department-by-department basis over the course of the following fiscal year" or (ii) that no one in the company is paid less than 500 or more than 20000.

More importantly, while the SELECT statement allows for a reasonable amount of specificity, as long as the fields are simple and the data types are standard, if one desired to subsequently analyze the relationships between a variety of fields with specific restrictions, SQL has no available constructs. Using the same COMPANYDATABASE example, above, we might want to investigate the relationship between HIREDATE and SALARY by DEPARTMENTNUMBER to see whether changes in policy at some date had any effect. This is not an example of wanting means or simple lists.

Compound Domain Restrictions

In complex cases, for example within the social sciences, domain restrictions are based on values of other attributes (and their domain restrictions). This type of operation is common and fairly convenient, not only with current statistical database management systems, but in the simple file management models of the statistical packages of the 1960s [e.g., Dixo 64; Couc 69; LAP 75]:

```
create_attribute colour_preference := nominal
  red for (colourcode > 0.001) & (colourcode <= 0.015),
  yellow for (colourcode > 0.015) & (colourcode <= 0.781),
  blue for colourcode > 0.781;
```

Another proposal to X3H2 [Celk 87] would extend SQL in the direction of the Entity-Relationship database model, isolating data integrity constraints into the column definitions, in much the same way one would SELECT with a CHECK on a view, as in the example below. Of course, if support were provided for enumerated set domains, this could be transformed into a simple domain definition constraint. Such "compound" domains are not provided by any currently active SQL2 proposal.

```
CREATE DOMAIN PERSON
  (NAME CHAR (30),
   SEX CHAR (1) CHECK ((SEX = "M") OR (SEX = "F")));
```

However, numerous problems still arise in attempting to address the components of a PERSON domain from, say, an EMPLOYEES table. Not only are the issues of domain value restrictions still largely unaddressed, but a majority of the standards body prefers to defer the topic of domains to some version of the standard beyond the proposal for SQL2 [Melt 88].

DATA ANALYSIS REQUIREMENTS AND THEIR IMPLICATIONS

A standard commercially-oriented database may require extensive reporting on its contents, usually for a few fields but for all (or many) records in the system. A typical example would be monthly reporting of payroll data or sales figures, following a standard formula and reporting format, for all employees in a company. While summary statistics may be of interest, (e.g., "aggregate sales by division"), such reports usually itemize individual data points and perform no analysis on that data. Summary statistics need not be saved back into the database (sums can always be re-computed, usually fairly efficiently) and large databases need not (indeed, generally should not) be reorganized, or subset and reconstructed, to produce the required reports.

By contrast, analysis of scientific and statistical data frequently requires complex reorganization of the data, often as part of the analytic and exploratory process. Data values derived during analysis must be saved back into the database. Datasets must be broken out based on analytic methods and results. Systems must be able to accommodate a variety of analytic techniques and database structures, and allow the results of one technique to be used by another, as is frequently required in the analysis of complex physical and social science data.

The nature of scientific and statistical databases is such that the data will often be used in, or to support, statistical analysis. Different data organizations are suggestive of, and consistent with, different statistical techniques and approaches. To say that one can represent all statistical data in "flat files" and convert them to the appropriate form from those files is to miss the point; the important issue is which organization as seen by the end user--the most "external" of schema--is optimal for the use to which the data will be put [Klen 83].

Qualitative Metadata

Adequate support for analysis often requires information about data quality, description of missing values and distinctions among different missing value types, aggregation and decomposition of data elements, and data about complex relationships between elements.

As examples, a physical or social scientific database might typically need to include information on the reliability of the instruments used, degrees of variability, changes in data collection methods, estimates of the "believability" of the data, and other metadata to qualify or describe the actual data values in the database. Similarly, a missing value in a statistical database may indicate a value below a certain threshold, no value measured at all, a trace value, that the question was not applicable, that data were not available, or that the respondent did not answer or didn't know. All of these examples have different and potentially significant meanings, depending on the data in question [Coom 67; Shos 83; IN22 87; Stew 88]. Analysis of statistical data frequently relies on keeping track of the complex relationships which can arise between data elements, including what the relationship's absence means. The "real world" is typically much more complex than a standard rectangular model allows, and information about that complexity must be preserved throughout the process [Shos 83].

Storing Values from Computations

SQL-based systems restrict the contexts in which computation of values that summarize several rows (entities) may be performed. They typically permit the results of those computations to be stored into tables (relations) only under very restricted circumstances. When stored, the bindings between the values (or computations stored in lieu of values) and the data from which they were computed is lost unless the user keeps track of it. For restricted cases [Gold 70], it is possible to push well beyond the strictures of SQL, but at the cost of considerable complexity; for other cases, it is not possible. Retention of these values, or computational specifications from which they can be dynamically constructed, and binding of the values to the data submodel from which they were computed, is critical for statistical use where the values will be used for summarization, scaling, data evaluation, or comparisons, as discussed below. The values produced will, in many cases, be from standard data type domains. In other cases, they can be represented by use of the extended type domains discussed above, such as array types. For these two groups of types, the statistical database requirement would be met if the values could be stored into the database and the appropriate bindings maintained (see below). Other data types, discussed after the binding section, may pose difficult challenges to the relational model--no relational system of which we are aware permits storing them as types with an appropriate complement of operators.

There have been several suggestions about how to record statistics computed from a database. They can be handled as separate relations [DeNs 83; Klen 83; Ozso 83] or stored as a special form within the relations ("at dataset level" [LAP 75]), or simply displayed (printed out) and not stored (e.g., the approach taken in Oracle and in Focus [Info 81] (typical, although not a relational system)). The latter, while a common solution, does not meet the requirement at all, but the other approaches have not been significantly more successful, at least within relational constraints.

The simple case of computing a mean can illustrate the difficulties of retaining derived statistics. It is relatively easy to talk about "the mean of the 'age' field", and to store it somewhere. At the same time, storing it violates integrity constraints unless the system can keep track of the relationship. It can, of course, be computed dynamically each time it is needed, but this involves serious performance degradation if the database is large. It is, of course, possible to keep track of whether or not the values of the field in question are modified, and to then either recompute the mean, or dynamically update it with the changed values (while algorithms are readily available for dynamically updating the mean, this is not the case for even slightly more complex statistics.) But, in reality, we often do not have "mean of the 'age' field". What we have is "mean of the 'age' field for well-educated men", "mean of the 'age' field for poorly-educated women", and so forth (see example). If one is going to try to keep track of what has changed, and update only conditionally, there are now three fields to keep track of, and only a subset of the "age" values.

In SQL, there are two approaches for this example. The following is the obvious analogy to the usual statistical database management system operation:

```
ALTER TABLE SURVEY ADD (AveAgeHiEdM NUMBER(2),
   AveAgeLoEdF NUMBER (2));
SELECT AVG(AGE) FROM SURVEY WHERE SEX = 'M' AND EDUCATION = 'HIGH';
SELECT AVE(AGE) FROM SURVEY WHERE SEX = 'F' AND EDUCATION = 'LOW';
UPDATE SURVEY SET AveAgeHiEdM = the output from the first SELECT,
   AveAgeLoEdF = the output from the second SELECT;
```

X3H2 [Hirs 88] suggests defining and using two views instead. For example, we might write:

```
CREATE VIEW AVGMALEHIGH (AGEM) AS SELECT AVG(AGE) FROM SURVEY
   WHERE SEX='M' AND EDUCATION='HIGH';
CREATE VIEW AVGFEMALELOW (AGEF) AS SELECT AVG(AGE) FROM SURVEY
   WHERE SEX='F' AND EDUCATION='LOW';
CREATE TABLE SURVEYSTATS
  (AveAgeHiEdM DECIMAL(3,2), AveAgeLoEdF DECIMAL(3,2));
INSERT INTO SURVEYSTATS (AveAgeHiEdM) SELECT AGEM FROM AVEMALEHIGH;
INSERT INTO SURVEYSTATS (AveAgeLoEdF) SELECT AGEF FROM AVGFEMALELOW;
```

In the first example, there is no automatic method for storing the results of the SELECT command, nor any way to dynamically update the values of the average ages. In the second, this problem is eliminated, but at the cost of potentially having to keep track of a separate view for every statistical inquiry: typically hundreds of them in exploratory use of a complex database. More generally, there is no obvious way to say, e.g., *UPDATE SurveyStats SET MEANAGE=AVE(AGE by sex and education).* There are some non-obvious ways, but they break down for more complex statistics and relationships. The difference between this hypothetical syntax and the first SQL example above illustrates an additional point: one really does not want to store the means as columns of "SURVEY"; the mean values are properties of aggregated subsets of the entities, not of the individual entities. The second SQL example uses a separate table to retain the derived values, eliminating this column problem, but adding a new "binding" complication (see below). At least one statistical DBMS can actually accomplish the process implied here through a variation on a join, using an explicit map ("relation" in the terminology of that system):

```
crd survey_stats := unique(age, sex in survey with relation=educsexr);
create_attribute meanage in survey_stats :=
   mean_thru_relation (age thru educsexr);
```

This example is not illustrative of the broader problem in another way: the mean is easy to compute, and perhaps not unacceptably expensive even if it must be computed each time it is asked for. Consider instead the results of a regression, or of a dimension-reducing procedure such as factor analysis or multidimensional scaling. These procedures can require significant computing resources, and may require user intervention in the estimation process [DeGW 80], or in adjusting the cases (records) to be considered [Bels 80], so recomputation cannot even be made transparent to the user. And, while computing a mean may require several selection fields, it only requires a single field on which to compute. These more complex (and more typical) procedures require several computational fields as well. Relational systems do not deal well with this set of problems, the SQL subset of them no better: new developments, or different approaches, are needed.

Binding Derived Values to Source Values

As discussed above, a common requirement of a scientific or statistical database management system is that it be able to represent and manage not only raw "statistical" data, such as listings of the characteristics of a population, but also that it be able to manage values representing the results of analyses or inferences about those data [Ande 72; Klen 81]. While there are exceptions, most of these values are identified differently from the original data values: they might represent summaries such as means, or the results of a fitting process (e.g., regression coefficients and accompanying residuals and estimates of variance accounted for), or projections onto another data space (e.g., factor scores and type of rotations performed), or complex tabulations. These types of secondary data raise two separate sets of problems for the relational database system:

(i) the logical bindings between these derived values and the sets of entities and attributes (records and fields) from which they come are different in character from the more traditional explicit or implicit bindings of the relational model, and

(ii) some of the results require data domains and storage types that are not obviously accommodated in the relational model.

Binding derived values to source values is critical to statistical and scientific databases. Failure to enforce those bindings can lead to integrity problems--these are not "just metadata".

Many of these binding difficulties are strictly consequences of the constraints of the relational model. Network systems with explicit mappings and linkages between data elements and sets of data elements could much more easily be extended to accommodate retention of the needed information. Similarly, if one constructed the internal structures of a database management system as a collection of linked list-like objects with property lists, many of the binding problems would become very simple (although others would appear).

Special Data Types Derived from Statistical Estimation

Both in the data management and, if separated, in the data analysis software, "suggestiveness" and the similar idea of "forc[ing] us to notice what we never expected to see" [Tuke 77] are important criterion. Data must be presented and [apparently] organized in a fashion that is consistent with the ways analysts think about the data and that facilitates developing and testing hypotheses. Optimizing external data organization and presentation for the convenience of the software chosen has the price of having to retrain analysts to think like programmers or of risking the loss of important inferences and insights. The following examples, while not comprehensive, illustrate this point.

Just as multidimensional arrays (discussed as data types, above) are the natural output of crosstabulation procedures and the natural inputs to multidimensional categorical analysis, many forms of analysis, e.g., factor and cluster analysis and multidimensional scaling, start from matrices of distances or similarities. These "matrices" are often represented in triangular form when the distances are undirected and in full form only with directed distances. They are the logical outputs of various inter-variable or inter-case (record) measuring processes and of some network analysis procedures. Like the multidimensional arrays discussed above, they should be representable and manageable within a statistical database management system. They raise similar problems: they are not ordinary "flat" relations, they must be carefully and closely bound to the records and fields from which they are computed, they are often computed from subsets of the data, complicating the binding process, and many different types of distance computations are possible--the type used must at least be stored as metadata, and the system must not preclude storing two different sets of distance computations derived from the same data.

SUMMARY AND CONCLUSIONS

We can now respond to the following two issues, restatements of the problem posed in the introduction:

(i) the suitability of SQL-based systems, and relational systems generally, at the current state of the art, for use with complex statistical database management problems and, conversely,

(ii) the modifications that would be required of a SQL-based system in order to permit it to accommodate a larger range of statistical database requirements.

As we have shown, if a database management system is to be designed to include complex statistical and quasi-statistical analysis capabilities, or to provide data management support for other software that provide those capabilities, SQL is unfortunately complex and tedious of expression in the better cases, and inadequate in the worse ones. Many of the inadequacies are consequences of the relational model itself: the obvious and natural ways to satisfy the requirements conflict with normalization constraints or the need to avoid navigational primitives and explicit inter-table mappings. Others could be accommodated within the relational framework, typically at the cost of considerable generalization of the domain-definition concept above that permitted in SQL (or in other commercially-oriented realizations of relational systems).

SQL is an international standard, and SQL2 will soon be proposed as an international standard. As such, the statistical and scientific database community can attempt to influence their content insofar as content changes, consistent with SQL's overall structure and approach, would improve our ability to manage databases of interest.

Some of the problems with SQL discussed above are imposed by its relational nature, others are the result of SQL's incomplete support for some aspects of the relational model. Posing system designs that would eliminate these problems is well beyond the scope of this paper. However, some ideas are suggested by our investigations that may be worth further investigation. The three most interesting of these appear to be

- introduction of extended domains, including array and ordered-set and unordered-set values,
- permitting explicit maps between tables and operations on those maps as well as operations on the tables and their components, and
- possible use of an external database representation based on lists of data values, rather than tables of them.

None of these ideas are new, but the fact that two of them appear to be contradictory to the relational approach suggests that the model, at least as generally understood, may not be the ideal one for statistical databases.

ACKNOWLEDGEMENTS

The authors would like to thank Dr. Ree Dawson for many helpful comments and suggestions on an early version of this paper. Mr. Alan Hirsch also reviewed the preliminary version and made several helpful corrections to our understanding of the current SQL standard developments. Remaining errors in interpretation or inference are, of course, the responsibility of the authors.

REFERENCES

[Ande 72] Anderson, R.E. and E. Coover, "Wrapping the package--critical thoughts on applications software for social data analysis", *Computers and the Humanities*, **7** (1972), pp. 81-95.

[Ande 77] Anderson, R.E. and F.M. Sim, "Data Management and Statistical Analysis in Social Science Computing", *American Behavioral Scientist*, **20**, 2 (January/February 1977), pp. 367-409.

[Asht 88] Mace, S., "Ashton-Tate, Microsoft Join Forces to Introduce SQL Database Server", *Info World*, **10**, 3 (January 18, 1988), p.1, and G. Abruzze, "Microsoft, A-T join forces to unveil new 'SQL Server'", *Computer & Software News*, **6**, 3 (January 18, 1988), p. 1.

[Bels 80] Belsley, D.A., E. Kuh, and R.E. Welsch, *Regression Diagnostics: Identifying Influential Data and Sources of Collinearity*, New York: John Wiley & Sons, 1980.

[Bish 83] Bishop, Y.M., and S.R. Freedman, "Classification of Metadata", *Proc., 2nd Int. Workshop on Statistical Database Management*, 1983, pp. 230-234.

[Brod 77] Brode, J., P. Werbos, and E. Dunn, *TSP in the Datatran Language*, Cambridge, MA, USA: MIT Laboratory of Architecture and Planning, 1977.

[Brow 83] Brown, V.A., S.B. Navathe, and S.Y.W. Su, "Complex Data Types and a Data Manipulation Language for Scientific and Statistical Databases", *Proc., 2nd Int. Workshop on Statistical Database Management*, 1983, pp. 188-195.

[Celk 87] Celko, J., ANSI X3H2 87-23, "Domains in SQL2", September 1987.

[Chan 83] Chan, P. et al., "Statistical Data Management Research at Lawrence Berkeley Laboratory", *Proc., 2nd Int. Workshop on Statistical Database Management*, 1983, pp. 273-279.

[Codd 70] Codd, E.F., "A Relational Model of Data for Large Shared Data Banks", *Communications of the ACM*, 13, 6 (June 1970), pp. 377-387. See esp. pp. 380-381.

[Codd 79] Codd, E.F., "Extending the database relational model to capture more meaning", *ACM Trans. on Database Systems*, **4**, 4 (1979).

[COMP 80] Barritt, M.M. and D. Wishart, *COMPSTAT 80: Proceedings in Computational Statistics*, Vienna: Physica-Verlag, 1980.

[Coom 67] Coombs, C.H., *A Theory of Data*, New York: Wiley, 1967.

[Couc 69] Couch, A.S., Armor, D.J., et al., *DATA-TEXT System: A computer language for social science research*, Cambridge, MA, USA: Harvard University Department of Social Relations, 1969.

[Date 87] Date, C.J., *A Guide to the SQL Standard: A user's guide to the standard relational language SQL*, Reading, MA, USA: Addison-Wesley, 1987. See especially Appendix E, "An Annotated SQL Critique".

[Denn 83] Denning, D., "A Security Model for the Statistical Database Problem", *Proc., 2nd Int. Workshop on Statistical Database Management,* 1983, pp. 368-390.

[DeGW 80] Dennis, J.E., D.M. Gay, and R.E. Welsch, "An Adaptive Nonlinear Least-Squares Algorithm", Technical Report #TR-20, Alfred P. Sloan School of Management, Cambridge, MA, USA: MIT, 1980.

[DeNS 83] Denning, D., W. Nicholson, G. Sande, and A. Shoshani, "Research Topics in Statistical Database Management", *Proc., 2nd Int. Workshop on Statistical Database Management,* 1983, pp. 46-53.

[Dixo 64] Dixon, W.J., ed., *BMD: Biomedical Computer Programs,* Los Angeles: Health Sciences Computing Facility, University of California, Los Angeles, 1964.

[Fran 76] Francis, I. and J. Sedransk, "Software Requirements for the Analysis of Surveys", *Proceedings of the International Biometric Conference,* 1976, pp. 228-253.

[Fran 81] Francis, I., ed., *Statistical Software, A Comparative Review,* New York: Elsevier North--Holland, 1981.

[Gold 70] Goldstein, R.C. and A.J. Strnad, "The MacAims Data Management System", *Proc. 1970 ACM SIGFIDET Workshop on Data Description and Access,* New York: ACM, 1970.

[Hirs 87] Hirsch, A.R., "Array Processing in SQL2", ANSI X3H2-87-294, October 26, 1987.

[Hirs 88] Hirsch, A.R., Personal communication.

[Info 81] Information Builders, Inc., "Sample MULTR Terminal Session", *FOCUS Users Manual,* pp. 8-18 - 8-20.

[IN22 87] Klensin, J.C., "Representation of trace, missing, and zero values in food data interchange", INFOODS Working Paper INFOODS/IS N22, Cambridge, MA, USA: MIT INFOODS Secretariat, 1987.

[ISO 87] ISO 9075-1987, Database Language--SQL, 1987 and ANSI X3.135-1986, ANSI Database Language--SQL, 1986.

[Klen 81] Klensin, J.C. and D.B. Yntema, "Beyond the package: A new approach to behavioral science computing", *Social Science Information,* **20**, 4/5 (1981), pp. 787-815.

[Klen 81a] Klensin, J.C., and R. Dawson, *The Consistent System, Program Summaries,* Cambridge, MA, USA: MIT Laboratory of Architecture and Planning, 1981.

[Klen 83] Klensin, J.C., "A Statistical Database Component of a Data Analysis and Modelling System: Lessons from eight years of user experience", *Proc., 2nd Int. Workshop on Statistical Database Management,* 1983, pp. 280-286.

[LAP 75] *The Consistent System, Janus Reference Manual,* Cambridge, MA, USA: MIT Laboratory of Architecture and Planning, 1975.

[Lotu 88] Mace, S. and E. Warner, "Lotus Choses Gupta Database Engine", *Info World,* **10**, 7 (February 15, 1988), p. 1.

[Malm 86] Malmborg, E., "On the Semantics of Aggregated Data", *Proc., 3rd Int. Workshop on Statistical and Scientific Database Management,* 1986, pp. 152-158.

[Mart 86] Martin, D. *Advanced Database Techniques,* Cambridge, MA: The MIT Press, 1986.

[McIn 70] McIntosh, S.D. and D.M. Griffel, "ADMINS Mark III--The User's Manual", Cambridge, MA, USA: MIT Center for International Studies, March 1970.

[Melt 87] Melton, J., ANSI X3H2-88-2, "ANSI SQL2 vs. SQL3 preliminary preferences (revised)", 22 December 1987 and X3H2-88-3, "Some criteria for discriminating between SQL2 and SQL3", 26 December 1987. Of the X3H2 members at the December 1987 meeting, 39% preferred to defer DOMAINs and 59% preferred to defer ENUMERATED DATA TYPEs until SQL3 or later.

[Melt 88] Melton, J., ISO Data Base Language CPH-2a, ANSI X3H2-88-127, ANSI X3H2 ISO/IEC JTC1/SC21/WG3 Data Base Languages, ISO-ANSI (working draft) SQL2, April 1988.

[Merr 83] Merrill, D., J. McCarthy, F. Gey, and H. Holmes, "Distributed Data Management in a Minicomputer Network: The SEEDIS Experience," *Proc., 2nd Int. Workshop on Statistical Database Management,* 1983, pp. 99-103.

[Mill 71] Miller, J.R. III, "Proposed user conventions and the kernel of a user language for the Janus (front-end) system", Cambridge Project working paper, April 8, 1971.

[Most 68] Mosteller, F., "Association and Estimation in Contingency Tables", *Journal of the American Statistical Association,* **63** (1968), pg. 1977.

[Mull 83] Muller, R.J., "Using Statistical Software with a Database Management Data Theory", *Proc., 2nd Int. Workshop on Statistical Database Management,* 1983, pp. 414-423.

[Mull 83a] Muller, R.J., *Data Organization: The integration of database management, data analysis, and software technology applied to the National Crime Survey,* unpublished Ph.D. dissertation, MIT Department of Political Science, 1983.

[Nie 75] Nie, N.H., C.H. Hull, J. Jenkins, K. Steinbrenner, and D.H. Bent, *SPSS Statistical Package for the Social Sciences,* New York: McGraw-Hill, 1975.

[Olke 86] Olken, F., D. Rotem, A. Shoshani, and H.K.T. Wong, "Scientific and Statistical Data Management Research at LBL", *Proc., 3rd Int. Workshop on Statistical and Scientific Database Management,* 1986, pp. 1-20.

[Orac 86] Oracle Corporation, *SQL*Plus Reference Guide* and *SQL*Plus User's Guide,* Belmont, CA: Oracle, 1986.

[Ozso 83] Ozsoyoglu, G. and Z.M. Ozsoyoglu, "Features of a System for Statistical Databases", *Proc., 2nd Int. Workshop on Statistical Database Management,* 1983, pp. 9-18.

[Part 82] Parta, R.E., J.C. Klensin, and I. de Sola Pool, "The Shortwave Audience in the USSR: Methods for Improving the Estimates", *Communications Research,* **9**, 4 (Oct 82), pp. 581-606.

[Pell 80] Pelletier, P.A. and M.A. Nolte, "Time and the Panel Study of Income Dynamics", in J. Rabin and A. Marks, eds., *Data Bases in the Humanities and Social Sciences,* 1980, pp. 53-60.

[Read 86] Read, B.J., "Scientific Data Manipulation in a Relational Database System", *Proc., 3rd Int. Workshop on Statistical and Scientific Database Management,* 1986, pp. 31-35.

[Rian 86] Riano, C., "Using Logic to Organize Statistical Databases", *Proc., 3rd Int. Workshop on Statistical and Scientific Database Management,* 1986, pp. 159-164.

[Shye 78] Shye, S., "Partial Order Scalogram Analysis", in Shye, S., ed., *Theory Construction and Data Analysis in the Behavioral Sciences,* San Francisco: Jossey-Bass, 1978.

[Sonq 77] Sonquist, J.A. and W.C. Dunkelberg, *Survey and Opinion Research: Procedures for Processing and Analysis,* New Jersey: Prentice-Hall, 1977, See especially pp. 20-40.

[SRCM 79] Survey Research Center, Computer Support Group, *OSIRIS IV: Statistical Analysis and Data Management Software System,* Ann Arbor, MI, USA: Institute for Social Research, University of Michigan, 1979.

[Stam 73] Stamen, J.P. and R. Wallace, "Janus: A Data Management and Analysis System for the Behavioral Sciences", *Proceedings of the 1973 Annual Conference of the ACM,* New York: ACM 1973.

[Stev 46] Stevens, S.S., "On the Theory of Scales of Measurement", *Science,* **103** (1946), pp. 677-680.

[Stew 88] Stewart, K. editorial, *Journal of Food Composition and Analysis,* Vol. 1, No. 2, (March 1988), p. 103.

[Ston 66] Stone, P.S., D.C. Dunphy, M.S. Smith, and D.M. Ogilve, *The General Inquirer: A computer approach to content analysis,* Cambridge, MA, USA: MIT Press, 1966.

[SuNa 83] Su, S.Y.W., S.B. Navathe, and D.S. Batory, "Logical and Physical Modeling of Statistical/Scientific Databases", *Proc., 2nd Int. Workshop on Statistical Database Management,* 1983.

[Teit 82] Teitel, R., "User Interface with a Relational Model of Data", *ACM SIGSOC Bulletin,* **13**, 2-3 (January 1982).

[Tuke 65] Tukey, J.W. and M.B. Wilk, "Data analysis and statistics: techniques and approaches", *Proceedings of the Symposium on Information Processing in Sight Sensory Systems,* Pasadena, CA, USA: California Institute of Technology, 1-3 November 1965.

[Tuke 77] Tukey, J.W., *Exploratory Data Analysis,* Reading, MA, USA: Addison-Wesley, 1977. See Preface, esp. p. vi.

[Ynte 70] Yntema, D.B., ed., "Summer Study Report", MIT: Cambridge Project, 1970. A summary of important points appear as Chapter 6 of Yntema, D.B., ed., *The Cambridge Project: Annual Progress Report June 1970-June 1971,* Cambridge, MA, USA: MIT Cambridge Project, 1971, available from USA Defense Documentation Center (Contract DAHC15 690347), July 1971.

THE REPRESENTATION OF A TEMPORAL DATA MODEL IN THE RELATIONAL ENVIRONMENT

Arie Segev† and Arie Shoshani‡

† School of Business Administration and
Computer Science Research Dept, Lawrence Berkeley Lab
The University of California
Berkeley, California 94720

‡ Computer Science Research Dept.
Lawrence Berkeley Laboratory
University of California
Berkeley, California 94720

An Invited Paper to the 4th International Conference on Statistical and Scientific Database Management

Abstract. In previous work, we introduced a data model and a query language for temporal data. The model was designed independently of any existing data model rather than an extension of one. This approach provided an insight into the special requirements for handling temporal data. In this paper, we discuss the implications of supporting such a model in the relational database environment. We show that the concept of a temporal relation needs to be introduced, and define a temporal normal form for it. We discuss several options for the representation of our model's constructs in the relational context, and explain why we chose a particular representation. We also suggest the concept of a temporal relation family as a virtual view to treat multiple temporal relations together.

1. INTRODUCTION

Temporal data are quite prevalent, yet very few data management systems (and especially commercially available systems) pay attention to the special needs of such data. Existing systems have been designed with the view that the task of a database system is mainly to keep the data in the database current. Typically, historical data will be lost by updates. In reality, there exist many applications where the time element is inherent. Such applications naturally include scientific and statistical databases (SSDBs), where physical experiments, measurements, simulations, and collected statistics are usually in the time domain. However, in many business applications, for which commercial systems have been designed, temporal data is also essential. Business applications, such as banking, sales, inventory control, and reservation systems, need to keep a complete history of transactions over the database. Furthermore, often this history needs to be statistically analyzed for decision making purposes.

This research was supported by the U.S. Department of Energy Applied Mathematics Sciences Research Program of the Office of Energy Research under contract DE-AC03-76SF00098.

In our previous work, we introduced a data model and a query language for temporal data. The model was designed independently of any existing data model rather than an extension of one. This approach provided an insight into the special requirements for handling temporal data. We have precisely characterized the properties of temporal data and defined operators over them without being influenced by traditional models which were not specifically designed to model temporal data.

Our approach differs from many other works which extend existing models to support temporal data. Examples of works that extend the relational model are [Ariav et al 84, Clifford & Crocker 87, Gadia & Yeung 88, Lum et al 84, Snodgrass 87, and Tansel 86]. Examples of works that extend the Entity-Relationship model are [Klopproge 81, Adiba & Quang 86].

In this paper, we discuss the implications of supporting our model in the relational database environment. We first review our work to date in section 2, and introduce those concepts that are relevant to this paper. In section 3, we show that the concept of a temporal relation needs to be introduced, and discuss several options for the representation of our model's constructs in the relational context. We explain why we chose a particular representation, which differs from most approaches proposed in the literature. In section 4, we introduce a new definition for a temporal normal form, and show how it differs from another definition suggested in the literature. In section 5, we characterize precisely the relational representation of our model's constructs. In section 6, we suggest the concept of a temporal relation family as a virtual view to treat multiple temporal relations together. We conclude with a short summary and a discussion of future work.

2. SUMMARY OF PREVIOUS WORK

In [Shoshani & Kawagoe 86] a framework was described for modeling temporal information. This work was followed by a detailed logical model and a query language [Segev & Shoshani 87]. The approach taken was to develop a temporal model that is independent of any existing logical data model (such as the relational or network models), so that our design can address itself solely to the semantics and operators of temporal data without being influenced by existing models. The reason is that existing models have not been designed to deal with temporal data, and simply extending existing models may unduly influence the design. This work resulted in a model that is based on the concepts of a *time sequence* (*TS*) and a *time sequence collection* (*TSC*), which we will explain shortly. Once this independent model (which we call the *TSC* model) was developed, our plan was to represent its structures and operations in specific logical models. Typically, this will require extensions or changes of the logical models, or perhaps will point out that some models have inadequate facilities for supporting temporal modeling. As mentioned previously, the purpose of this paper is to identify the requirements for representing the *TSC* model in the context of the relational model. In this section we describe those details of the *TSC* model that are needed to explain these requirements.

We are mainly interested in capturing the semantics of ordered sequences of data values in the time domain, as well as operators over them. Consequently, we define the concept of a Time Sequence (*TS*), which is basically the sequence of values in the time domain for a single entity instance, such as the salary history of an individual or the measurements taken by a particular detector in an experiment. We define the properties of the *TSs*, such as their type (continuous, discrete, etc.), their time granularity (minutes, hours, etc.), and their life span. We will elaborate on these properties below, but first we describe *TSs* and *TSCs* in more detail.

2.1. Time sequences and time sequence collections

Our model is based on the simple observation that temporal values are associated with a specific object, and are totally ordered in time; that is they form an ordered sequence. For example, the salary history of John forms an ordered sequence in the time domain. We call such a sequence a *time sequence* (*TS*). *TSs* are basic structures that can be addressed in two ways. Operators over them can be expressed not only in terms of the values (such as "salary greater than 30K"), but also in terms of temporal properties of the sequence (such as "the salary for the last 10 months", or the "revenues for every Saturday"). The results of such operators is also a *TS* whose elements are the temporal values that qualified.

Since all the temporal values in a *TS* are associated with a single surrogate value, it is convenient to view *TSs* graphically as shown in Figure 1. Consider the *TS* shown in Figure 1a, which may represent the cost of some item over time. In this example, the time point 1 has the value 50, time point 3 has the value 70, etc. However, in this case, these values extend to other time points of the

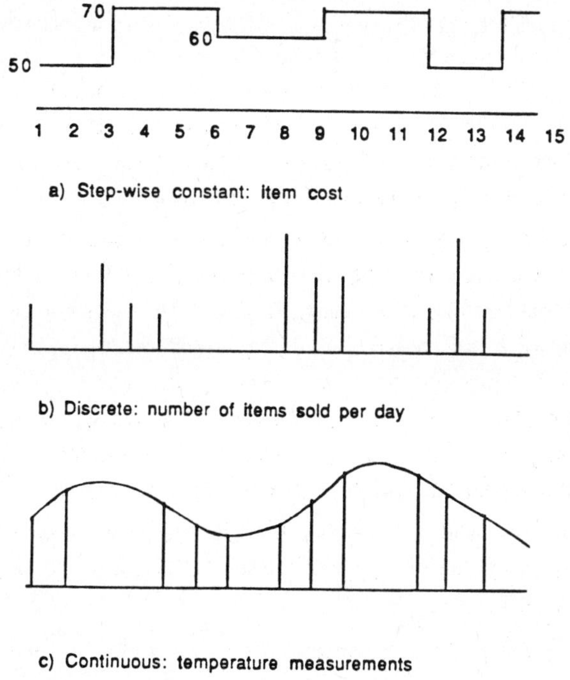

Fig. 1: Examples of different types of time sequences

sequence as shown (e.g. time point 2 has the same cost value as time point 1, etc.) We label such behavior of the *TS* "step-wise constant". In contrast, Figure 1b shows a *TS* of the number of items sold per day. Here the temporal values apply only to the days they are specified for. We call this property of the *TS* "discrete". A third example is shown in Figure 1c which represents temperature measurements. In this case, one can interpret the *TS* as being "continuous" in the sense that values in between the measured points can be interpolated if need be.

Note that in all these examples, the *TS* can be represented as a sequence of time-value pairs (e.g. (1, 50) (3, 70) (6, 60) ... for Figure 1a), and they differ only in the interpretation of their behavior, according to their type. The association of these properties to the *TSs* allow us the treatment of such sequences in a uniform fashion. First, we can define the same operators for *TSs* of different types, such as to select parts of a *TS* or to aggregate over its values. Furthermore, we can define operators between *TSs* of different types, such as multiplying a discrete *TS* with a step-wise constant *TS*. For example, multiplying the *TSs* of figures 1a and 1b, will generate a "revenues" *TS* = "number of items" *TS* X "cost" *TS*. Second, one can design the same physical structures for *TSs* that have different semantic properties, such as continuous or discrete. Two recent papers [Rotem & Segev 87, Gunadhi & Segev 88] describe the design of physical database structures for temporal data.

It is natural and useful to consider the collection of *TSs* for the objects that belong to the same class. For example, consider the collection of *TSs* that represent the salary histories for the class of all the employees in the database. We refer to such a collection as the *time sequence collection* (*TSC*). The usefulness of the *TSC* structure stems from the ability to address the temporal attributes of an entire class, and relate them to other (possibly non-temporal) attributes of the class. For example, we may be interested in the salary history of employees in the computer department for the last 6 months. Such operations over *TSCs* were discussed in [Segev & Shoshani 87].

In this paper we refer to two kinds of *TSCs*: simple and complex. A simple *TSC* is defined for a a single temporal attribute. A simple *TSC* can be described as a triple (S,T,A) where S, T, and A are the surrogate (or the identifier of the class), time, and attribute domains, respectively. A simple *TSC* can thus be viewed as the collection of all the temporal values of a single attribute for all the surrogates of a class. It is convenient to think of a simple *TSC* in a two-dimensional space as shown in Figure 2. In Figure 2a, the rows represent surrogates, columns represent times, and the temporal values are represented as points in the surrogate-time domain. Figure 2b, shows the equivalent graphical representation assuming that the *TSC* is of the type step-wise constant.

We note here that non-temporal values can be represented as a special case of the *TSC*. A non-temporal attribute has a single time point (which could be "current time"), and therefore its *TSC* will be reduced to a single column structure.

A complex *TSC* is a *TSC* whose attribute component A does not represent a single element. This situation exists when several attributes occur (or are measured) at precisely the same time points. For example, when collecting air pollution samples at regular intervals, several measurements are taken, such as carbon monoxide, nitrogen compounds, etc. We denote this case as $(S,T,\bar{A})$. Complex *TSCs* provide a concise way of representing together several simple *TSCs* that have the same temporal behavior. An important special case is representing non-temporal data as the degenerate *TSC* $(S,\bar{A})$.

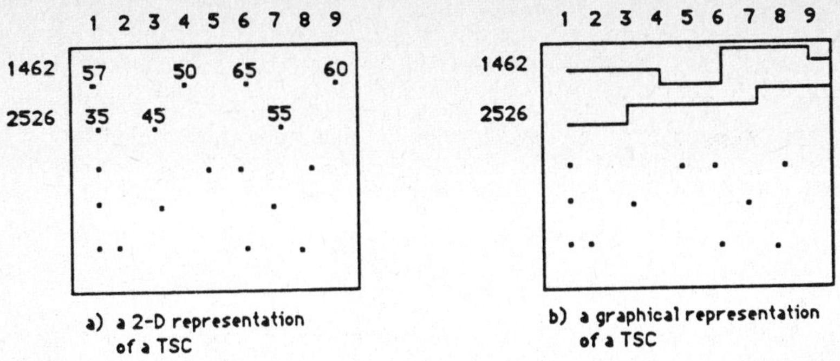

Fig. 2: A two-dimensional representation of a time sequence collection

where all the non-temporal attributes can be treated together in a single *TSC*.

Obviously, any combination of the above three cases can exist simultaneously. To simplify our discussion here, we only describe the properties for simple *TSCs*.

2.2. Properties of a *TSC*

In this section, we discuss in more detail the properties of a *TSC* that were mentioned briefly previously. They are *time granularity, lifespan, type,* and *interpolation rule.* The important point to note is that these properties apply to the entire *TSC*, and thus can be considered part of the metadata of that *TSC*. When we consider the representation in the relational context later, this point is crucial to the argument that we need a special kind of a relation to represent a *TSC*.

We distinguish between the *time points* and the *event points* of a *TSC*. The time points of a *TSC* are all the potential points in time that can assume data values. In contrast, the event points of a *TSC* are only the points that actually have data values associated with them. For example, suppose that the salary of an individual can change during any month of a certain year, but actual changes took place in April and October only. Then, only these two months are called the event points of that *TSC*. In general, the event points of a *TSC* are a subset of the time points. Often, the values of other time points can be inferred or interpolated from the event points. For instance, in the example above, the months of May through September will assume the same value as the event point for April. We will refer to data values of event points as "event data values".

Consecutive event points may have the same values, and in general the removal of these duplicate values results in a loss of information. This point will be discussed in more detail in the context of duplicate values and the temporal normal form in Section 4. Next, we define the properties of a *TSC*.

Time Granularity

This property specifies the granularity of the time points (t) of a *TSC*, i.e. the points in time that can potentially be event points. We allow for two time granularity representations - ordinal and calendar. The ordinal representation simply signifies that the time points are counted by integer ordinal position (1,2,3,...). The calendar representation can assume the usual calendar time hierarchy values: year, month, day, ..., second, etc.

Life Span

Each *TSC* has a life span associated with it. The life span is specified by a start_point and an end_point defining the range of valid time points of the *TSC*. The start-times and end-times are also represented as ordinal or calendar.

We are interested in three cases of a life span:

a) start_point and end_point are fixed.

b) start_point is fixed and end_point is current_time.

c) a fixed distance is defined between the start_point and the end_point. The end_point is "current_time" and the start_point is dynamically changed to maintain the fixed distance from the end_point.

Type

The type of a *TSC* determines the data values of the *TSC* for time points that are not event points (i.e. time points that do not have explicit data values.) In general, there is an interpolation function associated with each *TSC*. Some of the interpolation functions are very common, and therefore are given specific type names below.

We are interested in the following types of time sequences:

a) Step-wise constant: if (t_i,a_i) and (t_k,a_k) are two consecutive event points (and their data valies) in a *TSC* such that $t_i < t_k$, then $a_j = a_i$ for $t_i \le t_j < t_k$.

b) Continuous: a continuous function is assumed between two consecutive event points (t_i,a_i) and (t_k,a_k) which assigns a_j to t_j $(t_i < t_j < t_k)$ based on a curve fitting function.

c) Discrete: each event data value (a_i) in a *TSC* is not related to other values. Consequently, missing values cannot be interpolated.

d) User defined type: values for time points of a *TSC* that are not event points can be computed based on user defined interpolation functions.

In [Segev & Shoshani 87] temporal operators over *TSCs* have been discussed in detail, and a syntax defined for them. We do not discuss that subject here because it is not necessary as a background material. It is only important to note that the temporal operators we defined are over and between *TSCs*. For example, we defined an operation called COMPOSE, which can be best explained with an example such as the "revenue" *TS* mentioned previously. Recall that the "revenue" *TS* resulted from multiplying the two *TSs* of figures 1a and 1b (cost times number of items sold).

Suppose that we now have the corresponding "cost" *TSC* and "the numbers of items sold" *TSC*. Each *TSC* represents the set of *TSs* for many items. We would like an operation that will multiply the set of *TSs* from the two *TSCs* pairwise for *TSs* with matching surrogates. We call such operation COMPOSE, and allow different arithmetic operations over it. The significance of operators over *TSCs* to this paper is that we would like to have a structure in the relational context that is analogous to a *TSC*. This is indeed what we aim for in the next sections.

3. SELECTING A REPRESENTATION FOR TSCs IN THE RELATIONAL MODEL

As was discussed in the previous section, our starting point for modeling temporal data was independent of any specific data model. In trying to represent these independent concepts in a specific model, such as the relational model, the first question is whether the data structures of the target model are powerful enough to represent them. In the case of the relational model this question translates to "can we represent *TSCs* as relations?"

3.1. The Need for Temporal Relations

Certainly, one can choose some tabular representation for the time sequences. Many examples of that exist in the literature, as will be discussed later. The issue is whether it is sufficient for the chosen tabular representation to have the properties of regular relations as provided by the relational model, or do we need additional modeling power. We claim that the main concepts of a *TSC*, which include *type*, *time granularity*, *life span*, and *interpolation function*, are meta data concepts that have to be added to the relational model. We advocate the extension of the relational model to include *temporal relations* that have the above semantic properties associated with them. Thus, we would expect the extended relational system to provide interpolation for values that are not explicitly stored according to the type of the temporal relation within the granularity and lifespan semantic constraints.

To illustrate the above point consider, for example, the table in Figure 3a. Suppose that a request is made for the salary of E1 at time 11/87. In a relational system the answer would be "null" or "non-existent". However, given that this table is a temporal relation, the system would be expected to use the meta-data associated with the relation and return the value 30k using the information that this is a "step-wise-constant" relation with its implied interpolation function. In the case of Figure 3b, the request for the temperature in SFO at 4AM should result with a value between 67 and 73 because the type of the temporal relation is "continuous" and some 'curve-fitting' interpolation function will be used in that case. We will discuss more precisely the representation of *TSCs* as temporal relations in Section 5, including the implication of the lifespan. We continue here with a discussion of the alternatives for tabular representations.

3.2. Alternatives for Tabular Representation

There are several tabular representations that different authors have chosen. They all fall into one of four categories depending on two factors: 1) whether they deal with time points or time intervals, and 2) whether they chose a Non-First Normal Form (NFNF) or First Normal Form (FNF) representation. (Note that by FNF we include representations that may be in higher normal forms, such as third normal form.) The representation of time intervals can take different notations, such as square

E#	MONTH/YEAR	SALARY
E1	3/87	25k
E1	9/87	30k
E1	4/88	32k
E2	3/87	20k
E2	10/87	22k
E2	5/88	25k
.	.	.
.	.	.
.	.	.

(a)

LOCATION#	HOUR	TEMPERATURE
SFO	1AM	65
SFO	2AM	67
SFO	5AM	73
.	.	.
.	.	.
.	.	.
OAK	2AM	67
OAK	3AM	70
OAK	5AM	76
.	.	.
.	.	.
.	.	.

(b)

Fig. 3: Examples of *TSC* tables

brackets (e.g [3/87, 9/87]), or additional columns in the relations (e.g. labeled "time-start" and "time-end"). The NFNF representation allows for data values to have complex repeating values, permitting a time sequence to be represented as a single complex value. Let us denote these four categories as FP (FNF-points), FI (FNF-intervals), NP (NFNF-points), and NI (NFNF-intervals). The four categories, FP, FI, NP, and NI are shown in Figures 4a, 4b, 4c, and 4d, respectively, for the table of Figure 3a. Note that the generic title TIME is used for time points in category FP, START-TIME and END-TIME are used to denote an interval in category FI. Note also that the time is not denoted explicitly in categories NP and NI (but rather implicitly through the format of the complex values). The repeating values in the NP (point-value pairs) and NI (interval-value pairs) categories appear in consecutive lines for illustrative convenience, but they make up a single complex value for each surrogate.

As mentioned previously, the different representations which appear in the literature fall into one of these four categories. It is worth pointing out that the differences in representation are only in format or labels. [Snodgrass 87] uses both the categories FP and FI. For FP, the temporal column TIME is labeled "(at)", and for FI, the temporal columns START-TIME and END-TIME are labeled "(from)" and "(to)", respectively. [Navathe & Ahmed 87] use the category FI, where the columns START-TIME and END-TIME are labeled T_S and T_E, respectively. Others choose the NFNF representation. We illustrate the differences by showing the format used for the first salary entry of our example. [Clifford & Crocker 87] use the category NP, where the time-value pair notation used is "3/87 - 25k". [Ahn 86] uses the category NI, where the value-interval notation used is "25k [3/87, 9/87)". Note the use of the regular parenthesis at the end of the interval to denote "up

E#	TIME	SALARY
E1	3/87	25k
E1	9/87	30k
E1	4/88	32k
E2	3/87	20k
E2	10/87	22k
E2	5/88	25k
.	.	.
.	.	.
.	.	.

(a) FNF-points

E#	START-TIME	END-TIME	SALARY
E1	3/87	9/87	25k
E1	9/87	4/88	30k
E1	4/88	NOW	32k
E2	3/87	10/87	20k
E2	10/87	5/88	22k
E2	5/88	NOW	25k
.	.	.	.
.	.	.	.
.	.	.	.

(b) FNF-intervals

E#	SALARY
E1	(3/87, 25k)
	(9/87, 30k)
	(4/88, 32k)
E2	(3/87, 20k)
	(10/87, 22k)
	(5/88, 25k)
.	.
.	.
.	.

(c) NFNF-points

E#	SALARY
E1	(3/87 - 9/87, 25k)
	(9/87 - 4/88, 30k)
	(4/88 - NOW, 32k)
E2	(3/87 - 10/87, 20k)
	(10/87 - 5/88, 22k)
	(5/88 - NOW, 25k)
.	.
.	.
.	.

(d) NFNF-intervals

Fig. 4: Different representations of a *TSC*.

to, but not including 9/87". [Tansel 86] also fall into the NI category and uses a similar notation "{<[3/87, 9/87), 25k>}". Finally, [Gadia & Yeung 88] also use the NI category, except that the interval non-overlap is made explicit by having the next interval start at one time granularity unit after the end of the last interval. Thus for the example above, two consecutive intervals would look like "[3/87, 8/87] 25k" and "[9/87, 4/88] 30k".

The reason for the detailed discussion of the different representations which appeared in the literature is to point out that the notation and labeling is rather arbitrary, and that we need only be concerned with the four categories mentioned above. It may seem unimportant at first glance which category is selected, but there are advantages and disadvantages to each. We will discuss those in section 3.4 below, and will explain the reasons for our choice of the FP category for representing *TSCs*. Before doing so, we bring up additional representation problems.

3.3. Other representation issues

Another issue that effects the representation is how to put multiple attributes (representing, for example, salary history and manager history) together into the same relation. For FNF structures, this means having multiple temporal attributes match in time across the entire tuple. That is, every tuple has only a single time point (or interval) associated with all the attribute values of the tuple. This issue brings up the need to define a *temporal normal form*, i.e. some guidance as to when it is reasonable to combine multiple attributes into one relation and what are the conditions for doing so. This is the topic of section 4. Note that this issue is trivially avoided in NFNF representations, since an entire time sequence is considered a single complex value. Thus, multiple complex values (time sequences) can be put together into a single relation.

An interesting observation to make is that some authors have found it useful to include in their representation the lifespan associated with each surrogate [e.g. Clifford & Crocker 87, Gadia & Yeung 88]. [Clifford & Crocker 87] represent this as an additional column of the relation, called "lifespan", and [Gadia & Yeung 88] view the surrogate itself as a temporal attribute. The idea is that the temporal values associated with each temporal attribute should exist within the lifespan of that surrogate. For example, a lifespan can be associated with the employee surrogate (say "name"), then the integrity constraint can be made that salary history values (and any other temporal values) should exist only for times within that lifespan. We consider this capability important, and therefore discuss below how it would be represented in the context of a *TSC* model.

In our model a lifespan is associated with the entire *TSC* rather than each surrogate. In the context of the *TSC* model, we see the lifespan associated with each surrogate as a temporal property of the surrogate regarding its existence. Thus, a *TSC* representing the existence of the surrogate can be defined for that purpose. This view is closer to the lifespan column proposed by [clifford & Crocker 87]. Thus, the integrity constraint condition mentioned above (e.g. that salaries can exist only for times that the employee exists in the database) can be treated as any other temporal integrity constraint between attribute. For example, the temporal integrity constraint that an employee's manager can not exist for those times that a person is not assigned a department is just as valid. The advantage we see in our approach is that the existence of a surrogate is viewed as a property of the surrogate and need be expressed only once (in the corresponding *TSC*). Considering the surrogate as a temporal attribute, or associating the existence with the relation brings about the problem that when surrogates are used in different relations they must have the same existence values. This places a burden on the system to check that the conditions are met. Also, the existence conditions will have to be repeated in all the relations that the surrogate is used, which can be avoided when the existence is treated as a property of the surrogate as described above.

3.4. Reasons for selecting the FP category

We first consider the issue of FNF vs. NFNF. The main advantage of NFNF is that a time sequence can be represented directly as a single (complex) value. It is easy to visualize this encapsulated form. Multiple (temporal and non-temporal) attributes of each surrogate are simply considered a tuple of the NFNF relation. Of course, this representation requires a major extension to the relational language with the capability to express conditions that apply to the internal structure of the complex values.

In spite of these advantages, we have chosen to represent a *TSC* as a FNF relation mainly because of our desire to stay as close as possible to the relational model. There are three compelling reasons. First, the relational model is familiar and there exist many practical commercially available systems that support it. Second, the tabular representation where the values are simple (not complex) is straightforward to understand. This simplicity is in our view one of the main reasons for the popularity of the relational model and we wish to preserve it. Third, we would like to preserve the *TSC* paradigm as well as operations over *TSC*s (such as the COMPOSE operation discussed in the previous section), which is easier to do if each *TSC* is represented as a single temporal relation.

This choice brings about the problem of dealing with multiple temporal attributes. If each temporal attribute (e.g the salary history of all employees), corresponding to a *TSC*, is represented as a separate temporal relation, then a query that involves several temporal attributes of the same surrogate type would require join expressions between these relations. This is tedious and unnecessary. We therefore propose a way for maintaining the advantage of the NFNF view while staying within the FNF framework. We refer to a collection of FNF temporal and non-temporal relations that have the same surrogate type as a *family*. Naturally, it corresponds to the view of considering a collection of *TSCs* for the same surrogate as a *TSC* family. For example, all the *TSCs* that are associated with an employee, such as salary history, management history, and project history, could be considered as a single *TSC* family. We discuss further the concept of a *TSC* family and its implications in Section 6.

The second issue to consider is that of representing time points vs. intervals. Having made the choice of a FNF representation, our discussion is made in that context, although most of the points made apply to NFNF representation as well.

One of the advantages of explicit interval representation for FNF (see Figure 4b) is that each tuple is self contained. In the case of time point representation (Figure 4a) one has to look at two tuples in order to see the range where a value holds true. On the other hand, a time point representation is more suitable for event data, such as recording the time and magnitude of earthquakes, or the number of books sold per day. In general, interval representation is suitable for Step-Wise Constant type data only. This can be illustrated by considering an interval representation for the Continuous type data shown in Figure 3b. The corresponding interval representation is as shown in Figure 5, where the START-TIME and END-TIME values are identical. This is a confusing implicit representation that should be avoided. The same problem exists for Discrete type data. It is worth pointing out that even with Step-Wise Constant type data, the concept of the event exists. For example, in salary history data, the beginning of each interval corresponds to the event of a raise. Thus, events in interval representations have to be associated with the beginning or the end of intervals.

LOCATION#	START-TIME	END-TIME	TEMPERATURE
SFO	1AM	1AM	65
SFO	2AM	2AM	67
SFO	3AM	3AM	73
.	.	.	.
.	.	.	.
.	.	.	.
OAK	2AM	2AM	67
OAK	3AM	3AM	70
OAK	5AM	5AM	76
.	.	.	.
.	.	.	.
.	.	.	.

Fig. 5: Interval representation of a continuous *TSC*

For the sake of uniformity, in order to use a single representation for all types of time sequences, we prefer the time point representation. However, there is another advantage to this representation. One obvious property of time sequences is that each time point of a given surrogate can have only a single value. For example, we would not want to permit two different salary values for some employee to co-exist at the same time. This condition translates in the interval representation to saying that intervals cannot overlap for the a given surrogate. In order to enforce this condition, the system would have to check for overlaps. The point representation provides this capability in a trivial manner simply by declaring the time along with the surrogate as the key of the temporal relation. This point is discussed further in Section 5.

Even though we prefer the time point representation, we consider the concept of intervals for temporal databases important for the query language. When used in the query language, the interval has a more general meaning than "the period of time where the value is the same" as implied by an interval representation. The interval is a continuous period of time regardless whether the values associated with it are the same, or completely independent of values. For example, one can describe an interval as the "period of time where the temperature was greater than 50". This example also shows that the concept of a sequence of intervals is also useful (i.e. the sequence of intervals when the temperature was greater than 50). Intervals are also needed for expressing temporal conditions between temporal relations. For example, we may want to find the salary of some employee *while* he/she was in the a certain department. We observe that if the concept of an interval exists in the query language, it is not essential to explicitly represent it in the temporal relation.

4. TEMPORAL NORMAL FORMS

[Navathe & Ahmed 86] defined the concept of Time Normal Form (TNF). A relation is not in

TNF if, for at least one of its temporal attributes, there exist two tuples or more with consecutive† time intervals and equal data values. We will refer to such values as *duplicate values*. For example, the relation‡ in Figure 6 is not in TNF because duplicate values in consecutive intervals appear in the CITY and HOTEL attributes (duplicates in either one of them would have been sufficient to cause the relation to be non-TNF). Similarly, the relation in Figure 7 is not in TNF because of duplicate values in the SALARY and DEPARTMENT attributes. There are two problems with the above concept of temporal normal form. The first problem is that semantically the above definition is indiscriminate. For the SAL-DEPT relation of Figure 7, it makes sense (intuitively) to split it into two relations, one for salary and the other for department in order to get TNF relations. In contrast, the EXPENSE relation of Figure 6 cannot be split for semantic reasons; the COST value is dependent on the combination HOTEL, CITY, T_s, T_e. The last two tuples may represent the fact that the room rate at the San Francisco Hilton went up during the stay of E1. Moreover, we can lose useful information even if we merge intervals of a single temporal attribute. For example, the last two tuples of Figure 7 may imply

E#	HOTEL	CITY	COST	T_s	T_e
E1	HILTON	LA	950	1	5
E1	SHERATON	LA	900	6	12
E1	SHERATON	NYC	400	13	16
E1	HILTON	SF	300	17	18
E1	HILTON	SF	350	19	20

Fig. 6: The EXPENSE Relation

E#	SALARY	DEPARTMENT	T_s	T_e
E1	20K	ACCOUNTING	1	4
E1	20K	FINANCE	5	8
E1	25K	FINANCE	9	10
E1	25K	MARKETING	11	20
E1	25K	MARKETING	21	25

Fig. 7: The SAL-DEPT Relation

† Two intervals $[a,b]$ and $[c,d]$ are consecutive if $c-b=1$ (time unit) or $a-d=1$. Some papers use the term 'adjacent', but we think that given the directional nature of time 'consecutive' is more descriptive.

‡ In this section, we use the notation of [Navathe & Ahmed 86] for representing time intervals. T_s and T_e are the start and end times respectively.

that at time 21 there was an event where the employee was denied a raise. In general, permitting duplicate values in two or more consecutive event points† is a desirable property. This is quite obvious in scientific and statistical data, where consecutive measurements or statistics have the same value. Since the removal of duplicate values may cause the loss of semantic information it should be left as an option to the database designer.

A second problem with the TNF concept is that it disallows non-TNF relations that result from join operations. We believe that a data model and a language should allow multi-step queries; this implies that the result of a query has to be represented in the same construct as the one used by the query. Consequently, one cannot disallow non-TNF relations, as it is likely that joined relations will be in non-TNF even if the original relations were in TNF. Permitting non-TNF relations is consistent with traditional relational databases where non-normal form relations are allowed. A consistency with the relational model also requires that the retrieval power of the query language will apply to a relation regardless of its normal form. A syntax that was designed for TNF relations only, can generate logically incorrect answers. An example for such a case is using the syntax of [Navathe & Ahmed 87] to answer the query "get the times that employee E1 started working for the finance department" (applied to the SAL-DEPT relation of Figure 7); the result will be 5 and 9 rather than the correct answer, 5. This last point implies that the temporal query language should be able to handle duplicate values correctly (we elaborate on this point below). The foregoing discussion does not invalidate the definition of TNF, but present reasons why non-TNF relations have to be allowed and taken care of by the query language. TNF is a desirable property because putting together time sequence collections with different event points into the same relation can cause a proliferation of duplicate values. This is because a separate tuple will be needed for each event point of each attribute, and all the information for the other attributes have to be duplicated. On the other hand, there are situations, as was explained above, where relations with multiple attributes (for attributes that have semantic binding) are desirable. Thus, TNF relations should not be enforced.

In this paper we propose a definition of a temporal normal form (not related to Navthe & Ahmed's TNF) which is required rather than desirable. That temporal normal form is analogous to 1NF of the relational model and we refer to it as 1TNF. 1TNF means that for a given instance of surrogate and time point, each attribute has a single value (possibly null); in other words, a time-slice at point t has to result in a standard 1NF-relation. The relation in Figure 8 is not in 1TNF since time points 9 and 10 contain two salary values each. If time intervals are used to represent temporal relations, then the formal definition of 1TNF is as follows:

Definition 1: a relation with a schema $R(S, A_1, \cdots, A_n, T_s, T_e)$ is in 1TNF if there do not exist two tuples $r^1(s^1, a_1^1, \cdots, a_n^1, t_s^1, t_e^1)$ and $r^2(s^2, a_1^2, \cdots, a_n^2, t_s^2, t_e^2)$ such that $s^1 = s^2$ and the intervals $[t_s^1, t_e^1]$ and $[t_s^2, t_e^2]$ intersect.

† Recall that we use the term event point to mean a time point with a recorded (rather than interpolated) attribute value.

E#	SALARY	T_s	T_e
E1	20K	1	5
E1	25K	6	10
E1	27K	9	11

Fig. 8: The SAL Relation

Note that the above definition does not allow the last two tuples of Figure 8 to co-exist. Furthermore, it even disallows tuples with the same values to have intersecting intervals, such as the two tuples (E1, 25K, 6, 10) and (E1, 25K, 9, 11). In an actual implementation, one may choose not to reject tuples of the same surrogate with intersecting intervals, provided that $a_i^1 = a_i^2$ for all $i \in \{1, \cdots, n\}$. In this case the time intervals are unioned resulting in a single tuple. For example, the two SAL tuples above will result in the single tuple (E1, 25K, 6, 11).

In our model, we use time points, and 1TNF is defined as follows:

Definition 2: A temporal relation with a schema $R(S, T, A_1, \cdots, A_n)$ is in 1TNF if there do not exist two tuples $r^1(s^1, t^1, a_1^1, \cdots, a_n^1)$ and $r^2(s^2, t^2, a_1^2, \cdots, a_n^2)$ such that $s^1 = s^2$ and $t^1 = t^2$.

As mentioned in the previous section, enforcing 1TNF in the case of time points is as simple as enforcing a key constraint (recall that the key of the temporal relation is $<S,T>$.)

4.1. Query Language Implications

The discussion above argues in favor of permitting duplicate values. Consequently, it is necessary to introduce the concept of a *change point* for a non-TNF relation. Given an attribute of a temporal relation, the change points are defined relative to that attribute as the event points where the value of the attribute is different from the previous consecutive value. In the example of Figure 9, the first and third tuples represent change points for SALARY, while all three tuples represent change points for DEPARTMENT. In general, change points are a subset of event points. It follows that the syntax of the query language should enable reference to the change points of one or more of the attributes of a temporal relation. We will not introduce here a specific syntax, but rather use the example of Figure 9 to illustrate the required functionality. We use time intervals in the figure, but the required

E#	SALARY	MANAGER	DEPARTMENT	T_s	T_e
E1	20K	SMITH	ACCOUNTING	1	10
E1	20K	SMITH	FINANCE	11	20
E1	25K	SMITH	MARKETING	21	30

Fig. 9: The S-M-D Relation

syntactical properties are the same for both time-points and time-intervals representations. The representation of *TSCs* with time point relations will be discussed in the next section.

Consider the following three queries and their answers.

Q1: Find the points where the salary of employee E1 changed; list the salary values at these points.
A1: (20K,1); (25K,21)

Q2: Find the employees who did not have a change of either salary or manager during the time interval 1-20; list the tuple data during that period.
A2: (E1,20K,SMITH,ACCOUNTING,1,10); (E1,20K,SMITH,FINANCE,11,20).

Q3: Find the employees and time points where there was a change of salary or manager or department.
A3: (E1,1) (E1,11) (E1,21).

The above three queries illustrate the need to refer to time intervals (or points) associated with different combinations of temporal attributes. Query Q1 cannot use the tuple time interval, and need to associate time intervals with only the SALARY attribute. Query Q2 needs to associate time intervals with the combination of SALARY and MANAGER attributes. Q3 is a query that can utilize the tuple-interval syntax (such as TSQL of [Navathe & Ahmed 87]).

5. RELATIONAL REPRESENTATION OF A *TSC*

In the sequence-based temporal model, a *TSC* is a collection of time sequences with the same surrogate type, where each point is represented by the triplet $<S,T,A>$. We represent a *TSC* by a temporal relation† with a schema $R(S,T,A)$; this is illustrated in Figure 10. The key of the relation is $<S,T>$, and the default ordering is primary by surrogate and secondary by time. What makes the relation temporal is the meta-data associated with it. The lifespan meta-data is used in interpolating values at the start and end of the relation. As an example, consider the case of a single surrogate instance, say S_1, and a stepwise constant sequence. Let T_{start} and T_{end} define the lifespan of the relation and $A(t)$ be the attribute value at time t. The event points for the surrogate instance are $\{t_i\}$, $i = 1, \cdots, m$, where the event data values $a_i = A(t_i)$. Then the following conditions should hold.

For the start of the relation:

$t_1 \geq T_{start}$

$a_1 \neq NULL$

If $t_1 > T_{start}$, then $A(t) = NULL$ for $t < t_1$

---

† Whenever unambiguous we will use 'relation' to mean 'temporal relation'.

S	T	A
S_1	t_1^1	a_1^1
S_1	t_2^1	a_2^1
.	.	.
.	.	.
.	.	.
S_2	t_1^2	a_1^2
S_2	t_2^2	a_2^2
.	.	.
.	.	.
.	.	.

Fig. 10: A Temporal Relation

For the end of the relation:

$t_m \leq T_{end}$

If $t_m < T_{end}$, then $A(t) = a_m$ for $t > t_m$

The above definitions apply only to temporal relation of the type ''stepwise constant''. Similar definitions can be made for the ''discrete'' and ''continuous'' types.

Since we allow duplicate values in consecutive time points, we introduce the notion of *reduction*. A reduction of a relation means the removal (for a surrogate instance) of consecutive duplicate values (which results in the sequence of change points only.) Each type of a *TSC* can be reduced or unreduced. One should distinguish between compression (which may be done at the physical storage level) and reduction. In the case of compression, no information is lost and the process is reversible, that is, the original data can be reproduced. In the case of reduction, however, information is lost because of the distinction between event values and interpolated values. Figure 11 shows a time sequence and its reduced version. The figure shows the event values only (they are represented by the height of the vertical bars), the values at the other time points are generated by interpolation when necessary.

The foregoing discussion has been limited to a single temporal attribute. In the case of a stepwise constant sequence, placing all the temporal attributes of a given surrogate in a single relation amounts to unioning the event points and creating a tuple for each resulting time point. One can think of this as a special join operation, which we call *event join.* If a resulting time point is not an event point for some attribute, the attribute's value is derived by interpolation. In this case, the schema of the temporal relation is $R(S, T, \bar{A})$, where $\bar{A} = \{A_1, \cdots, A_n\}$. To simplify the discussion we will assume that all the attributes in $\bar{A}$ are temporal. For example, Figure 12 shows two temporal sequences and their representation in a single temporal relation (assume that the lifespan of the sequences is $[t_1, t_9]$).

Multiple temporal attributes can be present in the original relations or arise as a result of an event-join or other temporal operations (e.g. the COMPOSE operation mentioned in Section 2.) In the original relations, they usually correspond to storing multiple independent time sequences in a single

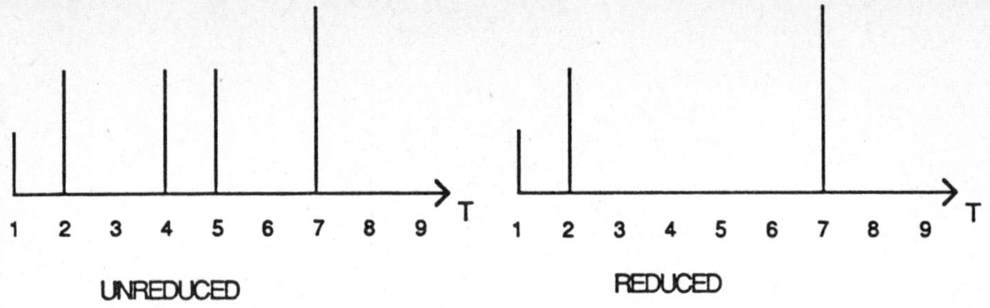

Fig. 11: A Reduction of a Time Sequence

relation, but sometimes there is an attribute that binds the time sequences together, e.g., the COST attribute in Figure 7. In both cases, the rules for $R(S, T, \overline{A})$ are basically the same as for $R(S, T, A)$

When multiple temporal attributes arise as the result of temporal operations, the information associated with duplicate values can be lost. To illustrate this point, consider Figure 13 where two single-attribute relations (a) are used to form a single relation (b). We assume here that the lifespans of the two original relations are the same; otherwise, null values have to be used in the relation of Figure 13b. If one would like to project the two original relations from the result relation, original consecutive duplicate values have to be identified. In Figure 13b, we used * to mark the fact that these values were stored (rather than interpolated) in the original relation. Note that there is no need to mark the first duplicate value since a change (in value) point must have been stored in the original relation.

6. *TSC* FAMILIES

A disadvantage of our temporal data model as presented in [Segev & Shoshani 87] is the number of *TSC* s generated for a single surrogate when it has many temporal attributes. This problem is carried over to a normalized relational representation of a *TSC*. In the case of traditional relational databases, the problem is similar but its scope is much smaller since normalization does not require the splitting of the entities' attributes (except for foreign key) into separate relations, while in temporal relations it is desirable to split all attribute into separate temporal relations. Consequently, a mechanism to view all the data of a surrogate as a single unit will be quite useful. We introduce the family construct for that purpose.

A family is a virtual complex object that enables a reference to all the data of a given surrogate. We illustrate it through the following example. Assume that the non-temporal attributes of an employee are E#, NAME, and ADDRESS, and the temporal attributes are SALARY, MANAGER, and DEPARTMENT. The schema of the resulting relations are: EMP(E#,NAME,ADDRESS); SAL(E#,T,SALARY); MGR(E#,T,MANAGER); DEPT(E#,T,DEPARTMENT). A family (say

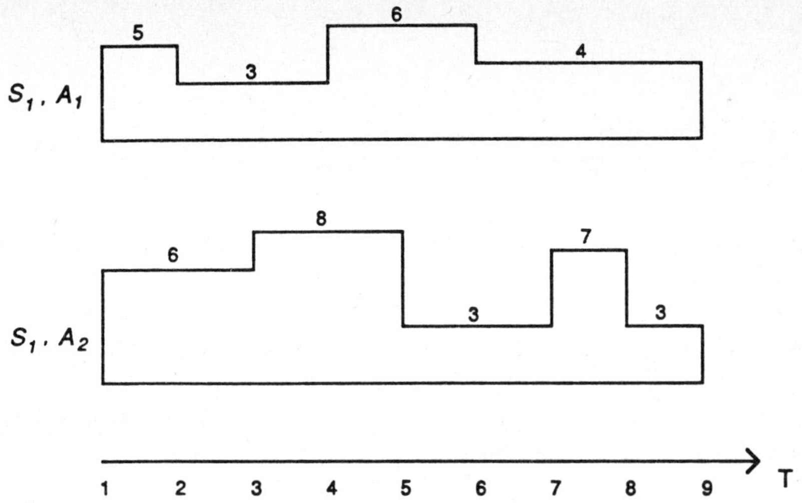

S	T	A_1	A_2
S_1	t_1	5	6
S_1	t_2	3	6
S_1	t_3	3	8
S_1	t_4	6	8
S_1	t_5	6	3
S_1	t_6	4	3
S_1	t_7	4	7
S_1	t_8	4	3

Fig. 12: A Temporal Relation with Multiple Attributes

S	T	A_1
S_1	1	1
S_1	2	5
S_1	4	10
S_1	7	10
S_1	8	10

S	T	A_2
S_1	1	2
S_1	3	4
S_1	5	5
S_1	6	7

(a)

S	T	A_1	A_2
S_1	1	1	2
S_1	2	5	2
S_1	3	5	4
S_1	4	10	4
S_1	5	10	5
S_1	6	10	7
S_1	7	10*	7
S_1	8	10*	7

(b)

Fig. 13: The Effect of Duplicate Values

EMP_F) of the employee relations is the object

EMP_F { EMP(E#,NAME,ADDRESS)
    SAL(E#,T,SALARY)
    MGR(E#,T,MANAGER)
    DEPT(E#,T,DEPARTMENT) }

Note that all the relations of a family must have the same surrogate. In principle, referencing a family should have the equivalent selective power of referencing the individual *TSCs* in the family. But, in addition, two more benefits should be provided: 1) it should provide a concise way of specifying only once conditions that apply to multiple *TSCs* in the family, and 2) it should provide a convenient way of specifying conditions between *TSCs*. The former may involve some surrogate conditions that apply to all *TSCs* (e.g. employee number between 100 to 200) or some temporal conditions (e.g. events that occurred in 1987 only). The latter includes interactions between *TSCs* (e.g. find the salary of Jones *while* his manager was Smith).

Another feature that should be available in referencing a *TSC* family, is the specification of what output is desirable once the selection of the surrogates of the family has been made. This is equivalent to a "project" operation, except that we need the capability of specifying which parts of each temporal attribute we wish to get in the output.

The above principles seem reasonable, but there are many details that have to be developed in future work. It is worth pointing out that a NFNF representation of *TSCs* would not eliminate the complexities needed to provided the features above; it only moves them to a level of dealing with complex values (representing individual time sequences). We would still need ways for specifying conditions to the complex values, conditions that apply between complex values, and the specification of output of parts or entire complex values.

7. CONCLUSIONS

Our approach to modeling temporal data differs from most approaches which suggest extensions to existing models. We have developed a model that is independent of existing models. Its development concentrated on the special properties and manipulation requirements of temporal data without being constrained by the modeling capabilities of existing models. In this paper, we explored the implications of representing our model and its requirements in the relational model. We concluded that a special temporal relation needs to be added to the relational model in order to support the different properties of the temporal data. We also considered the options for representing our model's structure in the relational model, and explained why representing time points is preferable to representing intervals. Another conclusion reached was that we have a pragmatic preference for structures that are in First Normal Form rather than Non-First Normal Form. We also considered the issue of a temporal normal form and its properties. We defined a temporal normal form and showed how it differs from another definition proposed in the literature.

In future work, we need to investigate the integration of traditional query capabilities of the relational model with the query capabilities of our temporal model. Since there are two kinds of relations, i.e. temporal and non-temporal, certain capabilities apply to one or the other or both. We need to explore how to integrate these capabilities into a single powerful query facility.

REFERENCES

[Adiba & Quang 86]

Adiba, M., Quang, N.B., Historical Multi-Media Databases, *Proceedings of the International Conference on Very Large Data Bases (VLDB)*, 1986, pp. 63-70.

[Ahn 86]

Ahn I., Towards An Implementation of Database Management Systems with Temporal Support, *Proceedings of the Third International Conference on Data Engineering*, pp. 374-381. February, 1986.

[Ariav et al 84]

Ariav, G., Beller, A., Morgan, H., A Temporal Data Model, Technical Report, New York University, December, 1984.

[Ariav 86]

Ariav, G., A Temporally Oriented Data Model, *ACM Transactions on Database Systems*, 11, 4(Dec. 1986), pp. 499-527.

[Bolour et al 82]

Bolour, A., Anderson, T.L., Dekeyser, L.J., Wong, H.K.T., The Role of Time in Information Processing: A Survey, *ACM-SIGMOD Record*, 12, 3, 1982, pp. 27-50.

[Clifford & Tansel 85]

Clifford, J., Tansel, A., On an Algebra for Historical Relational Databases: Two Views, *Proceedings of the ACM SIGMOD International Conference on Management of Data*, May 1985, pp. 247-265.

[Clifford & Croker 87]

Clifford, J., Croker, A., The Historical Relational Data Model (HRDM) and Algebra Based on Life Spans, *Proceedings of the Third International Conference on Data Engineering*, pp. 528-537. February, 1987.

[Gadia 86]

Gadia, S.K., Toward a Multihomogeneous Model for a Temporal Database, *Proceedings of the International Conference on Data Engineering*, 1986, pp. 390-397.

[Gadia & Yeung 88]

Gadia, S.K., Yeung C-S., A Generalized Model for A Relational Temporal Database, *Proceedings of the ACM SIGMOD International Conference on Management of Data*, June 1988, pp. 251-259. 1986, pp. 390-397.

[Gunadhi & Segev 88]

Gunadhi H., Segev A., Physical Design of Temporal Databases, *Lawrence Berkeley Lab* Technical Report LBL-24578, 1988.

[Klopproge 81]

Klopproge, M.R., TERM: An Approach to Include the Time Dimension in the Entity-Relationship Model, *Proceedings of the Second International Conference on E-R Approach*, 1981, pp. 477-512.

[Lum et al 84]

Lum, V., Dadam, P., Erbe, R., Guenauer, J., Pistor, P., Walch, G., Werner, H., Woodfill, J., Designing Dbms Support for the Temporal Dimension, *Proceedings of the ACM SIGMOD International Conference on Management of Data,* June 1984, pp. 115-130. March, 1986.

[McKenzie & Snodgrass 87]

McKenzie, E., Snodgrass R., Extending the Relational Algebra to Support Transaction Time, *Proceedings of the ACM SIGMOD International Conference on Management of Data,* May 1987, pp. 467-478.

[Navathe & Ahmed 86]

Navathe, S.B., Ahmed, R., A Temporal Relational Model and a Query Language, *UF-CIS Tech. Report TR-85-16, Univ of Florida,* April 1986.

[Navathe & Ahmed 87]

Navathe, S.B., Ahmed, R., TSQL- A Language Interface for History Data Bases, *Proceedings of Temporal Aspects of Information Systems,* North-Holland, May 1987, pp. 113-128.

[Rotem & Segev 87]

Rotem, D., Segev, A., Physical Organization of Temporal Data, *Proceedings of the Third International Conference on Data Engineering,* pp. 547-553. February, 1987.

[Segev & Shoshani 87]

Segev, A., Shoshani, A., Logical Modeling of Temporal Databases, *Proceedings of the ACM SIGMOD International Conference on Management of Data,* May 1987, pp. 454-466.

[Segev & Shoshani 88]

Segev, A., Shoshani, A., Modeling Temporal Semantics, in *Temporal Aspects in Information Systems,* Edited by C. Rolland, F. Bodart, and M. Leonard, North-Holland, 1988, pp. 47-58.

[Shoshani & Kawagoe 86]

Shoshani, A., Kawagoe, K., Temporal Data Management, *Proceedings of the International Conference on Very Large Databases,* August 1986, pp. 79-88.

[Smith & Smith 77]

Smith, J. M., and Smith, D. C. P., Database Abstractions: Aggregation and Generalization, *ACM TODS 2,* 2, June 1977.

[Snodgrass 84]

Snodgrass, R., The Temporal Query Language TQuel, *Proceedings of the Third ACM SIGMOD Symposium on Principles of Database Systems (PODS),* Waterloo, Canada, April 1984, pp. 204-213.

[Snodgrass & Ahn 85]

Snodgrass, R., Ahn, I., A Taxonomy of Time in Databases, *Proceedings of the ACM SIGMOD International Conference on Management of Data,* May 1985, pp. 236-246.

[Snodgrass 87]

Snodgrass, R., The Temporal Query Language TQuel *ACM Transactions on Database Systems,* 12, 2, June 1987, pp. 247-298.

[Studer 86]

Studer, R., Modeling Time Aspects of Information Systems, *Proceedings of the International Conference on Data Engineering,* 1986, pp. 364-373.

[Tansel 86]

Tansel, A.U., Adding Time Dimension to Relational Model and Extending Relational Algebra, *Information Systems*, 11, 4 (1986), pp. 343-355. *ACM Transactions on Database Systems,* 12, 2, June 1987, pp. 247-298.

KNOWLEDGE BASED SUPPORT FOR THE MANAGEMENT OF STATISTICAL DATABASES

Knut M. Wittkowski

Department of Medical Biometry, Eberhard-Karls-University
Westbahnhofstr. 55, D-7400 Tübingen, Fed. Rep. Germany

Requirements for database management systems (DBMS) are different for commercial and statistical applications. Some operations that are meaningful for commercial applications may be meaningless for statistical applications. Knowledge has to be taken into account to determine, which operations are semantically meaningful and which relations contain sufficient information for statistical analyses. The relevant knowledge is classified into six layers with different levels of abstraction. In the present paper, we concentrate on the intermediate SEMANTICS and STRATEGY layers, where the lattice structure of observational units (design knowledge) and relevant features of observational units and attributes (model knowledge) are represented. Although this type of knowledge has hardly ever been considered, it is demonstrated that a DBMS cannot be used for statistical applications, unless these knowledge bases are available.

INTRODUCTION

The term 'statistical database' refers sometimes to databases for statistical analysis and sometimes to databases of statistical summaries. In the present paper we will use the term predominantly in the former sense, although some problem concerning summary data will also be considered. In contrast to commercial applications, where both activity (changing the data) and processing (accessing and using the data) concentrate on observational units (cases), management of statistical databases is characterized by looking for relationships between attributes (variables) and by creating new attributes as combinations of others. We are also faced with a wealth of attributes which are observed in different observational units, which may be nested, (animal or person < farm or family < town < country), or crossed (time, exposition to certain risks, history of diseases etc.). Thus the data structures are more complicated than in commercial applications. On the other hand, few changes are typically applied to individual cases except for the purpose of data cleaning (CUBITT, WESTLAKE 1987b), so that concurrent access - a major problem in commercial applications - can be neglected. A summary of different requirements in commercial and statistical applications is given in Fig. 1

	Commercial	Statistical
Processing	replicative	innovative
User interface	batch	interactive
Processing	central	local
Updating	frequent/concurrent	very few
Missing/uncertain data	very few	frequent
Storage structures	fixed	varible
Data structure	scalars	vectors and matrices
Data compression	simple summaries	complex alrorithms
Meta-data	often not necessary	important

Fig 1: Requirements for database management systems (DBMS).

Knowledge based systems with applications to statistics differ from 'classic' expert systems in that different types of knowledge are to be handled: First the system will need some knowledge how to organize and optimize accessing the data. We have complicated data structures so that the knowledge about the structure within observational units has to be handled. We are further dealing with abstract models of the reality, so that the relevant features of attributes and observational units have to be represented. Knowledge about the domain (e.g. agriculture, medicine) might also help. Finally we have different types of users so that the system should have some knowledge on the user and the analysis path is often complicated so that the system might need to audit all actions.

The paper is organized as follows. We will first give a short data set that will be used as an example throughout the paper. We will then discuss the different layers starting at the bottom. Because approaches to knowledge based DBMS have so far concentrated on the bottom and the top of the hierarchy given in Fig. 2, we will concentrate on the new results concerning the SEMANTICS and STRATEGY layer.

U S E R

	INTERACTION	user-	
	AUDIT	dialogue-	
	APPLICATION	domain-	
LAYER:	STRATEGY	model-	knowledge
	SEMANTICS	design-	
	ACCESS	data-	
	EXECUTION		

Fig. 2: Layers of a knowledge based database management system

AN EXAMPLE

Consider a medical experiment (see Wittkowski 1985 for details), where a bypass operation is performed on 10 cardiac infarct patients either conventionally or with a laser (OPERTYPE = STEEL or LASER) and each patient is given four doses of an adjuvant medication of propranolol (BETADOSE). Thus the design is partially hierarchical with patients nested within OPERTYPE and crossed with BETADOSE. Bodyweight (BODYWGHT) was measured at entry in the study, vigor (ERGOMETR), subjective ranking of well-being (SUBJRANK), and type of heart failure (RHTHMERR) for each dose of propranolol. For statistical applications data is often represented by a relational model and processed by the associated relational algebra (CODD 1970). In the present paper we will use the relations given in Fig. 3 to demonstrate the need of extending the relational model by meta-data and to identify different layers of relevant knowledge.

OPERTYPE

OPER ID	OPER NAME
1	LASER
2	STEEL

ADJUVANT

TRTMT ID	BETA DOSE
1	10
2	20
3	50
4	100

PATBYDOS

PAT ID	TRTMT ID	ERGO METR	SUBJ RANK	RHTHM ERR
1	1	172	1	0
1	2	190	?	1
1	3	195	2	0
1	4	180	?	0
2	1	156	1	2
2	2	145	2	0
2	3	?	?	?
2	4	142	3	1
:	:	:	:	:
9	1	148	1	?
9	2	155	2	0
9	3	143	3	1
9	4	?	?	?
10	1	?	?	?
10	2	161	1	2
10	3	159	?	1
10	4	154	2	2

PATIENT

PAT ID	PAT NAME	BODY WGHT
1	MÜLLER M	74
2	DIETZ K	76
3	HOFMANN	52
4	MAU J	73
5	SCHULZ E	75
6	MEIER ER	70
7	SCHENZLE	72
8	WITTKOWS	69
9	SELBMANN	71
10	MÜLLER M	68

RANDOMIZ

OPER ID	PAT INGRP	PAT ID
1	1	1
1	2	2
1	3	3
1	4	4
1	5	5
2	1	6
2	2	7
2	3	8
2	4	9
2	5	10

Fig. 3: Relations used for numerical examples

EXECUTION AND ACCESS LAYER

At the bottom EXECUTION layer, we have the commercially available statistical analysis systems. We will use SAS as an example. The problems or features presented, however, are similar for most other systems.

In statistical applications we are often faced with a multitude of derived attributes, which may be results of numeric, boolean, statistical (c.f. FORTUNATO et al. 1986; GHOSH 1987), or time functions (c.f. BLUM

1982; BASSIOUNI, LLEWELLYN 1988). Corrections should be propagated to all derived attributes in a way similar to spread-sheet editors. On the other hand, only observed data may be edited. Whether or not derived attributes are permanently stored in the data base or computed while referenced is primarily a matter of efficiency (c.f. OLKEN 1986). It is also of little importance, whether statistical computations are performed by the DBMS (GHOSH 1988) or by external analysis systems. All requests for updating, aggregating, and retrieving data are handled at the ACCESS level. The data knowledge base contains the description of the network of causal links between attributes (see BLUM 1982 for an example).

SEMANTICS LAYER

Non-observational relations

The relational algebra offers propositions (carthesian product, union, intersection, difference) and predicates (restriction, projection, division and join). Available software, dominated by the needs of commercial data processing, often leads to erroneous or semantically meaningless results. In Fig. 4, the cases PATIENT (7,SCHENZLE,72) and OPERTYPE(2,STEEL) are disjunct in the lattice structure of observational units (WITTKOWSKI 1985). Thus the 'case' OBS=4 does not conform to the design knowledge. (The MERGE statement is the SAS join operator. SORT statements, which SAS requires prior to MERGE operations, are not shown.) The 'relation' ONEBYONE (Fig. 5a) may be at least useful for descritive purposes. The attributes OBS, OPERNAME, PATNAME, and BETADOSE, for instance, might be used for code transformation. However, relevant dependencies between data and observational units are ignored. OPERNAME=STEEL, for instance, refers semantically to OPERID=2 , although in ONEBYONE it is now in the same row (case) as OPERID=1 . Because these cases have no

```
DATA OPERPAT1;  SET PATIENT;  OPERID=MOD(PATID-1,2)+1;
DATA OPERPAT2;  MERGE OPERTYPE OPERPAT1;
```

OBS	OPER ID	OPER TYPE	PAT ID	PAT NAME	BODY WGHT
1	1	LASER	1	MÜLLER M	74
2	1	LASER	3	HOFMANN	52
3	1	LASER	5	SCHULZ E	75
4	1	LASER	7	SCHENZLE	72
5	1	LASER	9	SELBMANN	71
6	2	STEEL	2	DIETZ K	76
7	2	STEEL	4	MAU J	73
8	2	STEEL	6	MEIER ER	70
9	2	STEEL	8	WITTKOWS	69
10	2	STEEL	10	MÜLLER M	68

Fig. 4 Non-observational relation (Meaningless results are hachured.)

DATA ONEBYONE; MERGE OPERTYPE PATIENT ADJUVANT RANDOMIZ PATBYDOS;

OBS	OPER ID	OPER NAME	PAT ID	PAT NAME	BODY WGHT	TRTMT ID	BETA DOSE	ERGO METR	SUBJ RANK	RHTHM ERR
1	1	LASER	1	MÜLLER M	74	1	10	172	1	0
2	1	STEEL	1	DIETZ K	76	2	20	190	?	1
3	1		1	HOFMANN	52	3	50	195	2	0
4	1		1	MAU J	73	4	100	180	?	0
5	1		2	SCHULZ E	75	1		156	1	2
6	2		2	MEIER ER	70	2		145	2	0
7	2		2	SCHENZLE	72	3		?	?	?
8	2		2	WITTKOWS	69	4		142	3	1
9	2		3	SELBMANN	71	1		193	1	2
10	2		3	MÜLLER M	68	2		193	?	2
11	?		3		?	3		187	2	0
12	?		3		?	4		189	3	1
:	? ?		:		? ?	:		:	:	:

DATA JOINPAT; MERGE RANDOMIZ PATIENT PATBYDOS; BY PATID;
DATA JOINADJU; MERGE JOINPAT ADJUVANT; BY TRMTID;
DATA JOINBYID; MERGE JOINADJU OPERTYPE; BY OPERID;

OBS	OPER ID	OPER NAME	PAT ID	PAT NAME	BODY WGHT	TRTMT ID	BETA DOSE	ERGO METR	SUBJ RANK	RHTHM ERR
1	1	LASER	1	MÜLLER M	74	1	10	172	1	0
2	1	LASER	1	MÜLLER M	74	2	20	190	?	1
3	1	LASER	1	MÜLLER M	74	3	50	195	2	0
4	1	LASER	1	MÜLLER M	74	4	100	180	?	0
5	1	LASER	2	DIETZ K	76	1	10	156	1	2
6	1	LASER	2	DIETZ K	76	2	20	145	2	0
7	1	LASER	2	DIETZ K	76	3	50	?	?	?
8	1	LASER	2	DIETZ K	76	4	100	142	3	1
:	:	:	:	:	:	:	:	:	:	:

Fig. 5 Relations generated by different forms of MERGE statements

PROC TTEST DATA=ONEBYONE;
CLASS OPERID; VAR ERGOMETR;

VARIABLE = ERGOMETR

OPERID	N	MEAN	SD	SEM
1	5	178	15	7
2	4	169	29	14

VARIANCES	T	DF	P
UNEQUAL	0.6516	4.4	0.5476
EQUAL	0.6991	7.0	0.5071

FOR H0: VARIANCES ARE EQUAL
F'= 3.4 DF= 3/ 4 P=0.2651

PROC TTEST DATA=JOINBYID;
CLASS OPERID; VAR ERGOMETR;

VARIABLE = ERGOMETR

OPERID	N	MEAN	SD	SEM
1	18	169	23	5
2	17	168	19	4

VARIANCES	T	DF	P
UNEQUAL	0.1722	32	0.8644
EQUAL	0.1711	33	0.8652

FOR H0: VARIANCES ARE EQUAL
F'= 1.5 DF=17/16 P=0.4164

Fig. 6: Statistical analysis based on the relations ONEBYONE and JOINBYID

meaning, most statistical analyses based on ONEBYONE will be meaningless (Fig. 6a). In a knowledge based DBMS, the lattice structure of observational units (design knowledge) can be used on the one hand to identify whether or not a relation is observational. All non-observational relations should at least be flagged. On the other hand, design knowledge can also be used to provide defaults, so that the simple MERGE statement (Fig. 5a) might contain sufficient information for the system to generate a relation where data from the same oberservational unit is automatically related to the same case.

Dependent and replicated data

The relation JOINBYID is semantically meaningful as far as observational units are concerned. Thus it would be sufficient for most commercial applications. The following examples demonstrate that even relations conforming to the lattice structure of observational units may be useless for statistical applications. The t-test, for instance, is based on the assumption of independent observations. The four data observed from the same patient, however, are obviously not independent. Thus the results given in Fig. 6b are semantically meaningless. As a further example, consider the comparison of the LASER and STEEL group by BODYWGHT using the t-test. Although both relations contain the same information on the bodyweight of the 2·5=10 patients, we get different results (Fig. 7):

```
PROC TTEST DATA=PATIENT;
   CLASS OPERID; VAR BODYWGHT;

VARIABLE = BODYWGHT

OPERID   N  MEAN    SD   SEM
     1   5  70.0  10.1   4.5
     2   5  70.0   1.6   0.7

VARIANCES       T   DF       P
UNEQUAL   0.0000   4.2  1.0000
EQUAL     0.0000   8.0  1.0000

FOR H0:  VARIANCES ARE EQUAL
F'= 41.0  DF= 4/ 4  P=0.0033
```

```
PROC TTEST DATA=JOINBYID;
   CLASS OPERID; VAR BODYWGHT;

VARIABLE = BODYWGHT

OPERID   N  MEAN    SD   SEM
     1  20  70.0   9.3   2.1
     2  20  70.0   1.4   0.3

VARIANCES       T   DF       P
UNEQUAL   0.0000   20  1.0000
EQUAL     0.0000   38  1.0000

FOR H0:  VARIANCES ARE EQUAL
F'= 41.0  DF=19/19  P=0.0001
```

Fig. 7 Effect of join operations on test statistics

Analysis methods cannot determine, whether identical data have been observed from independent observational units or some data has been replicated by the DBMS to complete all cases. Thus relations generated by the join operator may cause serious problems for statistical applications, unless design knowledge (SEMANTICS layer) is considered. Universal relations may be used for the purpose of presentation, but not for statistical analyses, unless a description of the structure underlying the relation is also given to the procedure. Thus the DBMS must keep track of joins to ensure that only appropriate relations are used for analysis.

STRATEGY LAYER

So far we have considered problems that occur if knowledge on the lattice structure of observational units is not taken into account. We will now demonstrate that the decision, whether computation of statistical functions (e.g. POWER(x)AGGREGATE, GHOSH 1988) is meaningful cannot always be based on design knowledge. Thus we proceed to the next level of abstraction and discuss the relevance of knowledge on the model represented in the experimental design, i.e. on the observational units and attributes.

Scale level

The distinction of boolean, numeric and text attributes is often sufficient for commercial, but not for statistical applications. SUBJRANK, for instance, is an ordinal ranking and RHTHMERR is nominally scaled, so that neither the test statistics nor range and interquartil distances (Q3-Q1) in Fig. 8 can be meaningfully interpreted. With RHYTHMERR quantiles and boxplot should also be omitted.

```
PROC UNIVARIATE DATA=PATBYDOS;
   VAR SUBJRANK;

VARIABLE = SUBJRANK

N             26  SUM WGTS       10
MEAN         1.8  SUM          48.0
SD           0.8  VARIANCE      0.6
CV          42.5  SEM           0.1
T:MEAN=0    12.0  P>|T|      0.0001
SGN RANK   175.5  P>|S|      0.0001
NUM <> 0      26

FOR HO: DISTRIBUTION IS NORMAL
W = 0.797         P<W       <0.0100

100% MAX     3.0  RANGE         2.0
 75% Q3      2.3
 50% MED     2.0  Q3-Q1         1.3
 25% Q1      1.0
  0% MIN     1.0  MODE          1.0

S      3.0 000        6   B      |
T      2.7                O      |
E      2.3                X    +---+
M.L    2.0 00000     10   P    #---#
  E    1.7                L    | + |
  A    1.3                O    |   |
  F    1.0 00000     10   T    +---+
```

```
PROC UNIVARIATE DATA=PATBYDOS;
   VAR RHYTMERR;

VARIABLE = RHYTHMERR

N             31  SUM WGTS       31
MEAN         0.8  SUM          26.0
SD           0.8  VARIANCE      0.7
CV          97.8  SEM           0.1
T:MEAN=0     5.7  P>|T|      0.0001
SGN RANK    85.5  P>|S|      0.0001
NUM <> 0      18

FOR HO: DISTRIBUTION IS NORMAL
W = 0.778         P<W       <0.0100

100% MAX     2.0  RANGE         2.0
 75% Q3      2.0
 50% MED     1.0  Q3-Q1         2.0
 25% Q1      0.0
  0% MIN     0.0  MODE          0.0

S      2.0 0000       8   B    +---+
T      1.7                O    |   |
E      1.3                X    |   |
M.L    1.0 00000     10   P    #---#
  E    0.7                L    | + |
  A    0.3                O    |   |
  F    0.0 000000    13   T    +---+
```

Fig. 8 Statistical analysis of numerically coded attributes

Ranges

For commercial applications, data types are typically used only during to check consistency of the entered data. Searching and listing cases the most important processes in commercial application, are not affected by data types. For statistical applications, data types may determine the procedures that can be applied to the data. Because BODYWGHT cannot be less than zero, both the t-test and the signed rank test given in Fig. 9 are meaningless and should be avoided to reduce processing time, paper, and – last but not least – confusion of the user.

```
PROC UNIVARIATE DATA=PATIENT;  VAR BODYWGHT;

VARIABLE = BODYWGHT                         T:MEAN=0  32.4  P>|T|    0.0001
                                            SGN RANK  27.5  P>|S|    0.0059
N           10   SUM WGTS     10            NUM <>0       10
MEAN      70.0   SUM       700.0
SD         6.8   VARIANCE   46.7            FOR H0: DISTRIBUTION IS NORMAL
CV         9.8   SEM         2.2            W = 0.734         P<W   <0.0100
```

Fig. 9 Statistical analysis that contradicts to the data type (range)

Dependent and independent variables

In commercial applications, it is not necessary to distinguish between independent variables (factors and strata) and dependent variables (observations). If the data has to be analyzed with statistical methods, this distiction is very essential. Both analyses in Fig. 10 are meaningless (except for the number of cases), because neither PATID nor PATINGRP are random variables.

```
PROC UNIVARIATE DATA=RANDOMIZ;              PROC TTEST DATA=RANDOMIZ;
   VAR PATID;                                  CLASS OPERID; VAR PATINGRP;

VARIABLE = PATID                            VARIABLE = PATINGRP

N            10   SUM WGTS      10          OPERID   N  MEAN     SD    SEM
MEAN        5.5   SUM         55.0               1   5   3.0    1.6    0.7
SD          3.0   VAR          9.2               2   5   3.0    1.6    0.7
CV         55.0   SEM          0.9
T:MEAN=0    5.7   P>|T|     0.0003          VARIANCES        T    DF      P
SGN RANK   27.5   P>|S|     0.0059          UNEQUAL   0.0000   8.0  1.0000
NUM <> 0     10                             EQUAL     0.0000   8.0  1.0000

FOR H0: DISTRIBUTION IS NORMAL              FOR H0:  VARIANCES ARE EQUAL
W = 0.970         P<W       0.6780          F'=  1.0  DF= 4/ 4  P=1.0000
```

Fig. 10 Statistical analysis of deterministic variables

APPLICATION, AUDIT, AND INTERACTION LEVEL

Domain knowledge might be used e.g. to reduce manual input during acquisition of model knowledge. On of the first approaches to integrating domain knowledge into a data management system was the RX (BLUM 1982) system. Because domain knowledge is relatively independent on the lower knowledge bases, this level will not be discussed in the present paper.

Several approaches to audit the path of a statistical analysis and to draw conclusions have been recently published (see THISTED 1986 for a reference to current approaches). This task might become very important if the user might some day be able to ask for some sub-problems to be analysed by simply telling the system to repeat what has already been done to other attributes or observational units.

Man-computer-interfaces of statistical DBMS can be classified as natural-language, graphical, and form-based systems. Given the inherent limitations and advantages of these dialogue procedures, it is important that different levels of dialogue are handled within dedicated windows using different dialogue techniques. Natural language may be used for explanations, but for other purposes it might be a too clumsy dialogue procedure. Graphics are appropriate for bivarate data and simple tree structures. Where multivariate relations and lattice structures are to be represented, form-based dialogue techniques might be more efficient. These concepts have been discussed in WITTKOWSKI (1988).

SUMMARY

It is often indicated that the relational model should be extended to conform with the needs of statistical analyses (GHOSH 1988). The examples demonstrate that so-called 'relational models' in common DBMS may in some cases even be to rich. Thus it is not sufficient, to extend a DBMS for commercial applications by statistical operators. In some cases, it may even be necessary to restrict the set of operators. In order to check semantically integrity constraints, a DBMS for statistical applications needs knowledge on the lattice structure of observational units (design knowledge) and on the concepts underlying the observational units and attributes (model knowledge). This type of knowledge has so far been not considered, because it is of little importance for commercial applications. For statistical applications, however, it is necessary to avoid semantically meaningless analyses and to assist the user in performing analyses.

REFERENCES

BASSIOUNI MA, LLEWELLYN M (1988) Handling Time in Query Languages. In: RAFANELLI M, KLENSIN J, SVENSSON P (eds. 1988, in press)

BLUM RL (1982) *Discovery and representation of causal relationships from a large time-oriented clinical database: The RX project.* Berlin: Springer

BOARDMAN TJ (ed. 1986) *Computer Science and Statistics.* Washington,DC: ASA

CODD EF (1970) A relational model of data for large shared data banks. *Comm. ACM* **13**:377-387

CUBITT R, COOPER B, OZSOYOGLU G (eds. 1986) *Proceedings of the third international workshop on statistical and scientific database management.* CEPS/INSTEAD, Case Postale 2, L-7201 WALFERDANGE

CUBITT R, WESTLAKE A (eds. 1987a) Report on the third international workshop on statistical and scientific database management. Luxembourg, 22-23 July 1986. *Statistical Software Newsletter* **13**,3-27

CUBITT R, WESTLAKE A (1987b) Introduction. In: CUBITT R, WESTLAKE A (eds. 1987a)

ELLIMAN AD, WITTKOWSKI KM (1987) The impact of expert systems on statistical database management. In: CUBITT R, WESTLAKE A (eds. 1987a) 14-18

FORTUNATO E, RAFANELLI M, RICCI FL, SEBASTIO A (1986) An algebra for statistical data. In: CUBITT R, COOPER B, OZSOYOGLU G (eds. 1986) 122-134

GHOSH SP (1987) Category numerical relational operations for statistical database management. In: RAFANELLI M, KLENSIN J, SVENSSON P (eds. 1988, in press)

HAJEK P, IVANEK J (1982) Artificial intellicence and data analysis. In: CAUSSINUS H, ETTINGER P, TOMASSONE R (eds.) *COMPSTAT 1982 - Part I, proceedings in computational statistics.* Wien: Physica, 54-60

OLDFORD RW, PETERS SC (1986) Object-oriented data representations for statistical data analysis. In: DEANTONI R, LAURO N, RIZZI A (eds.) *COMPSTAT 86.* Heidelberg: Physica

OLKEN F (1986) Physical database support for scientific and statistical database management. In: CUBITT R, COOPER B, OZSOYOGLU G (eds. 1986) 44-60

PREGIBON D, GALE WA (1984) REX: an expert system for regression analysis. In: HAVRANEK T, SIDAK Z, NOVAK M (eds.) *COMPSTAT 84.* Wien: Physica, 242-248.

RAFANELLI M, KLENSIN J, SVENSSON P (eds. 1988) *Proceedings of the fourth international workshop on statistical and scientific database management.* (in press)

SAS INSTITUTE INC. (1982) *SAS user's guide: basics.*

SAS INSTITUTE INC. (1982) *SAS user's guide: statistics.* SAS Inst., Box 8000, Cary, NC 27511, USA

THISTED RA (1986) Tools for data analysis management. In: BOARDMAN TJ (ed. 1986) 152-159

WITTKOWSKI KM (1985) *Ein Expertensystem zur Datenhaltung und Methodenauswahl für statistische Anwendungen.* Stuttgart: Ph.D. Dissertation; Buchh. Hartinger, Xantener Str. 14, D-1000 Berlin 15

WITTKOWSKI KM (1986) Generating and testing statistical hypotheses: Strategies for knowledge engineering. In: HAUX R (ed.) *Statistical expert systems.* Stuttgart, FRG: Fischer 139-154

WITTKOWSKI KM (1986) An expert system for testing statistical hypotheses. In: BOARDMAN TJ (ed. 1986) 438-443

WITTKOWSKI KM (1988) Intelligente Benutzerschnittstellen für statistische Auswertungen. In: FAULBAUM F, UEHLINGER HM (eds.) *Fortschritte der Statistik-Software 1.* Stuttgart,FRG: Gustav Fischer 212-225 (in German)

A STUDY OF MODIFIED INTERPOLATION SEARCH IN COMPRESSED, FULLY TRANSPOSED, ORDERED FILES

Magnus Andersson
Per Svensson

Swedish Defence Research Establishment
S-102 54 Stockholm, Sweden

ABSTRACT. The CFTOF storage structure, introduced by one of the authors in 1979 and used in the data analysis system Cantor, and a modified interpolation search algorithm, presented in this paper, are combined to solve a conjunctive query search problem. The average complexity of the resulting method is analyzed, and the analytical performance model is compared with testbed experiments and an implementation in the Cantor system.

1. INTRODUCTION

A basic design requirement for a data analysis system is that successive data base transformations should not degrade query performance. Thus, derived data, i e, the result of a sequence of queries, and base data, i e, data loaded into the system, should be represented uniformly in the data base. In fact, in such systems, execution performance, secondary storage economy, and ergonomic considerations all discourage the use of auxiliary storage structures, such as indexes.

In the design of Cantor, this and other requirements [4] led to the development of a specialized storage structure, called compressed, fully transposed ordered files, CFTOF. In [14], a sequential search algorithm for the conjunctive query search problem was studied in the CFTOF framework.

Research interest in search methods has been stimulated by some important results on interpolation search algorithms [18, 16, 9, 17]. In this paper, we describe an improved algorithm for conjunctive query search problems over the CFTOF structure. The new algorithm, to be called modified CFTOF interpolation search, was obtained by combining interpolation search, sequential search, and binary search into a polyalgorithm which for each of a set of predefined subproblems dynamically selects the appropriate method. The performance of the algorithm is studied analytically and experimentally for an important subclass of the conjunctive query problem, called orthogonal range search.

Results from the evaluation of a probabilistic performance model are shown to conform well to performance data obtained by executing the algorithm in a testbed environment.

The probabilistic model was derived by modifying the model for sequential CFTOF search. The new model takes into account the log log N complexity for interpolation search over files with uniformly distributed keys [18] (here, as in the remainder of this paper, log denotes the base 2 logarithm). The testbed experiments, on the other hand, were carried out for two versions of the algorithm, differing in the value of a certain constant. The two values were chosen to approximate respectively the situation described by the probabilistic model, and the modified CFTOF structure used in Cantor, whose storage structure is a kind of blocked CFTOF (BCFTOF).

In the above-mentioned cases, the performance measure is the number of "probes", or data accesses, needed for the search.

We also present performance data obtained from an implementation of this algorithm in the Cantor system. Finally, performance data for the algorithm previously implemented in Cantor, similar to sequential CFTOF search except for relations with one-dimensional key, are given. Here, the performance measure is cpu time.

In our experiments only the search performance is considered. The resources needed to produce the resulting relation table has been disregarded, as well as query overhead such as grammatical analysis, optimization, and code generation.

The results are valid for orthogonal range queries over CFTOF respectively BCFTOF structures, satisfying the condition that attribute values are independent and uniformly distributed. Thus, they are relevant only for search over key attributes in a relation table organized as a (B)CFTOF. For search over non-key attributes, sequential fully transposed file (FTF) search (Batory [2]) may be used.

We show that modified CFTOF interpolation search significantly improves search performance over sequential CFTOF search in critical cases. It was shown in [14] that sequential CFTOF search, when applicable, represents a dramatic improvement over FTF search. Batory demonstrated in [2] that FTF search outperforms inverted file search (i e, search using an index for each search attribute) whenever the hit ratio, i e, the ratio between the number of satisfying tuples and the total number of tuples in the file, exceeds a few percent. CFTOF interpolation search as realized here reduces to FTF search in the worst case. Furthermore, it shows performance improvement over sequential CFTOF search in situations where the latter method compares unfavourably with inverted file search.

Thus, the availability of the new method strengthens the argument for avoiding index structures in a data base system designed for data analysis work. The only situation where inverted file range search retains a clear advantage is highly selective querying over non-key attributes.

General-purpose application of interpolation search must rely on some technique for avoiding the very poor worst-case performance of pure interpolation search. We have developed a method where interpolation search and binary search are used in combination. This method has an average complexity of approximately log log N for uniformly distributed keys and a worst-case complexity relatively close to that of binary search, in fact, 2 (floor(log N) + 2). This claim is supported by a combination of simple deduction and testbed experiments but a formal proof and quantification of the average complexity remains to be found. Previously published robust interpolation-binary search methods [17, 6] have the average complexity 2 log log N for uniformly distributed keys.

The rest of the paper is structured as follows. In Ch. 2, known results on interpolation search are summarized and our interpolation-binary search method is described. In Ch. 3, the modified CFTOF interpolation search method is informally described. A detailed description of its implementation is given in [1]. A probabilistic model of the average performance of CFTOF interpolation search is developed in Ch. 4, using results from [14]. In Ch. 5, the design and result of a performance measurement experiment is presented. Graphs showing search cost data vs. query permissivity are presented for a number of data sets and for the four different cost evaluation techniques mentioned above. For all these techniques, the performance of the new method is compared with that of the sequential CFTOF search method. Ch. 6 contains conclusions and suggestions for further work on search and sort methods for CFTOF structures.

2. A SUMMARY OF BASIC RESULTS ON INTERPOLATION SEARCH AND SOME OF ITS MODIFICATIONS

Interpolation search, described by Peterson already in 1957 [11, 5], can be characterized as a search technique which in a straight-forward way exploits sort order and a known and well-behaved key distribution to provide excellent single-key search performance.

Yao and Yao [18] showed that the expected number of probes is log log N to find a single key in a file satisfying these conditions. They also showed that log log N is a lower bound for the performance of any search algorithm under these conditions. To appreciate the implications of this result, recall that log log N increases monotonically from about 3.3 for $N = 10^3$ to 4.9 for $N = 10^9$ and 5.9 for 10^{18}; thus, one can expect that between three and six probes will on average suffice to find a single key in any real-world file satisfying the above-mentioned conditions.

However, using interpolation search on a file whose distribution does not satisfy the conditions, e g, whose key distribution is highly skewed when a uniform distribution was assumed in the algorithm, may require a disastrous worst case of N probes. Some modification of interpolation search is therefore required to obtain a robust algorithm.

One such method was suggested by Willard [17] and is further discussed and extended to batched search by Li and Wong, who call their method IBBIS [6]. This method alternates between (robust) binary search and interpolation search, resulting in a worst-case complexity of 2 (floor(log N) + 1) for any ordered file with unique keys, whereas the average number of probes for uniformly distributed keys is still quite good, 2 log log N.

In our work we have used a modification which, although obvious, has to our knowledge not been discussed previously. In this modification, following two initial interpolation search steps the reduction in the length I_n of the n:th search interval is compared to 2^{-n}, i e, the reduction resulting from n binary search steps. If after some interpolation search step we have $I_n > 2^{-n} I_0$, a binary search step is done before next interpolation search step is taken. As long as $I_n > 2^{-n} I_0$, the method alternates between interpolation and binary search steps, resuming pure interpolation search if and when $I_n \leq 2^{-n} I_0$ holds again. In this way, the worst case complexity is kept logarithmic in file size, never to exceed 2 (floor(log N) + 2), while the average number of probes for uniformly distributed files will be close to that of unmodified interpolation search, as Table 1, below, shows.

N	log log N	Average number intpol	Average number modif intpol	Maximum number of probes intpol	Maximum number of probes modif intpol
64	2.585	3.331	3.766	9	10
256	3	3.503	3.927	7	11
1024	3.322	4.241	4.957	9	14
4096	3.585	4.029	4.549	9	14
16384	3.807	4.736	5.465	9	16
65536	4	4.505	5.098	9	14

Table 1. Measured complexity (number of probes) in test runs of the ordinary and modified interpolation search algorithms for uniformly distributed single-key search. Each file was searched 10 000 times.

3. A MODIFIED INTERPOLATION SEARCH ALGORITHM FOR CONJUNCTIVE QUERIES IN CFTOF AND BCFTOF

3.1. The conjunctive query search problem

The class of conjunctive one-variable queries is defined as follows [2]:

$$(R): (\bigwedge_{i \in I} R_i \in W_i), \; I = \{i_1, i_2, \ldots, i_k\} \subseteq [1..d] \qquad (3.1)$$

where R is a record variable in the file F, consisting of unique records drawn from the sample space $V_1 \times V_2 \times \ldots \times V_d$, where V_i is the domain set of the i:th attribute. If $W_i \supseteq V_i$, then the "filter" conjunct $R_i \in W_i$ is trivially satisfied and may be excluded from the query.

We call the query complete when $\{i_1, i_2, \ldots, i_k\} = [1..d]$ and $W_i \subset V_i$, $i \in [1..d]$.

An orthogonal range query is the special case $W_i = [a_i, b_i]$, $i \in [1..d]$.

3.2. The CFTOF and BCFTOF concepts

The concept of a compressed, fully transposed, ordered file was introduced and formally defined in [14]. Instead of repeating this definition, we present the following informal description.

A fully transposed file (FTF) is a structure where each attribute in a relation table is stored as a separate entity, or *subfile*. It is assumed that each item in a subfile can be accessed individually via its ordinal number in the file.

A compressed, ordered FTF (CFTOF) can be formed for the key attributes of a relation table by sorting the table with respect to some permutation of the key and storing each key attribute in the result in a *run-compressed* subfile, i e, a subfile structure capable of representing any sequence of consecutive, equal values as a pair (nv, value), where nv is the number of items in the sequence. It is assumed that each subfile access yields not only the item value but also the two integers nv and iv, the latter being the position of the accessed item within its sequence of length nv.

In the data analysis system Cantor, a modification of this structure, called blocked CFTOF (BCFTOF) is used [4]. Here, the number of items stored together in each block is maximized by a constant B. Consequently, a long run will be distributed over more than one block and can not be accessed as a unit. This design decision was made to simplify the handling of updates to a relation table.

3.3. Applying interpolation search to the CFTOF structure

Conjunctive queries over fully transposed files can be processed by using interval-sequential search over one attribute subfile at a time [2, 15]. With this technique, input data for each search step consists of a set of tuple number intervals (TID intervals) in addition to the attribute subfile and its search filter. The output from each step, to be used as input to the next step or as final result, is a new set of TID intervals which defines a subset of the input set of TID:s. The initial set of TID intervals is [1, cardinality(F)], i e, the entire first subfile is searched sequentially.

In ordinary sequential FTF search [2] the attributes are processed in order of increasing *permissivity*, a term denoting that fraction of the attribute values which is on average admitted by the search filter associated with the attribute. We will use the term *hit* to denote an attribute value admitted by the filter.

In CFTOF search, defined for the key attributes of a CFTOF, these attributes are processed first, in sort key order. If filters on non-key attributes are present in the query, FTF search can then be used for the remaining set of TID intervals.

The performance improvement over FTF search which is obtained by introducing sequential CFTOF search is due entirely to the reduction in the number of file accesses caused by run length compression. The efficiency of sequential CFTOF search depends critically on permissivity and degree of compression for the most significant key attributes.

In the CFTOF structure each key attribute consists of subsequences of strictly ordered item values, or *suites*. The most significant key attribute consists of a single suite. At least when the attribute cardinality (and thus the average suite length) is high and the permissivity low it should therefore be possible to search a suite faster through the use of an algorithm that exploits ordering, such as interpolation search.

To use interpolation search over a suite, its endpoint TID:s must be known. Subordinate to any distinct value run v_{i-1} of the (i-1):st key attribute there is a suite in the i:th key attribute, if $1 < i \leq d$. The TID:s of the endpoints of such a suite are obtained immediately when v_{i-1} is accessed, as stated in Sec. 3.2, above. The corresponding attribute values can then be found using two direct access read operations (for the most significant key attribute the endpoint values are assumed known a priori). Starting from the suite endpoint "coordinates" thus obtained, interpolation search can be used to find the TID of the smallest hit within the suite (the smallest *suite hit*).

For the last key attribute, interpolation search may be used to find also the greatest suite hit. Otherwise, each suite hit has to be accessed, in order to find all necessary suite endpoints in the (i+1):st key attribute.

For an unblocked CFTOF, the unicity assumption for interpolation search holds, since this model assumes that each distinct value run is accessed as a unit and at the same fixed cost for all runs. However, for a blocked CFTOF where at most B items in an attribute can be accessed simultaneously, the unicity assumption does not hold. In the modified interpolation search algorithm BOXSEARCH [1] the TID value obtained by interpolation is therefore decreased by the sum of an estimate of the average length of a run and a constant S/2. In this way, the algorithm tries to lookup a value with a TID slightly less than that of the smallest suite hit, so as to increase the likelihood of finding the required point of value change after sequential reading of at most a few blocks.

The algorithm BOXSEARCH is designed for use in the Cantor system, in which the access time for an item value is highly dependent on the difference D between the TID to be accessed and the current TID. Access cost increases with this difference in a somewhat irregular manner from zero for items belonging to the same run in the same block to a maximum value when a new subfile lookup operation is needed to locate a distant item. For negative differences a lookup operation is required to find any item with a TID lower than the minimum TID in the current block. Due to effects caused by page buffering, access time increases with distance also in this case. The algorithm tries to exploit the low additional access cost for neighbouring values by switching to sequential search whenever $0 < D \leqslant S$. We will call S the *sequential search limit.*

Example. In [14], data from performance measurements made with a prototype of the Cantor storage and access system were presented. The experiment was carried out on a Dec-10 computer. The access cost for scanning an attribute subfile of 8500 items, measured as cpu time per item accessed, varied from about 3 μs for highly compressible key attributes to about 40 μs for non-compressible non-key attributes. The cpu time per item required to access a randomly selected subset of 64 items ranged between 0.6 and 2.5 ms.

To summarize, the modified interpolation search algorithm BOXSEARCH [1] employs the three basic search methods sequential search, binary search, and interpolation search in combination to solve the conjunctive query subproblem for the key attributes of a relation, organized as a compressed, fully transposed, ordered file. The search subproblem for the non-key attributes is then solved using ordinary FTF search. In both cases, the algorithm processes one attribute subfile at a time. For the CFTOF subproblem, sequential search is used for short search intervals, otherwise interpolation search is preferred, unless its convergence speed is slower than that of binary search.

4. PROBABILISTIC MODELS FOR SEQUENTIAL AND INTERPOLATION CFTOF SEARCH FOR ORTHOGONAL RANGE QUERIES

In [14], a probabilistic model was derived for the average search cost of the sequential search algorithm for complete conjunctive queries (and thus for orthogonal range queries) over CFTOF structures. The model assumes that attribute values are independent and uniformly distributed.

File assumption. Let V_i, $i \in [1..d]$, be strictly ordered finite sets of objects, called "attribute sets". The cardinality of V_i will be denoted c_i. Let N records be randomly drawn without replacement from the sample space $V = V_1 \times V_2 \times \ldots, \times V_d$. Let F be the CFTOF which corresponds to this set of records. Note that this model implies that the records represented by F are unique, i e, $[V_1, V_2, \ldots, V_d]$ is a key of F.

Query assumption. Let a complete conjunctive query

$$(R): \bigwedge_{i \in [1..d]} R_i \in W_i,\ W_i \subset V_i,\ i \in [1..d] \tag{4.1}$$

be characterized by a *permissivity vector* $\underline{\pi} = (\pi_1, \pi_2, \ldots, \pi_d)$, where $1/c_i \leqslant \pi_i < 1$, and π_i is defined by:

$$\pi_i = \left| W_i \cap V_i \right| / \left| V_i \right| \tag{4.2}$$

i e, on average the filter for the i:th attribute admits the proportion π_i of the attribute values.

Recurrence relation over the attributes. The derivation of the cost models is based on the following recurrence relation:

$$\begin{cases} C_d = g_d \\ C_i = g_i + \pi_i E_i C_{i+1},\ i = d-1, d-2, \ldots, 1 \end{cases} \tag{4.3}$$

where:

C_i is the accumulated cost after searching a suite in the i:th attribute subfile, including the cost of searching all its subordinate suites; E_i is the average cardinality of a suite in the i:th attribute subfile; g_i is the average cost of finding and when necessary accessing all hits within one suite of the i:th attribute.

The expression follows from the fact that the process of searching one suite in the i:th attribute search step consists of finding the TID:s of the suite hits, then, unless $i = d$, accessing each one of these hits to find its associated TID interval, and finally performing all following search steps for subordinate suites. There are on average $\pi_i E_i$ suite hits. Since the first attribute consists of a single suite, the accumulated search cost after searching this suite is also the total search cost of the query.

Average cardinality of a suite. In [14], the following expression was derived for the average cardinality E_i of a suite in the i:th attribute, i e, sequence of maximum length consisting of unique attribute values, subordinate to a single value in the (i-1):st attribute:

$$E_i = c_i * (1 - \binom{K_d - K_d/K_i}{N} / \binom{K_d}{N}), \quad i \in [1..d] \tag{4.4}$$

where $K_i = \prod_{s=1}^{i} c_s, \quad i \in [1..d]$

Sequential CFTOF search cost expression. The average sequential CFTOF search cost C_1 for a given complete conjunctive query is given by the expression:

$$C_1 = E_1 * (1 + \sum_{k=1}^{d-1} \prod_{l=1}^{k} (\pi_l E_{l+1})) \tag{4.5}$$

The formula follows from the observation that g_i in this case equals E_i, since each suite subordinate to a hit is scanned in its entirety. The closed form solution of (4.3) is immediately obtained.

Numerical values of C_1 can not be conveniently calculated from expression (4.5) for realistic parameter values. We therefore introduce a numerically tractable approximation from which the numerical values of C_1 in Diagrams 1-3 were computed.

By substituting Stirling's expression for the faculty function:

$$n! = \sqrt{2\pi n} * n^n * \exp(-n) * \exp(\theta/12n) \text{ for some } \theta,\ 0 < \theta < 1,$$

into the binomial coefficients in (4.4) and simplifying, one obtains:

$$\begin{aligned} E_i = c_i * (1 - {} & [1 + 1/(K_i K_d - K_d - K_i^2 N)]^{N+1/2} * \\ & [1 - K_i N/(K_i K_d - K_d)]^{K_d/K_i} * \\ & [1 - K_d/(K_i K_d - K_i N)]^N * \\ & \exp[\theta/6(K_d - N - K_d/K_i)]) \end{aligned} \tag{4.6}$$

Precise numerical bounds for E_i can be computed easily from (4.6).

<u>Cost expression for CFTOF interpolation search</u>. The expression (4.3) is valid also for CFTOF interpolation search. It remains to find an expression for g_i valid in this case. Now, whenever $E_i > 4$,

$$g_i = 2 + \log\log(E_i - 2) + \pi_i E_i \qquad (4.7a)$$

is the (approximate) average cost of finding and accessing all hits in a suite of cardinality E_i, when two accesses are required to find the attribute values at the suite endpoints, whose TID:s are known. For the first attribute in the CFTOF range search problem, the endpoint values are known, and the corresponding expression is:

$$g_1 = \log\log E_1 + \pi_1 E_1 \qquad (4.7b)$$

Since the expression (4.7a) is not defined for $E_i \leq 4$, and since interpolation search is not cost-effective for short suites, we use the following approximate expression for the average cost of a combined algorithm, where sequential search is used whenever the cardinality of a suite is $< L$, for some value of $L \geq 4$:

$$\begin{cases} C_d = \begin{cases} E_d & E_d \leq L \\ 2 + 2\log\log(E_d - 2) & E_d > L \end{cases} \\[2ex] C_i = \begin{cases} E_i + \pi_i E_i C_{i+1} & E_i \leq L \\ 2 + \log\log(E_i - 2) + \pi_i E_i * (1 + C_{i+1}) & E_i > L \end{cases} \quad i \in [2..d-1] \\[2ex] C_1 = \begin{cases} E_1 + \pi_1 E_1 C_2 & E_1 \leq L \\ \log\log E_1 + \pi_1 E_1 * (1 + C_2) & E_1 > L \end{cases} \end{cases} \qquad (4.8)$$

As above, Stirling's formula may be used to produce a computable approximation to these expressions.

5. AN EXPERIMENTAL COMPARISON OF SEQUENTIAL AND INTERPOLATION BOX SEARCH IN CFTOF AND BCFTOF

The techniques presented in Ch. 3 and 4, above, have been used to compare experimentally sequential and interpolation search in CFTOF and blocked CFTOF.

5.1. The experimental design

For each of two different sets of test files, the probabilistic models (Ch. 4) for sequential and interpolation search, two corresponding algorithms SEQBOXSEARCH and IPBOXSEARCH [1] in a testbed environment, and two versions of the Cantor system, were executed to provide performance data.

One of the Cantor versions used the new modified interpolation search algorithm outlined in Ch. 3, the other an old, entirely different implementation of the sequential search algorithm.

For the probabilistic models and the testbed environment, the cost measure is the number of accessed runs, whereas for the Cantor tests, the cost measure is cpu time.

In the interpolation search case, the main conceptual difference between the probabilistic model experiment and the testbed model experiment is that the latter uses a modified interpolation search algorithm, whereas the former estimates the performance of an almost pure CFTOF interpolation search algorithm.

In the sequential search case, on the other hand, there is no conceptual difference between the probabilistic model and testbed experiments. The two approaches should therefore produce similar results.

The different cost measures used for the testbed and Cantor experiments represent a difference in abstraction level; by using in the testbed the number of accessed runs as cost measure, the effect of blocking in the CFTOF structure was partially abstracted away. In experiment 1 (see below), the testbed measurements were run in two variations, one with a sequential search limit $S = 3$ (the smallest value permitted by the program logic), to simulate as closely as possible ideal CFTOF search, the other with $S = 80$, the value used in Cantor. The latter variation can be expected to somewhat better capture the relative improvement over sequential CFTOF search that can be achieved in a system like Cantor, where scanning is much faster than direct access. The ideal CFTOF model does not account for this phenomenon. Neither variation can however be said to constitute a quantitative model of system performance.

The two sets of test data were chosen to represent respectively a case advantageous for the new algorithm and the same experiment data as those one used in [14], a rather unfavourable case for CFTOF interpolation search.

Experiment 1. Three relations, each with two attributes and cardinality 50 000, were generated by random sampling without replacement from the sets:

$$V^{I} = [1..100] \times [1..1000]$$

$$V^{II} = [1..300] \times [1..300]$$

$$V^{III} = [1..1000] \times [1..100],$$

respectively.

Queries with permissivities 0.03, 0.1, and 0.3 were generated and the three different performance measures obtained.

As already mentioned, the testbed experiment was run in two variations, one with $S = 3$, the other with $S = 80$, the value used in the Cantor implementation.

Relations with two attributes were used because this is the minimum number which produces non-trivial results. Attribute cardinalities were varied since it is to be expected that the sort key order has a considerable effect on search performance, as the following calculation shows:

from (4.8), if $d = 2$ and $E_1, E_2 >> 1$,

$$C_1 = (3 + 2 \log \log E_2)\ \pi_1\ E_1 + \log \log E_1 + e$$

where e is a small error term. The first term in this expression will dominate unless $\pi_1 << 1/E_1$, which requires both a sparsely populated file and a low permissivity. Also, $3 + 2 \log \log E_2$ is a moderate number, between 7 and 13 for E_2 between 16 and 1 000 000 000.

As soon as $\pi_1 E_1 > 1$, C_1 is essentially proportional to this factor.

If one wants to minimize the average cost for orthogonal range queries in a system which uses CFTOF interpolation search, the sort key order should therefore in general be chosen according to increasing attribute cardinalities.

The results are shown in diagrams 1 and 2.1-2.3.

In diagram 1, the ratio between the cost estimate (number of probes) for the probabilistic model (4.8) and each of two versions of testbed measurements for interpolation

CFTOF search is shown. The three upper curves correspond to S = 80, the three lower to S = 3.

In diagrams 2.1-2.3, ratios between the cost estimate for interpolation CFTOF search and sequential CFTOF search are shown. In diagram 2.1, results of probabilistic models are compared. In diagram 2.2, interpolation CFTOF search (S = 80) and sequential CFTOF search testbed measurements are compared. In these diagrams, cost measure is number of probes. In diagram 2.3, Cantor measurements are compared. Cost measure is cpu time (Vax 11/750).

Experiment 2. Three relations, each with four attributes, were generated by random sampling without replacement from the set V = [1..10] x [1..10] x [1..10] x [1..10]. The three relations had cardinalities 947, 3253, and 5057.

Queries with permissivities 0.1 (0.1) 0.9 were generated and the three different performance measures obtained.

The testbed experiment was carried out for S = 80 only.

The results are shown in diagrams 3.1-3.3, analogous to 2.1-2.3.

5.2. The results

One should note that the data presented pertain to the search step of a query, disregarding the work needed to access the tuples that satisfy the query, as well as any query processing overhead that may be inherent in a system.

Since these costs are, respectively, proportional to the size of the output and independent of the searched data, for sufficiently large files and small permissivities they will in principle always be dominated by the search cost. Due to the high efficiency of interpolation search for uniformly distributed files, however, this situation may seldom be observed in practice for such files.

Diagram 1 shows a qualitative agreement between the results from the probabilistic model and the testbed measurements. In fact, in Experiment 1, within 10 %, the number of accesses is consistently 1.9 times as high in the testbed measurements with a sequential search limit of 3 as in the probabilistic model. We believe that the use of pure interpolation search in the testbed experiments would have removed this discrepancy.

Diagrams 2.1-2.3 show that ideally, interpolation CFTOF search would be between 6 and 50 times more efficient than sequential CFTOF search for these data, but that in the testbed measurement (S = 80), as well as in Cantor, the improvement was much less.

For the data in Experiment 2, diagrams 3.1-3.3 show that theoretically sequential and interpolation search are roughly equally efficient, that in the testbed measurements (S = 80) interpolation search consistently performs better, but that in Cantor the old program was on average somewhat faster.

In summary, the new algorithm does provide better system performance in critical situations, but the results also indicate that there might be room for further improvement by better tuning of the parameters of the implemented algorithm.

6. CONCLUSIONS AND SUGGESTIONS FOR FURTHER WORK

This study has shown that the use of a modified interpolation search technique can improve search efficiency for complete range queries, a subclass of the conjunctive queries, on compressed, fully transposed files, in comparison with the previously studied sequential search algorithm.

However, the improvement is significant only when the file is quite large and the query highly selective.

We have reached this conclusion by analyzing the algorithm's performance theoretically, by implementing it and running it in a testbed environment, and finally by measuring its performance as implemented in a specific database system environment, the data analysis system Cantor.

The use of compressed, ordered transposed files in a statistical and scientific dbms has been reported earlier to provide good secondary storage economy and to offer a good basis for fast "statistical" searches [4].

The efficiency of the new search algorithm further weakens the arguments for using inverted file structures (indexes) in connection with compressed transposed files.

Only for highly selective searches over non-key attributes will indexes provide a definite advantage.

We have also suggested a simple but, to our knowledge, new modification of the standard interpolation search algorithm, to make it robust with respect to the statistical distribution of its input data, without paying the cost penalty of previously reported modifications. The efficiency of this modification was demonstrated by a simulation experiment. A formal analysis of the complexity of this method remains to be carried out.

Further work on search performance in CFTOF structures should be done in the following areas:

- theoretical modelling and analysis of the effects of blocking

- modelling and analysis of the dependence of access cost on distance to accessed tuple and incorporation of this dependence in the CFTOF interpolation search model

- models for the number of disk accesses required by CFTOF search algorithms

- design and performance studies for adaptations of the basic search algorithms to the problems of join and key lookup searching

- studies of sort algorithms for CFTOF; in Cantor, a modified version of Quicksort works very well in practice, not least for very large files where virtual memory is heavily used.

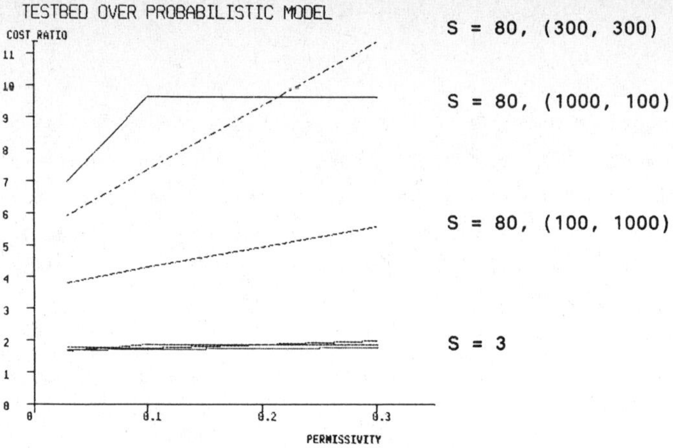

Diagram 1. Ratio between cost estimates (number of probes) from probabilistic model (formula (4.8)) and each of two versions of testbed measurements for interpolation CFTOF search. One version uses a sequential search limit of 80, the other a sequential search limit of 3. Three relations were used, each with cardinality 50 000. Two attributes with domain cardinality (100, 1000), (300, 300), and (1000, 100), respectively.

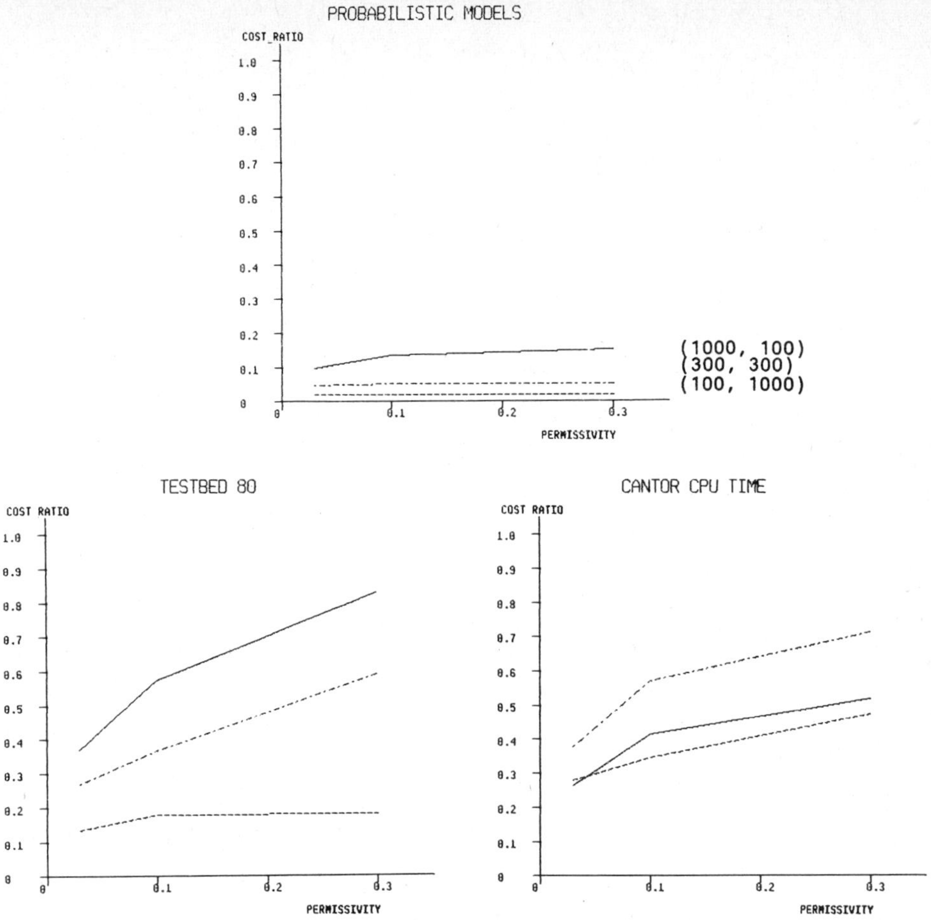

Diagrams 2.1-2.3. Ratios between cost estimates of interpolation CFTOF search and sequential CFTOF search obtained by three different methods. Same relation characteristics as in diagram 1.1.

In diagram 2.1, results of probabilistic models are compared. Cost measure is number of probes.

In diagram 2.2, interpolation CFTOF search (S = 80) and sequential CFTOF search testbed measurements are compared. Cost measure is number of probes.

In diagram 2.3, interpolation CFTOF search (S = 80) and sequential CFTOF search Cantor cpu time measurements (Vax 11/750) are compared.

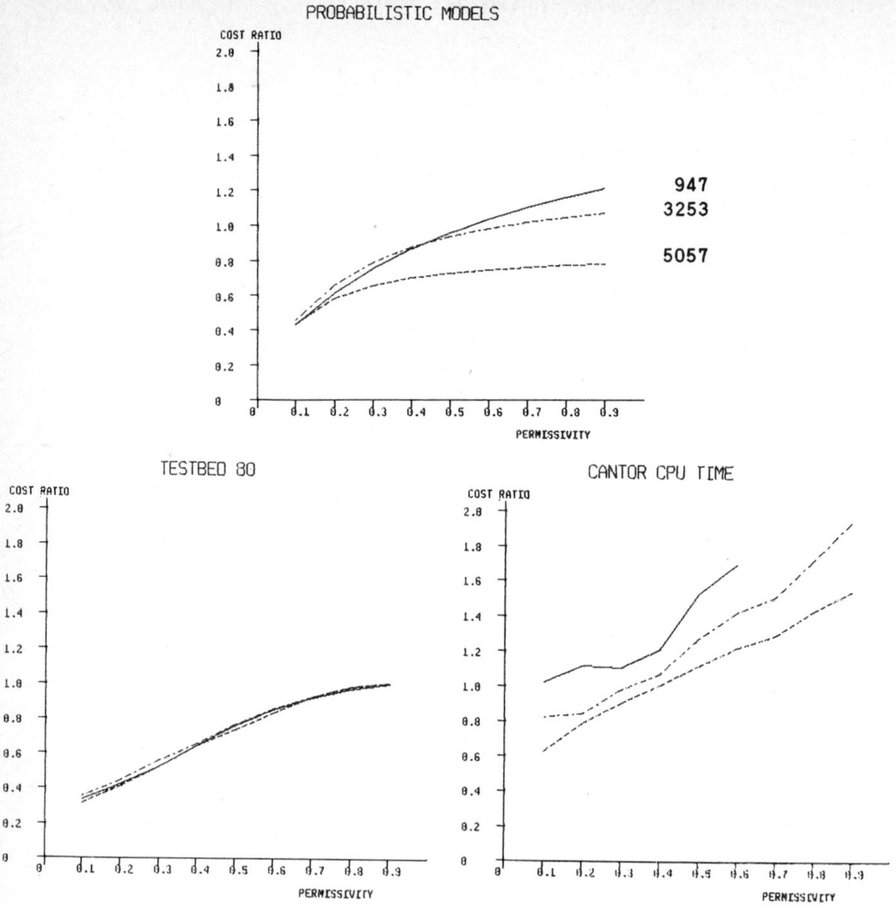

Diagrams 3.1-3.3. Ratios between cost estimates of interpolation CFTOF search and sequential CFTOF search obtained by three different methods. Three relations with cardinalities 947, 3253, and 5057. Four attributes, each with domain cardinality 10.

In diagram 3.1, results of probabilistic models are compared. Cost measure is number of probes.

In diagram 3.2, interpolation CFTOF search (S = 80) and sequential CFTOF search testbed measurements are compared. Cost measure is number of probes.

In diagram 3.3, interpolation CFTOF search (S = 80) and sequential CFTOF search Cantor cpu time measurements (Vax 11/750) are compared.

7. REFERENCES

1. Andersson M: Using interpolation search for box search problems in compressed fully transposed ordered files (in Swedish).
Swedish Defence Research Establishment (FOA), 1988.

2. Batory D S: On searching transposed files.
ACM Trans. on Database Systems, Vol 4, No 4, December 1979, pp 531-544.

3. Gonnet G H, Rogers L D & George J A: An Algorithmic and Complexity Analysis of Interpolation Search.
Acta Informatica, Vol 13, 1980, pp 39-52.

4. Karasalo I & Svensson P: The design of Cantor - a new system for data analysis.
Proc. Third Int. Workshop on Scientific and Statistical Database Management, Luxembourg July 24-26, 1986.

5. Knuth D E: The Art of Computer Programming, Vol 3, Sorting and Searching.
Addison-Wesley, Reading, Mass. 1973.

6. Li Z J & Wong H K T: Further Results on Interpolation Searching of Databases.
Lawrence Berkeley Laboratory Technical Report LBL-20708, February 1986.

7. Li Z J & Wong H K T: Batched Interpolation Searching on Databases.
Lawrence Berkeley Laboratory Technical Report LBL-22848, February 1987.

8. van der Nat M: On Interpolation Search.
Communications of the ACM, Vol 22, No 12, December 1979, p 681.

9. Perl Y, Itai A & Avni H: Interpolation Search - A Log Log N Search.
Communications of the ACM, Vol 21, No 7, July 1978, pp 550-553.

10. Perl Y & Reingold E M: Understanding the Complexity of Interpolation Search.
Information Processing Letters, Vol 6, No 6, December 1977, pp 219-222.

11. Peterson W W: Addressing for Random-Access Storage.
IBM Journal of Research and Development 1, 1957, pp 131-132.

12. Piwowarski, M: Comments on batched searching of sequential and tree-structured files.
ACM Trans. on Database Systems, Vol 10, No 2, June 1985, pp 285-287.

13. Santoro N & Sidney J B: Interpolation-Binary search.
Information Processing Letters 20, 1985, pp 179-181.

14. Svensson P: On Search Performance for Conjunctive Queries in Compressed, Fully Transposed Ordered Files.
Proceedings of the Fifth International Conference on Very Large Databases, IEEE Inc, New York 1979, pp 155-163.

15. Svensson P: Performance Evaluation of a Prototype Relational Data Base Handler for Technical and Scientific Data Processing.
FOA Rapport C-20281-D8, Swedish Defence Research Institute, December 1978.

16. Willard D E: Searching Unindexed and Nonuniformly Generated Files in Log Log N Time.
SIAM J. Comput. Vol 14, No 4, November 1985, pp 1013-1029 December 1978.

17. Willard D E: Surprisingly efficient search algorithm for non-uniformly generated files.
21st Allerton Conf on Communication, Control, and Computing, 1983, pp 656-662.

18. Yao A C & Yao F F: The Complexity of Searching an Ordered Random Table.
Proceedings of the Seventeenth Annual Symposium on Foundations of Computer Science 1976, pp 173-177.

DESIGN OF STATISTICAL INFORMATION MEDIA: TIME PERFORMANCES AND STORAGE CONSTRAINTS

G.Barcaroli[1], G.Di Battista[2], E.Fortunato[1], C.Leporelli[2]
[1] Istituto Centrale di Statistica (ISTAT), via C.Balbo 16 - 00184 Roma, Italia.
[2] Dipartimento di Informatica e Sistemistica, Universita' di Roma "La Sapienza" via Buonarroti 12 - 00185 Roma, Italia.

Abstract

A statistical database can be seen as a set of tables and a set of derivation functions; each function maps a set of tables into a new one. In order to optimize the time performance of the system it would be convenient to store the derived tables in secondary memory. However, a trade-off between storage resources and time performance arises: when the storage space is constrained, it is necessary to choose which derived tables have to be stored and which of them have to be computed on-line. In this paper such trade-off problem is investigated. We formulate it as an integer linear program, both for monadic and for polyadic derivation functions. In the first case we obtain a Simple Plant Location problem with a linear Knapsack constraint; in the second case the obtained program is equivalent to a Simple Plant Location problem with a submodular Knapsack constraint. Moreover we show that the problem is NP-complete and propose an efficient heuristic approach to solve it.

1. Introduction

In the last ten years statistical database applications have attracted the increasing interest of practitioners and researchers in statistics and computer science. A number of survey papers thoroughly explain the motivations of such interest (see for instance [S 82, SW 85, O et al. 86]). The scenario is rapidly evolving for several reasons; in the following three points we focus on the following three of them:

1. Technological opportunities: the availability of new technologies for low cost storage of large amount of data (e.g. optical disks) and of telecommunication networks, give

alternative ways to economically access large data collections; moreover such facilities provide opportunities for distributed analysis of statistical data on personal computers.

2. User requirements: the number of users of statistical data is rapidly increasing; their information needs are becoming more and more sophisticated and differentiated. For example a number of multi-way tables that result from different possible data views of a survey should be available to estimate characteristics and behaviors of the underlying population.

3. Management issues of statistical offices: because of costs and organizational complexity, the traditional distribution media cannot always meet the user requirements for information content and access flexibility.

An effective design for a statistical database application should be able to exploit new technological opportunities in order to take into account the new user requirements and management goals of the information providers. Thus, enhanced design methodologies are needed, and in particular a better integration is needed for the three different steps of the database design process: (1) conceptual design, in which the information content of the database is defined; (2) logical design, in which data are described in terms of the data model chosen for the implementation, and (3) physical design, in which a tuning is made on the system structures. In the following we give a contribution towards a possible integration of steps (1) and (2), showing how to optimize the time performance of the implemented database (under a storage constraint), starting from the data description given by the conceptual design activity and from a characterization of the user needs. For instance, this problem arises in the design of distribution media, like optical disks, or in the management of an on-line statistical database service. The inputs to the problem are:

- A formal description of the statistical tables involved in the application and of the logical links between them (we suppose that such description are produced during the conceptual design activity).
- A characterization of the frequency with which different tables are accessed.
- An estimate of the time that is required to compute a table starting from other tables.
- The storage constraint of the application.

The output consists of:

- The list of the tables to be stored, in order to optimize the time performance (under the mentioned storage constraint).
- The best derivation path for the tables that have to be computed on-line.

Since statistical databases are typically static [S 82], the redundant information introduced does not introduce a consistency violation problem; for the same reason, we can ignore the update costs in the optimization process.

In Section 2 we describe the conceptual and logical data models we have adopted and give some preliminary definitions and notations. An Integer Programming formulation of the problem is shown in Section 3 at different levels of generality; we prove also that the proposed optimization problem is NP-complete. In Section 4 the problem structure is exploited to give an efficient heuristic algorithm that guarantees a good approximation of the optimal solution. Finally, in Section 5, we present conclusions and outline future research topics.

2. Data models

Several data models have been proposed in the literature for the description of the information content of a statistical database: the Subject model [SC 80] describes data directories for large statistical applications; the Grass model [RR 83] is specifically tailored for describing simple statistical tables, the basic data structure for statistical databases; the Sam* model [Su 83], which is actually a conceptual model, can be used for describing both elementary and summary data; Csm [B et al. 86, DB 88] models summary data at different levels of aggregation (on the same topics see also [DG 87] and the references to metadata in [CCO 86]).

2.1 The conceptual model

For our purposes we refer to the following simple data description, which can be obtained using any of the above mentioned data models. We describe a <u>table</u> χ with a couple:

$$\chi = < f, C >$$

where C is a set of category attributes, and f is a function that maps each element of the cartesian product of the domains of the attributes of C into a summary value. For example, meaningful aggregations in an industrial production survey are:

χ_1. industrial production value by month, class of economic activity, and geographic area;
χ_2. industrial production value by year and branch of economic activity;
χ_3. index of the industrial production value by year.

For table χ_1, S_1 = {month, class of economic activity, geographic area} and f_1 associates to each triple < month, class , area > a value of industrial production.

We need also to describe how tables are derived from other tables. To do so we introduce a derivation mechanism: a table derivation maps a set $X = \{ \chi_1,...,\chi_n \}$ of tables into a new table χ using a computation procedure π. More formally a derivation δ is a triple:

$$\delta = < \pi, X, \chi >$$

where π is the computation procedure, X is the set of input tables and χ is the derived one. Referring to the above example χ_2 is derived from χ_1, and χ_3 is derived from χ_2. In both cases X is composed of a single element, therefore we can say that δ is a monadic derivation. Clearly in statistical applications polyadic derivations are also meaningful. As an example, consider tables:

χ_4. births by year and mother's age;
χ_5. population by year, age and sex;
χ_6. fertility ratios by year and mother's age.

χ_6 is derived from χ_4 and (part of) χ_5.

Our optimization process receives as input a description of a statistical application (produced during the conceptual design activity), expressed with the above data model and composed by a set of tables and a set of derivations.

2.2 The implementation model

We choose the relational data model for the implementation of the statistical tables, in which every table can be mapped into a relation. This assumption can be easily relaxed by observing that our approach is largely independent from the particular target model, as it can be used also if the tables are implemented as n-dimension arrays or using a mixed model, in which some tables are represented as relations and others as arrays. Moreover, our framework can be used to take into account the time required to derive different implementations of a given table.

3. Plant Location formulations

In this section we show how the problem, introduced in sections 1 and 2, can be formulated as an Integer Linear Program, in particular a Plant Location problem with a Knapsack constraint (see for instance [W 84] for a comprehensive bibliography on this class of problems). We describe the problem formulation through two refinement steps. The first proposed formulation (1) is for monadic derivations (covering a wide range of applications), and the second is for polyadic derivations (2). The proposed formulations are such that (2) is more general than (1).

3.1 Monadic derivations

A monadic derivation maps a table χ_j into another table χ_i. This table derivation process can be easily described using an acyclic digraph G(V,E), in which table χ_i is associated to vertex i and a directed edge from i to j means that χ_i can be generated starting from χ_j. We wish to minimize the weighted sum of access times to the tables or, equivalently, to maximize the access time savings with respect to a trivial solution. In the case of monadic operators the problem (P_m in the following) can be stated as an integer linear program:

$$z = \max_{x,y} \sum_{(i,j)\in E} s_{ij} x_{ij} =$$

$$= \sum_{i\in V} \left(n_i d_i^{max}\right) - \min_{x,y} \sum_{(i,j)\in E} \left(n_i d_{ij} x_{ij}\right)$$

subject to:

$$\sum_{j\in V} M_j y_j \leq M$$

$$\sum_{\{j \mid (i,j) \in E\}} x_{ij} = 1 \qquad \forall i \in V$$

$$x_{ij} \leq y_j \qquad \forall (i,j) \in E$$

$$y_j \in \{0, 1\} \qquad \forall j \in V$$

$$x_{ij} \in \{0,1\} \qquad \forall (i,j) \in E$$

where:

- M is the available storage;
- M_j is the storage space required for table j;
- $y_j =$ 1 if table i is stored, 0 otherwise;

- $x_{ij} = 1$ if table i is computed starting from table j, 0 otherwise;
- G(V,E) is the derivation graph;

$$d_{ij} = t_{ij} + T_j$$

$$d_i^{max} = \max_{\{j \mid (i,j) \in V\}} \left(d_{ij} \right)$$

$$s_{ij} = n_i \left(d_i^{max} - d_{ij} \right)$$

- T_j is the time required for accessing table j (if it is stored);
- t_{ij} is the time required to compute table i starting from table j; and
- n_i is the expected number of accesses to table i.

Notice that G is typically a transitive closure and that a self loop is associated with each vertex; we set $t_{ii} = 0$ to take into account that it is not necessary to compute stored tables.

At this point we can say that problem P_m is NP-complete. Since it has been formulated as an integer linear program, $P_m \in NP$. The completeness can be shown by restriction to the knapsack problem ([GJ 79]).

3.2 Polyadic derivations

Now we take into account the fact that in the generation process of a statistical application a table χ_i can be derived by a set of m_i tables $\chi_{j1}, ..., \chi_{jm_i}$, using a polyadic derivation. In this case the derivation process can be described by means of a directed acyclic hypergraph H(V,E), where each vertex of V is associated to a table and a hyperedge $(i; j_1, ..., j_{m_i})$ belongs to E iff χ_i can be generated by $\chi_{j1}, ..., \chi_{jm_i}$. Using hypergraph H the problem (we shall call it P_p) can be formulated as follows:

$$\min_{x,y} \sum_{(i;j_1,\dots,j_{m_i}) \in E} \left(n_i d_{i;j_1,\dots,j_{m_i}} x_{i;j_1,\dots,j_{m_i}} \right)$$

$$\sum_{j \in V} M_j y_j \leq M \qquad (1)$$

$$x_{i;\, j_1,\, ...,\, j_{m_i}} \leq y_{j_k} \qquad \forall (i; j_1, ..., j_{m_i}) \in E, \ 1 \leq k \leq m_i \qquad (2)$$

$$\sum_{\{j_1,\ldots,j_m \mid (i;j_1,\ldots,j_{m_i})\in E\}} x_{(i;j_1,\ldots,j_{m_i})} = 1 \qquad \forall i \in V$$

$$y_j \in \{0, 1\} \qquad \forall j \in V$$

$$x_{i;j_1,\ldots,j_m} \in \{0,1\} \qquad \forall (i;j_1,\ldots,j_m) \in E$$

with:

$$d_{j,i_1,\ldots,i_{m_j}} = t_{j,i_1,\ldots,i_{m_j}} + \sum_{k=1}^{m_j} T_{i_k}$$

where parameters are the same we have described for monadic operators and inequalities (2) say that in order to compute a table using a certain hyperedge it is necessary that all the starting tables of the hyperedge are stored.

3.3 Groups of tables with the same set of category attributes

For both monadic and polyadic derivations we can take into account the fact that it is often convenient to use a unique structure to store a group of tables with the same set of category attributes. For instance, if we have to store the following two tables:

χ_7. population by sex and marital status, and
χ_8. average age by sex and marital status,

two choices are possible:

- to use two relations; the first with attributes population, sex, marital status, and the second with attributes average age, sex, marital status;
- to use only one relation, with attributes population, average age, sex, marital status.

Clearly the second choice is more storage saving than the first in a relational environment. We can take into account this kind of problems substituting the following for inequality (1):

$$\sum_{i\in V} M_i y_i + \sum_{k\in F} \widetilde{M}_k z_k \leq M \qquad (1')$$

and adding constraints:

$y_i \leq z_k \quad \forall (i,k) \in B$

where

- the tables with the same set of category attributes are grouped into family k, F is the set of such families, and the couple (i,k) of B indicates that table i belongs to family k;
- $z_k =$ 1 if at least one table of family k is stored, 0 otherwise;
- $\widetilde{M}_k$ is the storage space required for the category attributes of family k; and
- M_i is the storage space required for the summary attribute of table i.

The proposed formulation does not take into account the increment of the access time due to the grouping of different summary attributes into a unique relation. However, we consider this increment as negligible.

It can be easily seen that this new problem is an instance of P_p.

4. Heuristic algorithms

Let $\mathcal{P}_S$ be the class of following problems:

$$\max \left\{ v(y) : \Sigma_j M_j y_j \leq M,\ y_j \in \{0,1\} \right\}$$

$v : \{0,1\}^n \to R$ is:

(i) submodular on $\{0,1\}^n$: $v(x)+v(y) \geq v(x \vee y) + v(x \wedge y)$ $\forall x,y \in \{0,1\}^n$, where $\vee$ and $\wedge$ are component by component logical OR and AND respectively; and
(ii) rational, piecewise linear, concave and nondecreasing function.

Wolsey ([Wol 82]) shows (using a previous result by Cornueljos et al. [C et al. 77]) that a greedy heuristic always attains at least 35% of the optimal value of problems in $\mathcal{P}_S$ and that P_m belongs to this class. More precisely, if Z is the optimal value of P_m and Z_H the value of the heuristic, then

$$Z_H \geq Z(1-e^{\beta}) \quad \text{where} \quad e^{\beta}=2-\beta$$

We restate here the greedy heuristic taking into account some details for P_m. For the sake of clarity, but without any loss of generality, we shall focus on the case in which all the tables of interest can be generated by a single base table j_0 with $n_{j_0} > 0$, and $d_{ij_0} \geq d_{ij}$ $\forall$ $(i,j) \in E$; in this case table j_0 must be stored and this first choice gives also a feasible solution of P_m; moreover, the value of this solution is 0, i.e. this is the trivial, worst solution that we use as a comparison in the statement of P_m.

Step 1: Set $k=1$, $J^* = \emptyset$, $j^k = j_0$, $M^0 = M$

Step 2: Let $M^k = M^{k-1} - M_{j^k}$

$J^* = J^* \cup \{j^k\}$

$J^\wedge = \{j \mid j \notin J^*, M_j \leq M^k\}$

$u_i^k = \max_{j \in J^*}(s_{ij})$ for $i \in V$

If $J^\wedge = \emptyset$, then go to Step 4

Step 3: Let $\rho_j(u^k) = (\Sigma_{i \in V} \max(0, s_{ij} - u_i^k)) \,/\, M_j$, for $j \in J^\wedge$, and

$$\rho_{j^{k+1}} = \max_{j \in J^\wedge}(\rho_j(u^k))$$

If $\rho_{j^{k+1}} > 0$, then set $k=k+1$ and go to Step 2;

else go to Step 4

Step 4: Stop: the greedy solution can be obtained as follows.

Set $y_j=1$ for $j \in J^*$, $y_j=0$ otherwise. For each i choose $j(i)$ such that $j(i) \in J^*$ and $s_{ij(i)} \geq s_{ij}$ for $j \in J$, set $x_{ij(i)}=1$ and $x_{ij}=0$ for $j \neq j(i)$. The value of the heuristic solution is

$$Z_H = \Sigma_{i \in V} s_{ij(i)} = \Sigma_{l \leq k-1} M_{j^{l+1}} \rho_{j^{l+1}}(u^l)$$

Each term $M_{j^{l+1}} \rho_{j^{l+1}}(u^l)$ represents the increment in the savings, with respect to the trivial solution, obtained by adding the table j^{l+1} to the first l stored tables. In order to take into account the scarcity of storage space, the heuristic selects, for each l, the table that yields the maximum improvement per unit of additional storage space.

The only difference between problem P_m and the polyadic problem P_p is the fact that each $x_{i;\, j_1, \ldots, j_{m_i}}$ is constrained in P_p by m_i different components of vector y. That means that we cannot improve, in general, the value of a solution of P_p with the addition of a single new table to those already selected. So, the previous heuristic cannot be used without changes. Moreover, if you call v(y) the optimal value of P_p when a particular value of the vector y is fixed, it can be easily seen that v(y) is no longer a submodular function, i.e. $P_p \notin \mathcal{P}_S$.

We shall sketch in the following a modified heuristic that , at each step, add to the current solution a set of tables sufficient to activate at least one additional hyperedge. To do so, we first give a slightly different representation of P_p.

Let $S=\{S_1,\ldots, S_k,\ldots S_p\}$ be the minimum collection of subsets of V such that:

$$(i; j_1, \ldots, j_{n_i}) \in E \Rightarrow \exists\, k,\ S_k \in S,\ S_k=\{j_1, \ldots, j_{n_i}\}$$

and let be

$$E'=\{(i,k) \mid (i; j_1, \ldots, j_{n_i}) \in E,\ S_k=\{j_1, \ldots, j_{n_i}\}\}$$
$$E''=\{(j,k) \mid j \in S_k\}$$

Now problem P_p can be stated as:

$$\min_{x,w} \sum_{(i,k)\in E'} (n_i d_{ik} x_{ik})$$

$$C(w) \leq M$$

$$x_{ik} \leq w_k \qquad \forall\ i,k \in E'$$

$$\sum_{\{k \mid (ik)\in E'\}} x_{ik} = 1 \qquad \forall i \in V$$

$$w_k \in \{0, 1\} \qquad k=1,\ldots, p$$

$$x_{ik} \in \{0,1\} \qquad \forall (i,k) \in E'$$

Here $w_k=1$ means that all the tables t_j such that $j\in S_k$ have been stored, while $x_{ik}=1$ means that the table t_i has to be computed using tables in S_k. Now the difference between P_m and P_p is that function C(w) replaces the linear knapsack constraint in giving the space necessary to store all the selected subsets of tables. It is easy to see that

$$C(w)=\sum_{j\in V} M_j \max_{k \mid (j,k)\in E''} w_k$$

is a submodular nondecreasing function. Moreover , if you call v(w) the optimal value of P_p when a particular value of the vector w is fixed, -v(w) is again a submodular nondecreasing function.

We are now able to give an informal description of the modified heuristic. We shall suppose, again, that feasibility is ensured first, by storing the necessary and sufficient set of tables.

Let K^* be the set of indexes of the already selected subsets S_k. and M the residual storage space.

Step 1: Compute, for each $k \notin K^*$, the additional storage space $\Delta_k(K^*)$ required to store all the tables in S_k.

$$\Delta_k(K^*)=C(w \vee e_k)-C(w)= \sum_{(j,k)\in E'',(j,l)\notin E'' \forall l \in K^*} M_j$$

Insert in K^* any k such that $\Delta_k(K^*) = 0$ and define $K^\wedge=\{k| \, k\notin K^*, \Delta_k(K^*) \leq M\}$, the set of indexes that can still be selected
If $K^\wedge=\emptyset$ then go to step 3.

Step 2:Define, for $k\in K^\wedge$, $K_k=\{k\}\cup\{ 1 \,|\, \Delta_l(K^*\cup\{k\}) =0, l\in K^\wedge\}$, the set of indexes that can be selected, if you choose k, using only the additional storage space required by k.
Compute, for each i, $u_i= \min_{k\in K^*}(n_i d_{ik})$ and, for each $k\in K^\wedge$, $\rho_k(u) =$
$(\Sigma_{i\in V} \min_{l\in K_k}(\min(0, d_{il} - u_i)) \,/\, \Delta_k(K^*)$,
$\rho_{k^*} = \min_{k\in K^\wedge}\rho_k(u)$
If $\rho_{k^*}<0$ then set $K^*=K^*\cup K_{k^*}$ and go to Step 1

Step 3: Stop: the solution can be computed storing all tables t_j such that $(j,k)\in E''$ and $k\in K^*$. Each table i shall be computed using tables in $S_{k(i)}$, where $d_{ik(i)} \leq d_{ik}$, for k(i) and k belonging to K^*.

5. Conclusions and future research topics

We have investigated the problem of storing derived data to improve the access time to a statistical database. This allows an efficient design of statistical information media such as optical disks distributed to end users or centralized data banks. The problem is also relevant for the design of memory management systems in interactive statistical analysis environments and for the choice among different physical storage models. It has been formally stated by means of an integer linear formulation that can be considered as a variant of a Simple Plant Location problem, characterized by a submodular resource constraint. We have also presented an heuristic to compute a good solution to the problem in polynomial time. The approach constitutes a basic step toward a methodology for logical design of statistical information media, or, more generally, of a statistical database. Future work will concern the developement of a complete methodology that embodies this contribution.

References

[B et al. 86] G. Barcaroli et al. - Csm: a Conceptual Statistical Model for Statistical Databases - Proc. of 7th Symposium on Computational Statistics (Compstat), Roma 1986.

[C et al. 77] G.Cornuejols, M.Fisher, G.L.Nemhauser - On the Uncapacitated Location Problem - Annals of Discrete Mathematics 1 1977.

[CCO 86] R.Cubitt, B.Cooper, and G. Ozsoyoglu (editors) - Proc. of the 3rd International Workshop on Statistical and Scientific Database Management, Luxembourg, 1986.

[DB 88] G.Di Battista, C.Batini - Design of Statistical Databases: a Methodology for the Conceptual Step - Information Systems, vol.13, no.4, 1988 (to appear).

[DG 87] A.Di Leva, P.Giolito - A Two Level Data Model for Distributed Statistical Databases - Dipartimento di Informatica, Universita' di Torino (manuscript) 1987.

[GJ 79] M.R.Garey, D.S.Johnson - Computer and Intractability - Freeman and co. 1987.

[O et al. 86] F.Olken, D.Rotem, A.Shoshani, and H.K.T.Wong - Scientific and Statistical Data Management Research at LBL - Proc. of the 3rd Intl. Workshop on Statistical Database Management, Luxembourg 1986.

[RR 83] M.Rafanelli and F.Ricci - Proposal for a Logical Model for Statistical Databases - Proc. of the 2nd Intl. Workshop on Statistical Database Management, Los Angeles 1983.

[S 82] A.Shoshani - Statistical Databases: Characteristics, Problems, and Some Solutions - Proc. VLDB, Mexico City, 1982.

[SC 80] A.Shoshani and P.Chan - Subject: a Directory Driven System for Organizing and Accessing Large Statistical Databases - Proc. of the 2nd International Conference on Very Large Data Base (VLDB), 1980, pp.553-563.

[SW 85] A.Shoshani and H.K.T.Wong - Statistical and Scientific Database Issues - IEEE Transactions on Software Engineering, vol. SE-11, N.10, October 1985.

[Su 83] S.Y.W.Su - SAM*: A Semantic Association Model for Corporate and Scientific-Statistical Databases - Information Sciences 29, 1983, pp. 151-199.

[Wol 82] L.A.Wolsey - Maximising Real-Valued Submodular Functions: Primal and Dual Heuristics for Location Problems - Mathematics of Operations Research, vol.7, no.3, 1982.

[W 84] R.T. Wong - Location and Network Design - in Combinatorial Optimization Annotated Bibliographies (M. O'hEigeartaigh, J.K. Lenstra, A.H.G. Rinnoy Kan eds.) J.Wiley, 1985.

Handling Time in Query Languages

M. A. Bassiouni

and

M. Llewellyn

Department of Computer Science

University of Central Florida

Orlando, FL 32816 USA

ABSTRACT

In this paper, an approach to handle time in relational query languages is outlined. The approach is based on extending Boolean and comparison operators by allowing their operands to be sets of intervals [BASS87]. The proposed temporal logic is shown to satisfy the properties of the normal Boolean logic. New syntax for retrieval statements is defined in order to separate the process of selecting entities (tuples) and the process of selecting required values of temporal attributes from the chosen entities. The extensions presented in this paper offer a good degree of flexibility in expressing different temporal requirements.

1. Introduction

Historical/temporal databases [AHN86, ARIA86, CLIF85, GADI85, NAVA87, NAVA88, SNOD86, SNOD87, TANS86] are designed to capture not only the present state of the modeled real world, but also its previous states. The capability to remember previous states is supported, in part, through a non-deletion policy. Once a tuple is entered into the database, it can never be deleted (except to correct for input errors).

For the purpose of this paper, we will consider a historical relational database in which each relation is viewed as a collection of related attributes: some of the attributes are constant valued while the others are temporal attributes (their values are time varying). A double time-stamping scheme (similar to that of Tansel's model [CLIF85]) for each time varying attribute is assumed. This means that time varying attributes within tuples are represented as sets of value/time-interval pairs. Although set-valued attributes complicate the physical implementation to some extent, the use of set-valued attributes is known to offer desirable advantages in the design of retrieval languages; some real query languages (e.g., the language GEM [ZANI83]) provide this feature. In addition, the scheme of time-stamping attributes has other attractive benefits as explained in [CLIF85].

Attribute time-stamping can be achieved by using either single stamping (e.g., Clifford's model [CLIF85]) or double stamping (e.g., Tansel's model [CLIF85] and the scheme used in [NAVA88]). We agree with the opinion expressed in [NAVA88] concerning the preference for double stamping: (1) the single time-stamping approach makes the implicit assumption of continuity of time, and (2) the double time-stamping approach can support point events since a point in time can be represented by a degenerate time interval in which the start and finish points are identical.

The notion of time normal form (TNF) has been proposed in [NAVA88] in order to avoid redundancy and retrieval/update anomalies. Although the model presented in our paper is based on attribute versioning (as opposed to the tuple versioning scheme used in [NAVA88]), the concept of TNF is applicable to our model. It is even pointed out in [NAVA88] that it is conceivable to implement the normalized single time-stamping scheme by clustering tuples by their time independent key and by physically grouping time-stamps with their corresponding attributes (thus virtually employing an attribute versioning scheme).

In the following discussion, we shall assume that time is represented by positive integers, and that the symbol **NOW** is used to represent the current value of time. The following relation will be used throughout the paper.

EMP (name , salary , dept , city , date_of_birth , manager)
name , date_of_birth : nontemporal attribute
salary , dept , city , manager : temporal attribute

The time-independent key of relation EMP is "name". The next sections will present our approach to provide a temporal retrieval capability using an extended logic for comparison and Boolean operators. The extended logic, developed by Bassiouni [BASS87], allows these operators to accept sets of intervals and return results in the form of sets of intervals. It is not the purpose of this paper to develop an exhaustive set of operations on time relations. Rather, our goal is to show the flexibility of the proposed approach and illustrate its usefulness in expressing temporal requirements. We will illustrate our approach and proposed operators using a syntax that generally resembles the languages QUEL [STON76] and GEM [ZANI83]. If the range variable e is defined over relation EMP, then "e.salary" denotes the set of salary/interval pairs corresponding to the tuple (employee) that variable e points to. For example, the set

{ (25000 , <2 , 4>) , (30000 , <5 , 8>) }

indicates a salary of 25,000 from time 2 to time 4 and a salary of 30,000 from time 5 to time 8. Unlike previous models of temporal queries, the proposed logic is built into the definition of Boolean and comparison operators. By extending the temporal logic to these operators, the proposed model offers a good degree of flexibility in expressing different temporal requirements.

2. Comparison Operators

Our first task is to extend the definition of comparison operators in order to allow for logical and temporal comparisons. For example, our model uses a logical "equal to" operator denoted by $=_L$ and a temporal counterpart denoted by $=_T$. We will adopt the notation that omitting the subscript will imply the logical operator (i.e., $=$ is the same as $=_L$). As an example, let e and m be two variables defined over relation EMP. The following expression

$$e.city \quad =_T \quad m.city$$

requires that both employees have lived in the same city at the same time (i.e., the corresponding time intervals must overlap). The expression

$$e.city \quad = \quad m.city$$

requires that the two employees have lived in the same city, but not necessarily at the same time. Thus logical comparison operators compare values without regard to their time intervals, while temporal operators can only compare values that belong to the same time (i.e., values that have overlapping time intervals).

To summarize this section, our model uses the logical comparison operators

$$= \qquad > \qquad < \qquad \geq \qquad \leq \qquad \neq$$

and their temporal counterparts

$$=_T \quad >_T \quad <_T \quad \geq_T \quad \leq_T \quad \neq_T$$

If the comparison condition is not met, the empty set NIL is returned (by both logical and temporal operators). Otherwise, the value returned would be the universal interval $<0\ , NOW>$ for logical operators, or the set of intervals common to both operands (and satisfying the comparison condition) for temporal operators. For example, the value of the expression

$$\{(10, <3,5>)\ , (2\ , <6,9>)\ , (8\ , <10,12>)\} >_T \{(5\ , <4,11>)\}$$

is the set { <4,5> , <10,11> }. As shall be seen later, the interval selector operators can be used to override the returned non-NIL values.

The rule for handling operands that are constants or nontemporal attributes is simple: such operands are automatically given the universal interval $<0\ , NOW>$. Thus the comparison $7 >_T 5$ returns the universal interval, the comparison $5 >_T 7$ returns NIL, and the composite comparison $(5 >_T 7) = NIL$ returns the universal interval.

3. Temporal Boolean Operators

Our next step is to extend the temporal model by including temporal counterparts of the Boolean operators. In our model, there are two types of the operator AND: the logical operator AND_L and the temporal operator AND_T. In a real implementation, the notations LAND and TAND may be used to replace AND_L and AND_T, respectively. Also, if the subscript is omitted, the logical operator is assumed (i.e., AND is the same as AND_L). To see the difference between the two operators, let us assume that the range variable e has been declared on relation EMP. The expression C1 given by

$$e.salary >_T 50000 \quad AND_T \quad e.dept =_T ADM$$

requires that the (selected) employee makes more than 50,000 while (i.e., at the same time when) working in the "ADM" department. The expression C2 given by

$$e.salary >_T 50000 \quad AND_L \quad e.dept =_T ADM$$

requires that the selected employee has received a salary greater than 50,000 and has worked in the "ADM" department, but the two events don't have to occur simultaneously. From the definition of AND_L, it will be clear that replacing the temporal comparison operators in C2 by their logical counterparts would give the same result. Both AND_L and AND_T operate on sets of time intervals (associated with attribute values) and they return a set of time intervals (possibly empty). For example, if a specific employee gained more than 50,000 during three different intervals, say

$$(51000\ ,\ <3, 4>\ )\quad ,\quad (54000\ ,\ <6, 8>\ )\quad ,\quad (56000\ ,\ <10, NOW>\ )$$

and that same employee worked in the "ADM" department during two intervals, say

$$(\ ADM\ ,\ <7, 8>\ )\quad ,\quad (\ ADM\ ,\ <9, NOW>\ )$$

Then both conditions C1 and C2 are satisfied for that employee. As explained in Section 2, the evaluation of the operand "$e.salary >_T 50000$" in C1 gives the three intervals <3,4> , <6,8> , <10, *NOW*> while the evaluation of "$e.dept =_T ADM$" gives the two intervals <7,8> , <9, *NOW*>. Finally, the operator AND_T will return the two intervals <7, 8> and <10, *NOW*>. These are exactly the intervals in which the employee has a salary rate of more than 50,000 while working in the "ADM" department.

When the two sets of intervals representing the two operands of AND_T don't have overlapping intervals, the result returned by AND_T is the empty set denoted by *NIL* (which corresponds to the value FALSE in the binary Boolean logic). Otherwise, AND_T returns the set of intervals common to both of its operands. In our model, any nonempty set of intervals corresponds to the Boolean value TRUE. On the other hand, the operator AND_L returns the value NIL only if one (or both) of its operands evaluates to NIL. If neither operand is NIL, the value returned by AND_L is the universal interval <0 , *NOW* >. The universal interval returned by AND_L indicates that both conditions are satisfied, and since the time intervals of the two conditions are not relevant, the universal interval is returned. This choice makes it easy to include AND_L as a subexpression within larger expressions. For example, the following expression can be used if we are interested in the "service periods in the ADM department" of those employees who earned more than 35,000 and who lived at some point of time in the city of Orlando (we do not require that the events of earning the stated salary and residing in Orlando occur at the same time, and furthermore we are not interested in the time periods of these two events).

$$e.dept =_T ADM \quad AND_T \quad (e.salary > 35000 \quad AND_L \quad e.city = Orlando)$$

In the above expression, the right-hand operand of AND_T will evaluate to the universal interval

$< 0 , NOW >$ only if both operands of AND_L are satisfied. Notice that whenever one of the operands of AND_T is the universal interval, the value returned by AND_T will be the set of intervals of its other operand. As explained later, the interval $<0 , NOW >$ returned by AND_L can be changed to the intervals of one (or both) of its operands.

Our model uses similar extensions for the operators NOT and OR. The unary operator NOT_L returns the value NIL if its operand is a nonempty set of intervals. Otherwise it returns the universal interval. The unary operator NOT_T returns NIL if its operand is the universal interval. Otherwise, it returns the set of intervals which is the complement of its operand (with respect to the universal interval). For example the expression

$$NOT_T \ \{ <2,2> , <6,8> , <7,9> \}$$

returns the set $\{ <0,1> , <3,5> , <10,NOW> \}$.

Both the logical and temporal versions of NOT can be very useful in expressing temporal conditions. The following expression returns the value NIL if the employee has lived in Orlando (at any time) and returns the universal interval if he has never lived in Orlando.

$$NOT_L \ (e.city{=}Orlando)$$

Similarly, the following expression gives the set of intervals during which the employee has lived outside Orlando.

$$NOT_T \ (e.city =_T \ Orlando \ )$$

Notice that, as expected, the expression $NOT_T \ (e.city =_T Orlando)$ gives the same result as the expression $e.city \neq_T Orlando$.

The operator OR also has two versions. The binary operator OR_L returns the value NIL only if both of its operands are empty (NIL). Otherwise, it returns the universal interval. The binary operator OR_T returns NIL if both of its operands are NIL. Otherwise, it returns the union of the intervals of its operands. Thus the expression

$$e.manager =_T John \quad OR_T \quad e.manager =_T Mark$$

returns the set of intervals during which the employee worked for either John or Mark.

4. Properties of Proposed Logic

It is easy to see that both the temporal version and the logical version of the proposed logic satisfy the properties of the normal Boolean logic [BASS87]. The following lemmas give some examples of these properties. Proof of the lemmas are not given in this paper.

Lemma 1: DeMorgan's Law

The expression

$$NOT_T \quad (C1 \ \ AND_T \ \ C2)$$

is equivalent to the expression

$$(NOT_T \ \ C1) \quad OR_T \quad (NOT_T \ \ C2)$$

Similarly, the expression

$$NOT_T \quad (C1 \ \ OR_T \ \ C2)$$

is equivalent to the expression

$$(NOT_T \ \ C1) \quad AND_T \quad (NOT_T \ \ C2)$$

Similar equivalences can be established for the corresponding logical operators.

Lemma 2: Distributivity

The following expression

$$C1 \quad AND_T \quad (C2 \ \ OR_T \ \ C3)$$

is equivalent to

$$(C1 \ \ AND_T \ \ C2) \ OR_T \ (C1 \ \ AND_T \ \ C3)$$

Similarly, the expression

$$C1 \quad OR_T \quad (C2 \ AND_T \ C3)$$

is equivalent to the expression

$$(C1 \quad OR_T \quad C2) \ AND_T \ (C1 \quad OR_T \quad C3)$$

Similar equivalences can be established using the corresponding logical operators.

5. Interval Selectors

When the value returned by AND_T or AND_L is non-NIL, the interval selection operators "R." or "L." can be used to override the default result and to select the set of intervals corresponding to the right or left operand, respectively. For example, the expression C3 given by

$$(e.salary >_T 50000 \ \ L.AND_L \ \ e.dept{=}ADM) \quad AND_T \quad e.city =_T Orlando$$

is the same as the expression C4:

$$(e.salary >_T 50000 \ \ AND_T \ \ e.city =_T Orlando) \quad L.AND_L \quad e.dept{=}ADM$$

Notice that in C3, the composite operator $L.AND_L$ performs the logical AND operation but returns either the empty set NIL or the set of intervals during which the salary exceeded 50,000. The value returned by this composite operator is used as the left operand for the temporal operator AND_T. Both C3 and C4 require that the selected employee has received a salary rate greater than 50,000 while living in Orlando and that he has worked in the "ADM" department at any arbitrary time. Furthermore, both C3 and C4 return the value NIL if the above condition is not met. Otherwise, they return the set of time intervals during which the employee lived in Orlando and made more than 50,000. The expressions C3 and C4 are two equivalent expressions (performing the same task). Writing the same query in more than way is usually possible in our model (as is the case in most other languages).

6. The Syntax of Temporal Queries

Unlike previous temporal models, our approach distinguishes between the qualification needed to select entities (tuples) and the qualification needed to select temporal values (output) from the chosen entities. The syntax of a retrieval statement is as follows.

retrieve (*target list*)
[**for output** *output−qualification*]
[**where** *selection−qualification*]

where keywords are shown in boldface and the notation [...] denotes optional syntax. The selection qualification gives the condition that determines whether an entity (or a combination of entities) should be selected for output. The output qualification is used to select required values of temporal attributes from the selected entities. If the "for output" phrase is omitted, the set of time intervals returned by the selection qualification is used to select the output (i.e., to replace the output qualification). If the "where" clause is omitted, a default interval of $<0, NOW>$ is used to replace the selection qualification. The keyword **SELQUAL** can be used in the output qualification to designate the set of time intervals produced by the selection qualification.

Before proceeding to give examples that illustrate the above syntax, we will first clarify the nature of the operands of temporal operators such as AND_T. Basically, operands of AND_T are expected to be sets of time intervals. For example, the following expression

$$(e.salary >_T 50000) \quad AND_T \quad < 0 , NOW >$$

has a right operand equal to the universal interval (treated as a singleton set), and a left operand equal to the set of time intervals during which the salary exceeded 50,000 (notice that since the right operand is the universal interval, the result of the above expression is always equal to the left operand). We will also permit an operand to be a set of value/interval pairs. In this case, the intervals contained in such a set will be used as the real operand. Thus in the following expression

$$e.salary \quad AND_T \quad \{ < 5 , 7 > , < 9 , 10 > \}$$

the left operand is a temporal attribute (a set of value/interval pairs), and we will assume that only the time intervals of this operand will be used to evaluate the result of AND_T. Notice that it is possible to introduce an operator to explicitly extract intervals from a set of value/interval pairs. For simplicity, however, we will assume that there is a mechanism to automatically cast each operand into the type required by the operator.

Example 1

The following query gives the salary at time 10 of any employee who, while living in Orlando, worked in the ADM department and earned a salary of more than 50,000.

range of e is EMP

retrieve (e.name , e.salary)

for *output e.salary* AND_T <10,10>

where e.city $=_T$ *Orlando* AND_T *e.dept* $=_T$ *ADM* AND_T *e.salary* $>_T$ 50000

Notice that the selection qualification (which follows the keyword "where") is used to determine whether a given employee is a candidate for output (the selection qualification must evaluate to non-NIL for such an employee). The output qualification (which follows the "for output" keyword) is used to extract the salary at time 10 of the selected employee (salaries outputted must give a non-NIL value when used to evaluate the output qualification). If a selected employee has no salary at time 10, he will have no output. Notice that if the output qualification (the "for output" phrase) is omitted, the query would return the salaries corresponding to the time intervals returned by the selection qualification (obviously any such salary must be greater than 50,000).

In the above example, the operator AND_T in the output qualification acts like an "overlap" function. In fact the Boolean overlap function used in many temporal models is a special case of the AND_T operator defined in this paper. Using the keyword "OVERLAPS" would be a better choice than AND_T in the output qualification of Example 1. We will therefore add the redundant operator "OVERLAPS" to our syntax.

The OVERLAPS Operator

This redundant operator is equivalent to AND_T: it operates on two sets of intervals and its result is a set of intervals (possibly empty). It is included in the syntax in order to replace AND_T whenever this replacement makes the meaning clearer. Notice that the generalized Boolean logic of section 3 eliminates the need to use an overloaded "OVERLAP" operator (to deal separately with Boolean and time-interval expressions).

Example 2

The following query gives the salary at time 10 for those who earned more than 20,000 at time 5.

range of e is EMP
retrieve (e.name , e.salary)
for *output e.salary OVERLAPS* <10,10>
where (e.salary $>_T$ 20000) *OVERLAPS* <5,5>

In the above example, the selection qualification is used to determine whether an employee satisfies a required condition, whereas the output qualification is used to select the salary at a certain time. Combining the two qualifications into a single one would not be possible in this example since each qualification requires different time intervals.

Example 3

For every employee that worked in the ADM department under manager "Fred", print his salary at that time and his current city.

range of e is EMP
retrieve (e.name , e.salary , e.city)
for *output (e.salary OVERLAPS SELQUAL)* AND_L
(e.city OVERLAPS < NOW , NOW >)
where e.dept $=_T$ *ADM* AND_T *e.manager* $=_T$ *Fred*

The selection qualification returns the set of time intervals during which the employee worked under Fred in the ADM department. This set of intervals replaces the keyword SELQUAL in the output qualification. Any salary/city combination printed in the output must give a non-NIL result when used to evaluate the output qualification.

To conclude this section, we shall give one more example to show that the model gives good degree of flexibility in expressing different temporal requirements.

Example 4

We are interested to collect some information about any employee who has made more money than any of his managers (while serving under that manager). The data to be collected are: 1) name of employee, 2) name of the manager and salary of employee at that time, i.e., the time when the employee earned more than the manager, and 3) name of the current manager of the employee.

range of e , m *is EMP* /* e=*employee* , m=*manager* */
retrieve (*e.name* , *e.salary*, *m.name* , *e.manager*)
for *output* (*e.salary OVERLAPS SELQUAL*) AND_L
(*e.manager OVERLAPS* < *NOW* , *NOW* >)
where *e.salary* $>_T$ *m.salary* AND_T *e.manager* $=_T$ *m.name*

Notice that the temporal operator $>_T$ in the selection qualification ensures that the compared salaries belong to the same time (i.e., are not from nonoverlapping time intervals). Notice also that the operator $=_T$ has a right operand (m.name) of type nontemporal, and as explained before such operand is given the universal time interval. Now suppose that we want to change the query so that the information is collected for any employee who ever earned more than his manager, but not necessarily during the time he served under that manager (the two salaries must still belong to the same time albeit that time is now arbitrary). A simple change in the selection qualification of the above query will achieve the new requirement. The new selection qualification will be

where *e.salary* $>_T$ *m.salary* $L.AND_L$ *e.manager* = *m.name*

The composite operator $L.AND_L$ is used to ensure that entity m has been the manager of entity e and that during some interval(s) of time, the salary of entity e has exceeded that of entity m. Furthermore, the operator $L.AND_L$ will return the set of intervals of its left operand (assuming the right operand is non-NIL), i.e., the set of intervals during which the salary of entity e has exceeded that of entity m. This set is used to replace the SELQUAL keyword as explained before.

7. Other Operators

It is possible to build a set of useful operators to be used in a query language based on the proposed logic. Since this paper is not intended to present an exhaustive set of operations on time relations, we shall only present some examples of possible operators (detailed discussion of these and other operators will be given in a forthcoming publication).

The following three unary operators are used to extract information from their single operand. They can be used in the target list as well as in the output and selection qualifications. To explain the meaning of each operator, we will use the following value/interval set as the single operand.

$$S_1 = \{ (A, <1,2>) , (B, <4,5>) , (A, <6,8>) , (C, <7,9>) , (D, <11, NOW>) \}$$

CURRENT

This function returns the current values (corresponding to the time NOW) contained in its operand. Thus $CURRENT(S_1)$ gives the set $\{ D \}$ which is the set of values corresponding to the time intervals specified by: S_1 *OVERLAPS* $<NOW , NOW>$.

START

This function returns the starting times of intervals contained in its operand. Thus $START(S_1)$ gives the set $\{1, 4, 6, 7, 11\}$.

FIRST_START

This function returns the earliest starting time contained in its operand. Thus $FIRST\_START(S_1)$ is the same as $Min\{START(S_1)\}$ and hence returns the value 1.

Example 5

Give the name and current salary of any employee who never lived in Orlando and who first joined the ADM department before time 10.

range of e is EMP
retrieve (e.name , CURRENT(e.salary))
where (FIRST_START(e.dept $=_T$ *ADM) < 10)* AND_L
((e.city=Orlando) = NIL)

Notice that the use of the function CURRENT in the target list replaced the condition *e.salary OVERLAPS* $< NOW , NOW>$ that would be otherwise needed in the (omitted) output qualification.

The comparison operators presented in Section 2 are primarily designed to compare values. It is also possible to define operators that are intended to compare sets of value/interval pairs. The following is one example of a binary operator that compares two sets of value/interval pairs.

COVERED_BY

This function operates on sets of value/interval pairs. It returns the universal interval only if for every value/interval pair (v, I_1) in the left operand, there is a value/interval pair (v, I_2) in the right operand such that the interval I_1 overlaps the interval I_2.

Example 6

Give the name of any employee who co-worked with Fred in every department that Fred served.

range of *e* , *f* *is EMP*
retrieve (*e.name*)
where *f.name=Fred* AND_L (*f.dept* *COVERED_BY* *e.dept*)

8. Conclusion

In this paper, we outlined an approach to handle time in relational query languages. The approach is based on extending Boolean and comparison operators to handle operands that are sets of intervals. The proposed temporal logic satisfies the properties of the Boolean logic. The operators presented in this paper are not meant to be exhaustive. Other useful operators as well as syntax/semantics for update queries can be easily incorporated into the proposed logic. Several examples have been given in this paper in order to show that the proposed model can give good degree of flexibility in expressing different temporal requirements.

References

[AHN86] Ahn, I. "Towards an implementation of Database management systems with temporal support" Proc. IEEE Conf. on Data Engineering, 1986, pp. 374-381.

[ARIA86] Ariav, G. "A temporally oriented data model" ACM Trans. of Database Systems, Vol. 11, 1986, 499-527.

[BASS87] Bassiouni, M. "Set-valued Boolean and comparison temporal operators" Technical Report, Department of Computer Science, University of Central Florida, 1987.

[CLIF85] Clifford, J. and Tansel, A. "On an algebra for historical relational databases: two views" Proc. ACM SIGMOD Int. Conf. on Management of Data, 1985, pp. 247-265.

[GADI85] Gadia, S. and Vaishnav, J. "A query language for a homogeneous temporal database" Proc. ACM Symp. on Principles of Database Systems, 1985.

[GADI86] Gadia, S. "Towards a multihomogeneous model for a temporal database" Proc. IEEE Int. Conf. on Data Engineering, 1986, pp. 390-397.

[MCKE87] McKenzie, E. and Snodgrass, R. "Extending the relational algebra to support transaction time" Proc. ACM SIGMOD Conf. on Management of Data, 1987, pp. 467-478.

[NAVA87] Navathe, S. and Ahmed, R. "TSQL- A language interface for temporal databases" Proc. of Temporal Aspects of Information systems, North-Holland, 1987.

[NAVA88] Navathe, S. and Ahmed, R. "A temporal relational model and a query language" to appear in International Journal of Information Sciences.

[SEGE87] Segev, A. and Shoshani, A. "Logical modeling of temporal data" Proc. ACM SIGMOD Conf. on Management of Data, 1987, pp. 454-466.

[SNOD86] Snodgrass, R. and Ahn, I. "Temporal databases", Computer, Vol. 19, No. 9, 1986, 35-42.

[SNOD87] Snodgrass, R. "The temporal query language TQUEL" ACM Trans. on Database Systems, Vol. 12, No. 2, 1987, pp. 247-298.

[STON76] Stonebraker, M. et al "The design and implementation of INGRES" ACM Trans. on Database Systems, Vol. 1, No. 3, 1976, pp. 189-222.

[TANS86] Tansel, A. et al "A query language for historical relational databases" Proc. 3rd Int. Workshop on Statistical and Scientific Databases, 1986.

[ZANI83] Zaniolo, C. "The database language GEM" Proc. ACM SIGMOD Int. Conf. on Management of Data, 1983, pp. 207-218.

Software and Hardware Enhancement of Arithmetic Coding

M. A. Bassiouni

N. Ranganathan

Amar Mukherjee

Department of Computer Science

University of Central Florida

Orlando, Florida 32816 USA

Abstract

Arithmetic coding utilizes the skewness of character distribution by assigning larger intervals (code ranges) to characters having higher probabilities of occurrence. In this paper, we present a scheme to enhance arithmetic coding by utilizing the locality of character reference, i.e., the tendency of consecutive characters to fall within the same type (e.g., alphabets, digits, trailing blanks, successive zeros). The proposed scheme effectively increases the code ranges of individual characters by splitting the interval assignment into different groups. This will decrease the rate of interval narrowing and hence improve the compression efficiency. The paper is concluded by giving the VLSI design of the modified arithmetic coding scheme. Successful VLSI compression chips would be a significant enhancement to the technology of data encoding and would greatly contribute to reducing the cost of data transmission and data access within statistical database systems.

1. Introduction

Data compression is becoming increasingly important as a useful tool for the reduction of the cost of data transmission and storage. Data compression is particularly important to the design of high-speed vector machines. Some supercomputers (e.g., Fujitsu VP-200) use a compress/ expand mode as one of the available vector instruction modes. Numerous data compression algorithms have been proposed [e.g., HUFF52, EGGE80, EGGE81, BASS87, BASS86, WELC84, etc.] Recently, arithmetic coding has received an increased interest as a technique that gives high compression efficiency for a variety of data types [WITT87] as well as being amenable to efficient software/hardware implementations. In this paper, we present a modification of the arithmetic coding technique to further improve its compression efficiency. VLSI design of the modified scheme is also described.

2. Arithmetic Coding

The idea of arithmetic coding is to map a message into an interval of real numbers between 0 and 1. In practical implementation, an extra special symbol [WITT87] can be appended as an end-of-message (EOM) indicator. This EOM character enables the mapping of a message into a unique real number between 0 and 1. The following discussion illustrates the mechanism of arithmetic coding and prepares for our modification based on correlational data properties.

Consider a simplified set of three characters whose probabilities of occurrence are given in Table 1. It is assumed that the character "#" will be used only to indicate the end of an encoded portion of data (and cannot occur within the input data).

Table 1. A Simplified Character Set

Character	Probability
A	0.4
1	0.4
#	0.2

In arithmetic coding, characters are assigned nonoverlapping intervals whose union is the interval [0.0,1.0). Table 2 shows one possible assignment for the characters of Table 1.

Table 2. Interval Assignment

Character	Interval
A	[0.0,0.4)
1	[0.4,0.8)
#	[0.8,1.0)

The idea of arithmetic coding is to start with the initial interval [0.0,1.0) and then narrow it repeatedly (as characters are processed) such that each interval is totally contained in the preceding one. As an example, suppose that we want to encode the two-character message "1A". Upon receiving the first character, the initial interval I_0=[0.0,1.0) is transformed into the new interval I_1=[0.4,0.8) which corresponds to the code range of character "1" in Table 2. Upon encountering the second character, the interval I_1 is narrowed once again to give the new interval I_2=[0.4,0.56) which represents a subinterval of I_1 corresponding to the range of character "A" in Table 2. To see this, notice that Table 2 implies that character "A" represents the first 40% of a given interval. Accordingly, interval I_2 is obtained by extracting the first 40% portion of I_1. Also, the length of interval I_2 is equal to 0.56-0.40=0.16 which is 40% of the length of I_1.

In general, If $I_k = [a_k, b_k)$ is the current interval, and the range of the next character is given by $[s, f)$, then the next interval $I_{k+1} = [a_{k+1}, b_{k+1})$ is computed as follows.

$$a_{k+1} = a_k + s * (b_k - a_k)$$

$$b_{k+1} = a_k + f * (b_k - a_k)$$

This assures that I_{k+1} is a subinterval of (i.e., totally contained in) the interval I_k. Furthermore, the following relationship holds true.

$$length(I_{k+1}) = (f - s) * length(I_k)$$

where $length(I_k) = b_k - a_k$.

Returning back to the encoding of the string "1A", the interval I_2=[0.4,0.56) is transformed to the final interval I_3=[0.528,0.56) by appending the EOM character, i.e., the string to be encoded becomes "1A#". Any number within the interval I_3, say 0.53, can be used to represent the transmitted message.

The decoding process of the above message uses primarily the same logic of the encoding process. When the decoder receives the value 0.53, it knows that the first character is "1" since the value 0.53 is contained in the interval [0.4,0.8) corresponding to that character in Table 2. The decoder therefore knows that after processing the first character, the encoder narrowed the initial interval I_0=[0.0,1.0) to the interval I_1=[0.4,0.8). Next, by examining I_1 and the value 0.53, the decoder concludes that the next character is "A" since this the only character that can transform I_1 into an interval containing the value 0.53 (and this new interval is [0.4,0.56) as explained before). The decoder continues in this fashion until it finds the EOM character.

3. Modified Scheme for Arithmetic Coding

In the previous example, the message "1A" gave the final interval I_3=[0.528,0.56) and any number in that interval could have been used to encode this message (the value 0.53 was randomly chosen in the above example). In practice, the process of computing successive intervals is normally carried out in binary and the final value chosen is the one with minimal number of bits. For example, the binary representation of the interval I_3=[0.528,0.56) truncated to 7 fractional bits is given by I_3=[0.1000011,0.1000111). Therefore the binary value 0.10001 (equivalent to the decimal value 0.53125) is the number within I_3 that can be represented by the smallest number of bits. It is easy to see that when the length of the final interval (i.e., the difference between its upper and lower limits) gets larger, less number of bits are needed to encode the corresponding message. To see this consider an interval $I=[V_1,V_2)$. As the difference V_2-V_1 gets smaller (corresponding to narrowing the code range), more top bits of V_1 and V_2 become the same. On the other hand if V_2 is much larger than V_1, only fewer top bits of V_1 and V_2 would agree and hence fewer bits are needed to represent the interval $[V_1,V_2)$.

The above discussion reveals that the precision (number of bits) required to represent the final interval grows with the length of the message, and at some point it might not be possible to continue encoding a long message. One possible solution is to divide the message into portions (not necessarily of the same length) such that each portion, when padded with EOM character, can be encoded into a fixed-length field (say 64 bits). Thus the encoder starts with an initial interval I_0=[0.0,1.0) and continues to process characters and narrow this interval until it judges that the precision limit has been reached (after padding EOM). A new portion would then be started with an initial interval I_0=[0.0,1.0) as before. The number of characters that can be encoded in a fixed-length field depends on how fast the initial interval gets narrowed. It is easy to see that characters with higher probabilities of occurrence (and hence with larger intervals) have slower effect on narrowing the interval than characters with smaller probabilities. Assigning larger intervals to the most frequent characters would therefore increase the compression efficiency since it enables encoding more characters (on the average) in the same fixed-length field. The arithmetic coding scheme therefore utilizes the distributional properties of data (i.e., the skewness of character frequency) in achieving tighter compression rate. We next present a scheme that also utilizes the correlational properties of data, i.e., the property that consecutive characters tend to be of the same type. The scheme is based on the observation that many data files and commercial/business data records exhibit a strong locality behavior of character reference, i.e., consecutive characters tend to be of the same type (alphabets, digits, trailing blanks, successive zeros, etc.) We first illustrate the basic idea by an example.

Consider again the simplified set of Table 1 whose intervals are given in Table 2. Suppose that we know that character "A" (representing alphabets) tends to occur in groups, and so does character "1" (representing digits). Thus a typical string may be of the form "AAAA1111". Table 3 shows the process of coding this string using the intervals of table 2.

Table 3. Coding of AAAA1111

Character	Interval	Length of Interval
	[0.0,1.0)	1.0
A	[0.0,0.4)	0.4
A	[0.0,0.16)	0.16
A	[0.0,0.064)	0.064
A	[0.0,0.0256)	0.0256
1	[0.01024,0.02048)	0.01024
1	[0.014336,0.018432)	0.004096
1	[0.0159744,0.0176128)	0.0016384
1	[0.01662976,0.01728512)	0.00065536
#	[0.017154048,0.01728512)	0.000131072

Now suppose instead of using Table 2, we employ a scheme of two separate interval assignments as shown in Tables 4 and 5. In each of these two tables, an extra character is introduced. This character, denoted by the symbol "@", is an imaginary character that is used to indicate a switch from one locality to another. When encoding the string "AAAA1111", for example, the encoder would use table 4 for the first 4 characters then would discover a change of locality. The encoder would then apply the interval of the imaginary character "@" (from Table 4) and switch into Table 5 for the encoding of the remaining characters. When the decoder processes the coded message and finds the symbol "@" it knows that a switch of locality occurred at this point and hence uses the other table for the decoding process.

Table 4. Interval Assignment: Alphabet

Character	Interval
A	[0.0,0.8)
@	[0.8,0.9)
#	[0.9,1.0)

Table 5. Interval Assignment: Digit

Character	Interval
1	[0.0,0.8)
@	[0.8,0.9)
#	[0.9,1.0)

Notice that the code range of character "A" in Table 4 is larger than that of the original scheme (Table 2). This means that character "A" in the new scheme will have a slower effect on narrowing the interval than in the original arithmetic coding scheme. Similar remarks apply to character "1". In practice the range assigned to the special characters "@" and "#" should be based on their expected probabilities of occurrence. The expected probability of "@" depends on the average length of consecutive alphabets (in Table 4) or digits (in Table 5), The probability of "#" depends on the expected number of characters that can be encoded (on the average) into the chosen fixed-length field.

Using the new scheme, the process of encoding the string "AAAA1111" is shown in Table 6 (assuming control starts at table 4).

Table 6. New Encoding for AAAA1111

Character	Control	Interval	Length of Interval
	Table 4	[0.0,1.0)	1.0
A	Table 4	[0.0,0.8)	0.8
A	Table 4	[0.0,0.64)	0.64
A	Table 4	[0.0,0.512)	0.512
A	Table 4	[0.0,0.4096)	0.4096
@	Table 4	[0.32768,0.36864)	0.04096
1	Table 5	[0.32768,0.360448)	0.032768
1	Table 5	[0.32768,0.3538944)	0.0262144
1	Table 5	[0.32768,0.34865152)	0.02097152
1	Table 5	[0.32768,0.344457216)	0.016777216
#	Table 5	[0.3427794944,0.344457216)	0.0016777216

Notice that the length of the final interval obtained in Table 6 is approximately 13 times the length of the interval obtained in Table 3. Thus the extra narrowing effect produced by the character "@" (at each locality switch) is more than offset by the increase in code ranges of individual alphabet and digit characters (which resulted from splitting them into separate tables).

Numerical Results

Our preliminary numerical tests have shown that the modification improves the compression efficiency of arithmetic coding by an amount that varies depending on the type of data. Improvements of up to 40% is obtainable for files showing good degree of locality (.e.g., inverted files and business records). Good improvement has also been obtained for image and text data. Improvement for files containing sound data has been very small (but arithmetic coding performed better than other schemes on these files; the LZ method for example actually expanded sound files).

4. VLSI Design for Modified Scheme

Due to the complexity of most compression methods, past implementations of data compression techniques have been mostly restricted to software. Few hardware designs using associative memory [LEA78], and microprocessor based systems [HAWT82] have been reported. A proposed design for compression by textual substitution is given in [GONZ85]; no implementation was done. A brief discussion about the hardware design (Sperry proprietary) of the LZW algorithm (using hash tables) is given in [WELC84]. A fast VLSI implementation of the Huffman's scheme is given in [MUKH87]. In this section, we present a high-level description of of a VLSI chip for the implementaion of the modified arithmetic coding scheme.

Fig. 1 gives a hardware implementation of the arithmetic coding technique. The hardware realization of the modified arithmetic coding scheme is given in Figure 2. The arithmetic coding scheme consists of arithmetic operations like addition and multiplication and hence can be easily implemented in hardware. The codeword is an interval represented as (C, C+A) where C is the current low point and A is the width of the current interval. Say, for character i, $p(i)$ is the symbol probability and $P(i)$ is the cumulative probability. Then while encoding character i, the new low point is calculated as $C + A * P(i)$ and the new width is calculated as $A * p(i)$. The new values of C and C+A will represent the interval soon after the encoding of character i. The RAM stores for each character, the corresponding symbol probability and the cumulative probability. We assume integer arithmetic since it has been shown [WITT87] that arithmetic coding can be implemented using integer arithmetic. During the first cycle $P(i)$ is input to the multiplier and the value $A*P(i)$ is calculated. During the second cycle, the product is input to the adder and is added to C. Also, during the second cycle, the product $A*p(i)$ is computed to obtain the new width. A 2:1 multiplexer is used to multiplex between $P(i)$ and $p(i)$ as input for the multiplier. The boxes labeled New A, Cur A, New C and Cur C are registers to hold the new and current values of A and C.

For the modified arithmetic coding scheme, the probabilities are stored in the RAM by separating the characters in each group. Also, the probabilities corresponding to the group switch characters are stored. The circuit works as follows: The next input symbol is loaded into a register from the buffer, which is used by the Group Switch Logic GSL to determine if there is a change of group. If there is no change in the group,

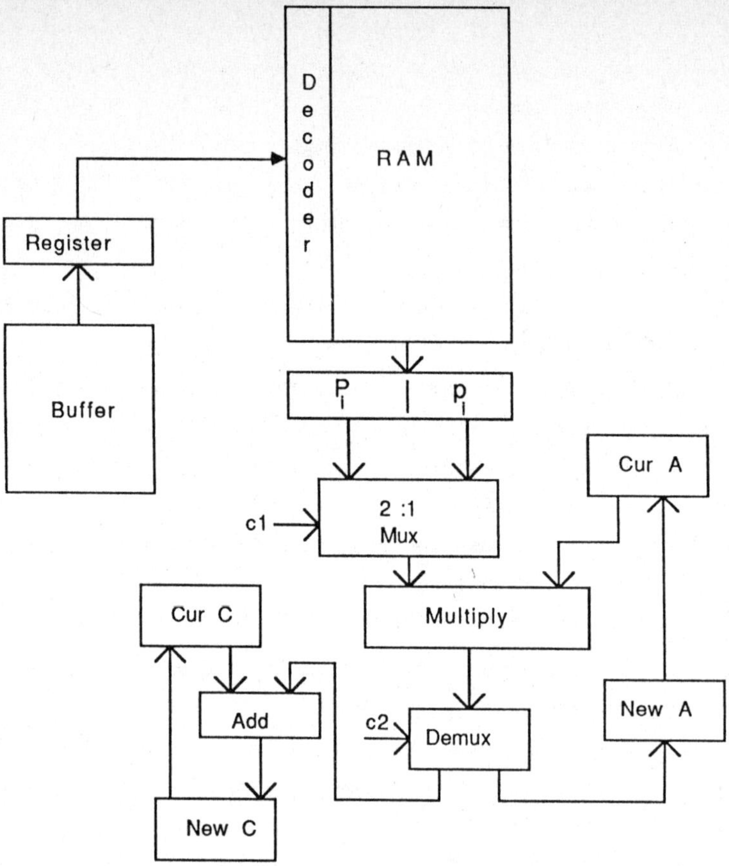

Figure 1. Hardware implementation of arithmetic coding

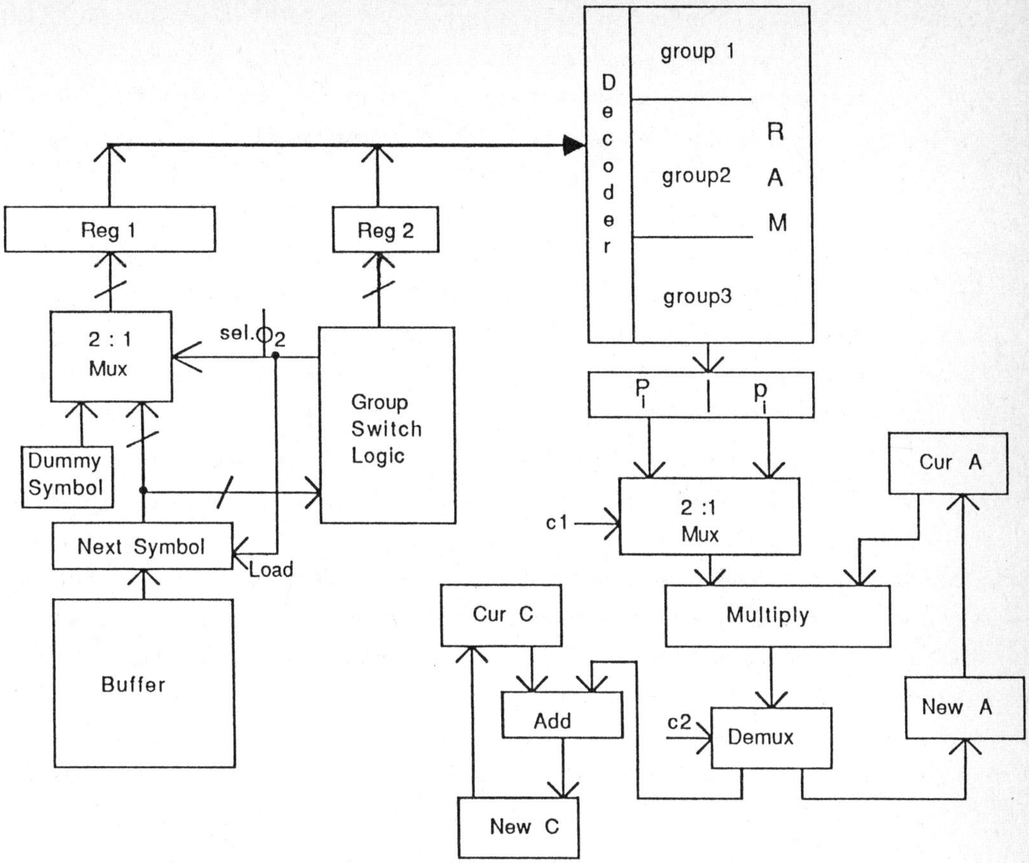

Figure 2. Hardware implementation of modified arithmetic coding

then GSL activates the $sel.\phi_2$ signal which passes the symbol to the register Reg1 to be decoded. In this case GSL loads the other register Reg2 with a fixed binary value. The bits in Reg2 are appended to those in Reg1 before they are input to the decoder in the RAM. These extra bits from Reg2 will be used in selecting the current group. Also $sel.\phi_2$ signal loads a new character into the next symbol register from the buffer. If there is a change in the group, the next symbol is made to wait and the $sel.\phi_2$ loads the dummy symbol into Reg1. Thus the $sel.\phi_2$ signal acts as a control input to the 2:1 multiplexor which selects between the next symbol and the dummy symbol. Again this dummy symbol corresponds to the situation where no symbol from any group is selected. Simultaneously, GSL loads a value in Reg2 which will select the group switch character and the corresponding probabilities will be outputted from the RAM. The size of Reg1 will be 8 bits in order to hold the ASCII symbols, while the size of Reg2 will depend on the number of groups. If there are three groups, then Reg2 will need 3 bits in order to represent six possible group switches.

The group switch logic is given in Figure 3. For simplicity, the logic is worked out for an example where we have three groups as follows: the original codes in the first group are assumed to be equal to or less than '11', the codes in the third group to be above or equal to the limit '13' and the second group falling in between. Say, group 3 is alphabets, group 1 is digits and group 2 is the blank. The group switch logic consists of mainly three comparators, a 1's complement adder and two registers. As soon as a new symbol is loaded into the symbol register from the buffer, the two comparators connected to the register along with the bubbled AND gate produce and load register Reg3 with a binary value of 100, 010 or 001 depending on whether the symbol is an alphabet, digit or a blank character. The state that corresponds to the previous symbol is stored in register Reg4. It is compared with the contents of Reg3 and if a match occurs, the control for the 2:1 Mux is enabled in order to pass on the symbol to be decoded. If there is a mismatch, the control signal is set to low and the adder is enabled. The adder performs a 1's complement subtraction of the contents of Reg3 from Reg4. The result is the output of the group switch logic which is to be loaded into Reg2 in Figure 2. The switch logic for the real ASCII codes has been designed using appropriate extensions/modifications of the scheme described in Fig. 3.

5. Conclusion

Efficient data encoding schemes are greatly needed to cope with the problems associated with data proliferation and data transmission. Schemes to improve compression efficiency or reduce compression overhead will significantly contribute to reducing the cost of data transmission and data access/storage within statistical database systems. In this paper, we presented a scheme to improve the efficiency of arithmetic coding by utilizing one of the observed properties of data, namely the locality of character reference. The proposed scheme effectively increases the code ranges of individual characters by splitting the interval assignment into separate groups. This will decrease the rate of interval narrowing and hence improve compression efficiency.

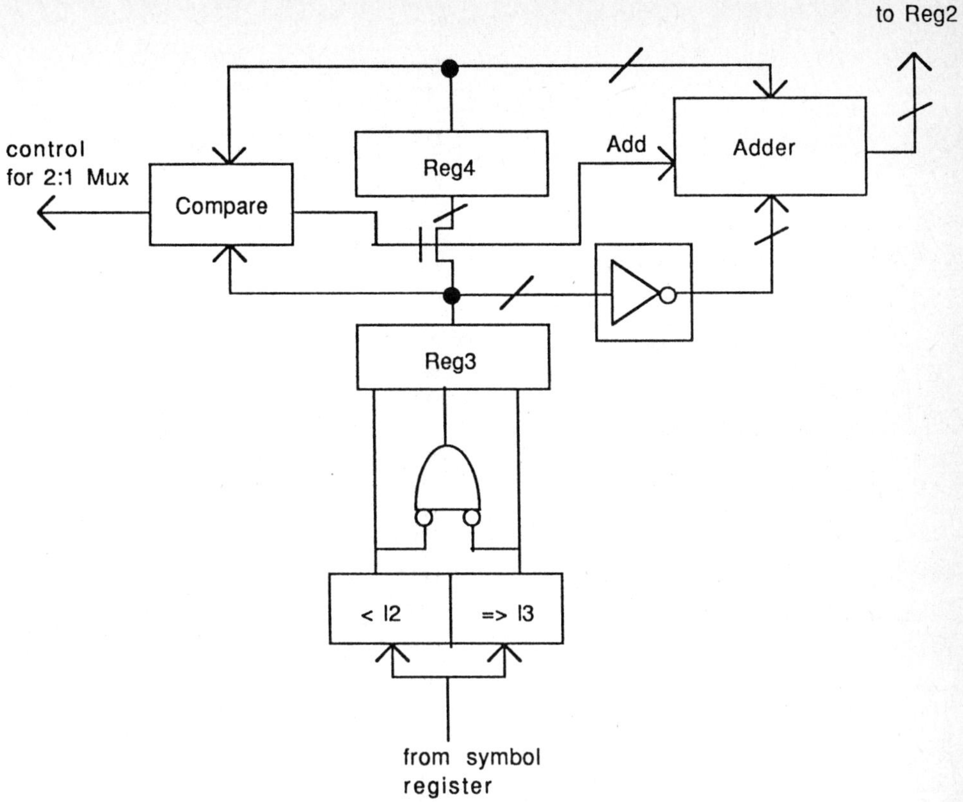

Figure 3. Group switch logic

Hardware assistance for data compression would be of valuable help to many information systems. The paper has discussed an approach for the VLSI realization of the modified arithmetic coding.

References

[BASS87] Bassiouni, M. and Mukherjee, A. "Supercomputer algorithms for data transmission and encoding" Proc. 2nd International Conf. on Supercomputing, 1987.

[BASS86] Bassiouni, M. and Ok, B. "Double encoding- a technique for reducing storage requirement of text" J. Information Systems, Vol. 11, No. 2, 1986, pp. 177-184.

[EGGE80] Eggers, S. and Shoshani, A. "Efficient access of compressed data" Proc. VLDB, 1980, pp. 205-211.

[EGGE81] Eggers, S.; Olken, F. and Shoshani, A. "A compression technique for large statistical databases" Proc. VLDB, 1981, pp. 424-434.

[GONZ85] Gonzalez-Smith, M. and Storer, J. "parallel algorithms for data compression" JACM, Vol. 32, No. 2, April 1985, pp. 344-373.

[HAWT82] Hawthorn, P. "Microprocessor assisted tuple access decompression and assembly for statistical database systems" Proc. VLDB, 1982, pp. 223-233.

[HUFF52] Huffman, D. "A method for the construction of minimum redundancy codes" Proc. IRE, Vol. 40, 1952, pp. 1098-1101.

[LEA78] Lea, R. "Text compression with associative parallel processors" The Computer Journal, Vol. 21, No. 1, 1978, pp. 45-56.

[MUKH87] Mukherjee, A. and Bassiouni, M. "A VLSI chip for efficient transmission and retrieval of information" Proc. 10th ACM SIGIR International Conf. on Research and Development in Information Retrieval, June 1987, pp. 208-216.

[MUKH86] Mukherjee, A. *Introduction to nMOS and CMOS VLSI Systems Design.* Prentice-Hall, Englewood Cliffs: N.J., 1986.

[WELC84] Welch, t. "A technique for high-preformance data compression" Computer, Vol. 17, No. 6, 1984, pp. 8-19.

[WITT87] Witten, I.; Neal, R. and Cleary, J. "Arithmetic coding for data compression" Communications of ACM, Vol. 30, No. 6, June 1987, pp. 520-540.

ORTHOGONAL RANGE RETRIEVAL USING BUCKET ADDRESS HASHING

C. C. Chang* and C. Y. Chen**

* C. C. Chang is with the Institute of Applied Mathematics
National Chung Hsing University, Taichung, Taiwan 40227
Republic of China

** C. Y. Chen is with the Department of Electronics
Feng Chia University, Taichung, Taiwan 40724
Republic of China

ABSTRACT

This paper is concerned with the multi-attribute file system design problem for orthogonal range retrieval. A formula which can be used to determine the optimal number of bits of the bucket address to assign to each attribute of a record is derived, assuming that all the queries are not equally likely. Also, an algorithm which can be applied to produce optimal integer solutions is presented. It will be seen that an optimal file structure for partial match queries does not guarantee the optimal file structure for orthogonal range queries.

1. INTRODUCTION

The multi-attribute file system design problem for partial match queries (PMQs) is to design a multi-attribute file system for which the average number of buckets that need to be examined, over all possible PMQs, is minimized. A multi-attribute file is a file in which each record is characterized by more than one attribute. By a PMQ, we mean a query of the following form: Retrieve all records satisfying $A_{i_j}=a_{i_j}$ for $1 \le j \le h$ and $A_{i_j}=*$ for $h+1 \le j \le N$, where A_{i_j} represents the i_j-th attribute, a_{i_j} denotes a value belonging to the domain of A_{i_j}, and * represents a don't care condition. Because of the importance of the PMQs in information retrieval system, much attention has been devoted in recent years [Aho and Ullman 1979, Bolour 1979, Burkhard 1976a, Burkhard 1976b, Chang 1984a, Chang 1984b, Chang, Du and Lee 1984, Chang, Lee and Du 1980, Chang, Lee and Du 1982, Chang and Su 1987, Du 1985, Lee and Tseng 1979, Lin, Lee and Du 1979, Liou and Yao 1977, Rivest 1976, Rothnie and Lozano 1974, Tang, Buehrer and Lee 1985]. Unfortunately, a solution of this problem is at large. Tang, Buehrer and Lee [1985] showed that this problem is NP-hard. Thus many heuristic methods, like the bucket-address hashing (BAH) method etc., had been proposed.

Aho and Ullman [1979] concerned a scheme in which the records were stored in a set of buckets with fixed capacity. For a query, they computed the addresses of the buckets that need to be examinted, assuming that the probability of one attribute being specified in a query depends on the attribute itself and is independent of the other attributes. We call this scheme the BAH method. Since in a query either all or none of the bits of an attribute are specified, to minimize the average number of buckets that need to be examined for all queries, the real issue is knowing how to determine the number of bits of the bucket address to allocate to each attribute of a record.

The concept of the BAH method can be explained by considering a file of telephone subscribers, with each subscriber record consisting of five attributes: first name, middle name, last name, street address and town. Suppose these five attributes are attached with hypothetical probabilities as shown in the second column of Table I. Suppose that there are 2^{20} records and that each bucket nominally contains 16 records. Then there are totally $2^{20}/16=2^{16}=65536$ buckets. If we select 16 bits of the last name to determine the address of the bucket, then the average number of buckets examined is $(0.9)\times 1+(0.1)\times 65536=6555$ buckets. If, however, we choose to use 8 bits from the first name and 8 bits from the last name to specify a bucket, we can do better. The average number of buckets would then be $(0.8\times 0.9)\times 1+(0.2\times 0.9)\times 256+(0.8\times 0.1)\times 256+(0.2\times 0.1)\times 65536=1378$ buckets. We can do even better than this, on the average, however.

Attributes	Probabilities	b_i	$\bar{b}_i$
First name	0.8	5.20	5
Middle name	0.1	0.03	0
Last name	0.9	6.37	7
Street address	0.2	1.20	1
Town	0.5	3.20	3

Table I

An important formula derived by Aho and Ullman [1979] to calculate the average number of buckets which must be examined to satisfy a query is

$$ANB_{PMQ}=2^{B}\cdot p\cdot \prod_{i=1}^{N}(1+r_i\cdot 2^{-b_i}), \tag{1.1}$$

where $B=\sum_{i=1}^{N} b_i$ represents the total number of bits to be used, b_i is the number of bits from the i-th attribute used to compute a bucket address, $P=\sum_{i=1}^{N}(1-P_i)$, P_i is the probability that the i-th attribute is specified, $r_i = \frac{P_i}{1-P_i}$, and N is the number of attributes in a record.

By minimizing (1.1), the optimal number of bits to allocate to attribute i is

$$b_i = \frac{B}{N} + \log\frac{P_i}{1-P_i} - \frac{1}{N}\sum_{j=1}^{N}\log\frac{P_j}{1-P_j}, \quad i=1,2,\ldots,N. \tag{1.2}$$

The third column of Table I shows the optimal values of the b_i's calculated from (1.2).

Unfortunately, the b_i's are not integers in general. A method which was also proposed by Aho and Ullman [1979] to obtain an optimal integer solution is as follows. Let $\bar{b}_i$ be the optimal integer approximation of b_i. Then $\bar{b}_i$ is b_i, rounded either up or down. Let FP be the sum of fractional parts of all b_i's (Note that since $b_1+b_2+\ldots+b_N=B$, FP must be an integer). Select those FP b_i's having the largest fractional parts and round them up; round the others down. In Table I the fractional parts of the b_i's sum to 1. We therefore select b_3, the one with the largest fraction and round it up; the other b_i's are rounded down. These $\bar{b}_i$'s are listed in the last column of Table I.

In this paper, we consider the BAH method for orthogonal ragne queries (ORQs) instead of PMQs. By an ORQ, we mean a query of the following form: Retrieve all records staisfying $\ell_{i_j} \leq A_{i_j} \leq u_{i_j}$ for $1 \leq j \leq h$ and $A_{i_j}=*$ for $h+1 \leq j \leq N$, where ℓ_{i_j} and u_{i_j} are lower and upper limits of the i_j-th attribute, and if $\ell_{i_j}=u_{i_j}$ for j=1,2,...,h, then it is reduced to a standard PMQ.

In Section 2, we derive a performance formula of the BAH method for ORQs. In Section 3, a formula is derived for computing the optimal number of bits of the bucket address to assign to each attribute of a record. We shall show that an optimal file structure based upon the BAH method for PMQs does not guarantee an optimal file structure based upon the BAH method for ORQs. Conclusions will appear in Section 4.

2. PERFORMANCE ANALYSIS OF THE BAH METHOD FOR ORQs

Let there be N attributes $A_1,A_2,\ldots,A_N$ for each record. Let P_i be the probability that A_i is specified and let $P = \prod_{i=1}^{N} (1-P_i)$, $r_i = \frac{P_i}{1-P_i}$. Since queries are not always random, we assume that P_i depends on A_i itself and is independent of the other attributes. Let b_i be the number of bits from the i-th attribute used to compute a bucket address, $m_i = 2^{b_i}$, $B = \sum_{i=1}^{N} b_i$ be the total number of bits used, and $NB = 2^B$ be the total number of buckets. We denote $Q_{i_1 i_2 \ldots i_h}$ as the set of queries for which $\ell_{i_j} \leq A_{i_j} \leq u_{i_j}$, $1 \leq j \leq h$, and $A_{i_j} = *$, $h+1 \leq j \leq N$, for $0 \leq h \leq N$. Meanwhile, for h=0, we denote $Q_{i_1 i_2 \ldots i_h}$ as Q_0 which means to list the entire set of records.

In the following, we shall analyze the performance of the BAH method for ORQs.

(1) (a) For $q \in Q_0$, the expected number of buckets need to be examined is NB.

(b) The probability for q is P.

(2) (a) For $q \in Q_{i_1}$, the expected number of buckets need to be examined is $\frac{NB}{m_{i_1}} \cdot$(expected value of $(u_{i_1} - \ell_{i_1} + 1)$) $= \frac{NB}{m_{i_1}} \cdot e_{i_1}$, where $e_{i_1} = \frac{m_{i_1}+3}{4}$ represents the expected value of $u_{i_1} - \ell_{i_1} + 1$ which is found by assuming that the probabilities of ℓ_{i_1} being k, $1 \leq k \leq m_{i_1}$, are all identical to $\frac{1}{m_{i_1}}$. Since $u_{i_1} - \ell_{i_1} + 1$ may be varied from 1 to $m_{i_1} - k + 1$, where $1 \leq k \leq m_{i_1}$, therefore, the expected value of $u_{i_1} - \ell_{i_1} + 1$ equals

$$e_{i_1} = \frac{1}{m_{i_1}} \left[\frac{1}{m_{i_1}} \cdot (1+2+\ldots+m_{i_1}) + \frac{1}{m_{i_1}-1} \cdot (1+2+\ldots+(m_{i_1}-1)) + \ldots + \frac{1}{m_{i_1}-k+1} \cdot (1+2+\ldots+(m_{i_1}-k+1)) + \ldots + \frac{1}{1} \cdot 1 \right]$$

$$= \sum_{t=1}^{m_{i_1}} \sum_{y=1}^{t} \frac{y}{m_{i_1} \cdot t}$$

$$= \frac{1}{m_{i_1}} \cdot \sum_{t=1}^{m_{i_1}} \frac{(t+1)}{2}$$

$$= \frac{m_{i_1}+3}{4}.$$

For the remainder of this section, we shall use e_{i_j} $(=(m_{i_j}+3)/4)$ to denote the expected value of $u_{i_j}-\ell_{i_j}+1$, $1 \le j \le N$.

(b) The probability for q is $P_{i_1} \cdot \prod_{\substack{j=1 \\ j \ne i_1}}^{N} (1-P_j) = \frac{P_{i_1}}{1-P_{i_1}} \cdot \prod_{j=1}^{N} (1-P_j) = r_{i_1} \cdot P.$

(3) (a) For $q \in Q_{i_1 i_2}$, the expected number of buckets need to be examined is $\frac{NB}{m_{i_1} \cdot m_{i_2}} \cdot e_{i_1} \cdot e_{i_2}$.

(b) The probability for q is $P_{i_1} \cdot P_{i_2} \cdot \prod_{\substack{j=1 \\ j \ne i_1, i_2}}^{N} \cdot (1-P_j) = r_{i_1} \cdot r_{i_2} \cdot P.$

(4) (a) For $q \in Q_{i_1 i_2 \ldots i_h}$, $3 \le h \le N$, the expected number of buckets need to be examined is $\frac{NB}{m_{i_1} \cdot m_{i_2} \cdot \ldots \cdot m_{i_h}} \cdot e_{i_1} \cdot e_{i_2} \cdot \ldots \cdot e_{i_h}$.

(b) The probability for q is $r_{i_1} \cdot r_{i_2} \cdot \ldots \cdot r_{i_h} \cdot P$.

Therefore, the average number of buckets need to be examined over all possible ORQs is

$$ANB_{ORQ} = \sum_{h=0}^{N} \sum_{\substack{i_1 < i_2 < \ldots < i_h \\ \{i_1, i_2, \ldots, i_h\} \subseteq \{1,2,\ldots,N\}}} \cdot \frac{NB}{m_{i_1} m_{i_2} \ldots m_{i_h}} \cdot e_{i_1} \cdot e_{i_2} \cdot \ldots \cdot e_{i_h} \cdot r_{i_1} \cdot r_{i_2} \cdot \ldots \cdot r_{i_h} \cdot P$$

$$= NB \cdot P \cdot \prod_{i=1}^{N} \left(1 + \frac{e_i r_i}{m_i}\right)$$

$$= 2^B \cdot P \cdot \prod_{i=1}^{N} \left(1 + \frac{r_i}{m_i} \cdot \frac{m_i+3}{4}\right)$$

$$= 2^B \cdot P \cdot \prod_{i=1}^{N} \left(1 + \frac{r_i}{4} + \frac{3r_i}{4} \cdot 2^{-b_i}\right)$$

$$= 2^B \cdot P \cdot \prod_{i=1}^{N} \left(1 + \frac{r_i}{4}\right) \cdot \prod_{i=1}^{N} \left(1 + r_i' \cdot 2^{-b_i}\right), \qquad (2.1)$$

where $r_i' = \frac{3r_i}{4+r_i}$, $1 \le i \le N$.

Comparing to (1.1), we find that the above formula is very similar to (1.1) since (2.1) can be reexpressed as

$$ANB_{ORQ} = 2^B \cdot P' \cdot \prod_{i=1}^{N} (1 + r_i' \cdot 2^{-b_i}),$$

where $P' = P \cdot \prod_{i=1}^{N} (1 + \frac{r_i}{4})$ = a constant.

3. OPTIMAL BUCKET DESIGN AND OPTIMAL INTEGER SOLUTIONS

From (2.1), our file system based upon BAH method design problem can be stated as follows: Find the values of the b_i's to minimize $\prod_{i=1}^{N} (1 + r_i' \cdot 2^{-b_i})$ subject to the constraint $\sum_{i=1}^{N} b_i = B$.

By applying the method of Lagrange's multipliers to minimize $\prod_{i=1}^{N} (1 + r_i' \cdot 2^{-b_i})$, we have a similar result as (1.2). That is

$$b_i = \frac{B}{N} + \log r_i' - \frac{1}{N} \sum_{j=1}^{N} \log r_j', \quad 1 \le i \le N. \tag{3.1}$$

Substituting $\frac{3r_j}{4+r_j}$ for r_j' in (3.1), we have that

$$b_i = \frac{B}{N} + \log \frac{P_i}{4-3P_i} - \frac{1}{N} \sum_{j=1}^{N} \log \frac{P_j}{4-3P_j}, \quad 1 \le i \le N. \tag{3.2}$$

Generally, b_i's are not always integers. Now, the algorithm proposed in [Aho and Ullman 1979] can be applied to determine the optimal integer approximations. That is, $\bar{b}_i = b_i$ (rounded either up or down). Let FP be the sum of fractional parts of all b_i's. Select those FP b_i's having the largest fractional parts and round them up; round the others down.

Example 3.1.

Consider the file of telephone subscribers as shown in Table I for ORQs. In this case, N=5, B=16, P_1=0.8, P_2=0.1, P_3=0.9, P_4=0.2, and P_5 =0.5. The values of $\log \frac{P_i}{4-3P_i}$ for the various values of P_i are given in the fourth column of Table II. The optimal values of the b_i's, computed from (3.2), are listed in the fifth column. Since FP=0.80+0.73+0.26+0.74 +0.47=3.00, we select b_1, b_4 and b_2, the three with the largest fractional parts and round them up; the others are rounded down. This leaves the optimal integer approximations as listed in the last column.

i	P_i	$P_i/(4-3P_i)$	$\log(P_i/(4-3P_i))$	b_i	$\bar{b}_i$
1	0.8	0.50	-1.00	4.80	5
2	0.1	0.03	-5.06	0.73	1
3	0.9	0.69	-0.54	5.26	5
4	0.2	0.06	-4.06	1.74	2
5	0.5	0.20	-2.32	3.47	3

Table II

Since an ORQ can be decomposed into a set of PMQs assuming that each attribute domain is finite and discrete, Du [1985] made the conjecture that an optimal file structure for PMQs is also likely to be "good" for ORQs. However, from (1.2) and (3.2), or from the results of Table I and Table II, we conclude that the optimal file structure based upon the BAH method for PMQs does not guarantee optimality for ORQs.

4. CONCLUSIONS

The multi-attribute file system design problem for PMQs is a special kind of the multi-attribute file system design problem for ORQs. Since, in [Tang, Beuhrer and Lee 1985], it was shown that the former is NP-hard, we can easily infer that the later is also an NP-hard problem.

In this paper, the BAH method for ORQs is concerned. We have analyzed the performance of the multi-attribute file systems based upon the BAH method for ORQs. A formula for the optimal number of bits associated to each attribute is presented. An algorithm to compute optimal integer number of bits for each attribute is also introduced. Furthermore, we have pointed out that the optimal file structure based upon the BAH method for PMQs does not imply that it is optimal for ORQs.

REFERENCES

[1] Aho, A. V. and Ullman, J. D., (1979): "Optimal Partial-Match Retrieval When Fields are Independently Specified," ACM Transactions on Database Systems, Vol. 4, No. 2, pp. 168-179.

[2] Bolour, A., (1979): "Optimality Properties of Multiple Key Hashing Functions," Journal of the Association for Computing Machinery, Vol. 26, No. 2, pp. 196-210.

[3] Burkhard, W. A., (1976a): "Hashing and Trie Algorithms for Partial Match Retrieval," ACM Trans. Database Syst., Vol. 1, No. 2, pp. 175-187.

[4] Burkhard, W. A., (1976b): "Partial Match Retrieval," BIT, Vol. 16, No. 1, pp. 13-31.

[5] Chang, C. C., (1984a): "Optimal Information Retrieval When Queries Are Not Random," Information Sciences, Vol. 34, pp. 199-223.

[6] Chang, C. C., (1984b): "Optimal Partial Match Retrieval When The Number of Buckets is a Power of Prime," Proceedings of International Computer Symposium, Taipei, Taiwan, Dec. 1984, pp. 807-813.

[7] Chang, C. C., Du, M. W. and Lee, R. C. T., (1984): "Performance Analyses of Cartesian Product Files and Random Files," IEEE Transactions on Software Engineering, Vol. SE-10, No. 1, pp. 88-99.

[8] Chang, C. C., Lee, R. C. T. and Du, H. C., (1980): "Some Properties of Cartesian Product Files," Proceedings of ACM-SIGMOD 1980 Conference, pp. 157-168.

[9] Chang, C. C., Lee, R. C. T. and Du, M. W., (1982): "Symbolic Gray Code as a Perfect Multi-attribute Hashing Scheme for Partial Match Queries," IEEE Transactions on Software Engineering, Vol. SE-8, No. 3, pp. 235-249.

[10] Chang, C. C., and Su, D. H., (1987): "Performance Analyses of Multi-attribute Files Based upon Multiple Key Hashing Functions and Haphazard Files," Journal of the Chinese Institute of Engineers, Vol. 10, No. 1, pp. 99-105.

[11] Du, H. C., (1985): "On the File Design Problem for Partial Match Retrieval," IEEE Transactions on Software Engineering, Vol. SE-11, No. 2, pp. 213-222.

[12] Lee, R. C. T. and Tseng, S. H., (1979): "Multi-key Sorting," Policy Analysis and Information Systems, Vol. 3, No. 2, pp. 1-20.

[13] Lin, W. C., Lee, R. C. T. and Du, H. C., (1979): "Common Properties of Some Multi-attribute File Systems," IEEE Transactions on Software Engineering, Vol. SE-5, No. 2, pp. 160-174.

[14] Liou, J. H. and Yao, S. B., (1977): "Multi-dimensional Clustering for Data Base Organizations," Information Systems, Vol. 2, pp. 187-198.

[15] Rivest, R. L., (1976): "Partial-match Retrieval Algorithms," SIAM Journal of Computing, Vol. 14, No. 1, pp. 19-50.

[16] Rothnie, J. B. and Lozano, T., (1974): "Attribute Based File Organization in a Paged Memory Environment," Communications of the Association for Computing Machinery, Vol. 17, No. 2, pp. 63-69.

[17] Tang, T. Y., Buehrer, D. J. and Lee, R. C. T., (1985): "On the Complexity of Some Multi-attribute File Design Problem," Information System, Vol. 10, No. 1, pp. 21-25.

AUTOMATIC DRAWING OF STATISTICAL DIAGRAMS

Giuseppe Di Battista

Dipartimento di Informatica e Sistemistica
Universita' di Roma "La Sapienza"
Via Buonarroti, 12 - 00185 Roma

Abstract

Statistical diagrams, obtained using data models like SUBJECT, GRASS, SAM*, or CSM, are widely used in statistical databases, both for design purposes and as iconic representations for user friendly interfaces. In this paper an efficient layout algorithm is proposed that allows the automatic drawing of statistical diagrams, according to a set of aesthetics. The algorithm receives as input a graph, representing a statistical schema, and produces a drawing of the corresponding diagram, through an incremental specification of its features.

1. Introduction

Methodologies for information systems design suggest wide use of diagrams as production tools (see for instance [OSV 82] [TRC 79]): in the design phase diagrams are an effective documentation means and represent a common language for the designer and the user to express the requirements of the application; in the usage phase, diagrams provide a friendly interface for query expression and for manipulation of objects stored in the database.

Several data models have been proposed in literature for describing a statistical database application: Shoshani and Chan [SC 80] proposed SUBJECT that has been designed to describe large statistical databases at the logical level, Rafanelli and Ricci [RR 83] defined GRASS, designed to describe tables (the basic data structure in statistical applications). Other two data models are SAM* and CSM, proposed respectively by Su [S 83] and Di Battista, Ferranti, and Batini [DFB 86]; the aim of these two proposals is to describe data at the conceptual level, without any concern with the specific language or model chosen for the implementation. Schemas produced with the above mentioned statistical models are usually represented by diagrams that have typically a hierarchic structure (see in figure 1 an example of GRASS diagram).

Such diagrams are usually drawn manually or using a graphic editor, but in both cases placement of symbols and routing of connections is the responsibility of the designer. From this point of view the availability of an automatic layout facility reduces costs involved in drawing and in successive maintenance activities.

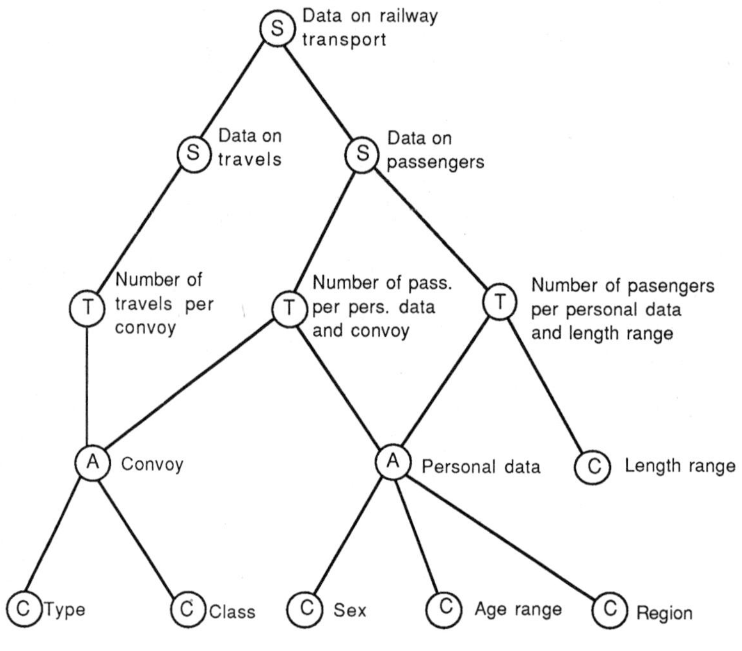

Figure 1: A diagram defined with the GRASS model.

Several database design tools have been implemented in the last ten years, some of these tools are provided with a graphic editor, but such editors have at least two limitations that can be overcome only with the aid of an automatic layout tool:

1. If the user must integrate partial products of the design activity, no means is provided to automatically produce an integrated diagram;

2. For systems that manage both text and graphic documentation, no means is provided to automatically update diagrams when the textual part has been updated.

In statistical databases such need for an automatic layout facility is especially felt, in fact statistical database applications are typically defined incrementally, producing at each step a diagram describing a new aggregation that has to be integrated into a global diagram. Moreover the importance of graphical user interfaces for statistical applications has been pointed out by several researchers (see for instance [C et al. 87]).

The problem of automatically producing a tidy and therefore readable drawing of a diagram has been deeply investigated in the literature and is gaining a growing interest (a recent survey of this field appears in [TDB 88]). The main results can be taxonomized referring to the graph underlying the diagram (for standard terminology on graphs see [E 79]): they concern the drawing of trees, of flat (undirected) graphs, and of directed graphs (in the following digraphs).

Concerning trees, the main result is described in [RT 81], where the drawing is produced with a divide and conquer strategy. For undirected graphs, general results are in [T 87] and [TDB 88]; in [BTT 84] and [BNT 86] two algorithms are presented to produce respectively Entity- Relationship and Data Flow diagrams.

Due to their hierarchic structure, the field of acyclic digraphs seems to be the best related one to the problem of drawing a statistical diagram (SD in the following). In this area all the proposed algorithms produce drawings in which the edges all flow in the same direction (e.g. from top to bottom or from left to right). A theoretical result is in [DT 88] in which the planarity problem is investigated, with the above mentioned constraint. In [W 77], [C 80], and [STT 81] algorithms are proposed to draw diagrams with a hierarchic structure, and therefore similar to SD's. However they present at least two limitations:

- The Warfield algorithm ([W 77]) deals only with the problem of minimizing the number of crossings between edges in the drawing and the Carpano algorithm ([C 80]) is substantially an improvement to the Warfield approach. Both of them do not give any bound on the number of crossings obtained.
- The third algorithm ([STT 81]) defines an interesting framework for the layout problem, which we partially reuse here, but nothing is said about the performance of the proposed approach.

We have also noticed that no analysis is provided about time complexity of the three algorithms and that they are probably well suited for dense graphs (i.e. graphs in which the number of edges is close to the maximum) but their performance with sparse diagrams (i.e. diagrams with a number of edges that is approximately linear in the number of vertices) is not well known.

In this paper a new automatic layout algorithm is described, specifically tailored for SD's; the algorithm takes advantage of the observation that SD's are typically are sparse acyclic digraphs with only one source. In section 2 we identify a set of aesthetics to capture the concept of a tidy and readable SD; section 3 is devoted to define a mathematical model for diagrams; then (section 4) we outline the algorithm, showing its properties in terms of performances and time complexity. Conclusions are in section 5.

2. Aesthetics for Statistical Diagrams

In producing drawings of SD, symbols are usually represented with circles and edges with polygonal lines. We adopt the same graphic standard here, with the following two additional requirements:

- circles are placed on a grid,
- polygonal lines representing edges are composed of sequences of segments connecting grid points.

The above graphic standard allows diagrams to be produced that have great regularity and modularity.

We have found that aesthetics typically adopted in real life applications (see for instance the experiments that have been performed at the Italian National Bureau of Statistics and during the design of the Italian National Transportation Information System) of SD's are as follows (in the following each aesthetic is identified by an acronym):

VERT - Verticality in the structure, i.e. each parent has to be placed above all its children;
CROSS - Minimization of the number of crossings between polygonals;
BENDS - Minimization of the global number of bends in polygonals;
LENGTH - Minimization of the global length of polygonals;
CENTER - Each parent has to be placed in a barycentric position with respect to its children.

Semantic features of a schema can be expressed by means of constraints, that can be explicitly provided to the algorithm by means of additional inputs. The mathematical model that we propose in this paper is suitable for several types of constraints; as an example in the following we consider how to impose constraints of type:

ISO-LEVEL - Vertices of a certain set must have the same ordinate in the drawing (e.g. a meaningful ISO-LEVEL constraint on a GRASS schema is to require that all T-nodes are represented on the same level, in order to better understand which tables have been defined in the statistical application);

The above mentioned aesthetics and constraints are typically not compatible: in figure 2 two drawings are shown of the same schema, in which, due to aesthetic VERT, edges are monotonically decreasing in the vertical direction: the first (figure 2(a)) minimizes the global length of the lines and the second (figure 2(b)) minimizes the number of crossings. A way to solve this kinds of conflicts is to establish a priority ordering between aesthetics. We impose this ordering in the next section and we show also how this ordering is naturally induced by the adopted graphic standard.

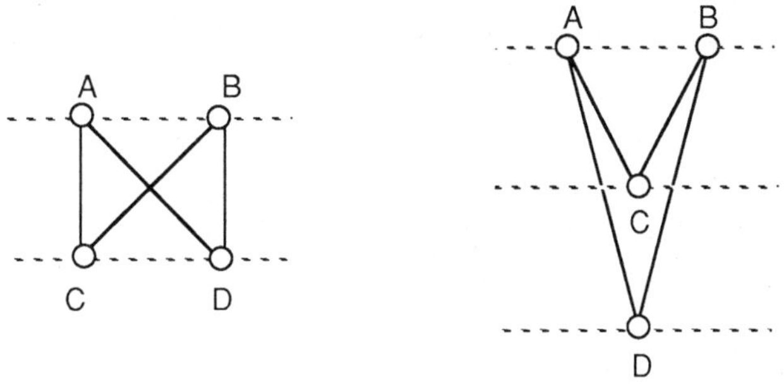

Figure 2: A conflict between aesthetics.

3. A mathematical model for SD's drawings

A statistical schema can be formally represented with a directed acyclic graph G(V,E), where V is the set of vertices and E is the set of directed edges. Notice also that G has typically only one source.

The aesthetics that we have pointed out in section 2 refer to different properties of a drawing: VERT refers to the vertical metric, CROSS to topology, BENDS to shape, CENTER to the horizontal metric, LENGTH to both the horizontal and vertical metrics. Concerning constraints, ISO-LEVEL refers to vertical metrics. In SD's the VERT aesthetics seems to be the most relevant, and then topology can be considered. These considerations suggest the stepwise generation of the drawing described in figure 3.

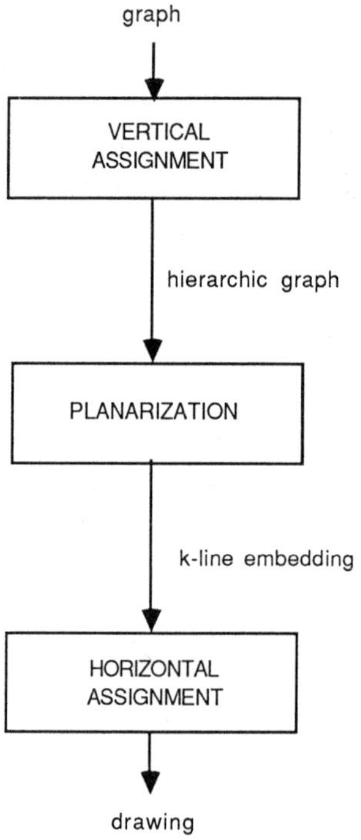

Figure 3: The steps of the proposed algorithm

Vertical assignment step takes into account aesthetics VERT and LENGTH, determining the vertical metrics of the drawing; it receives as input an acyclic digraph G(V,E) with only one source s and produces a hierarchic graph. A hierarchic graph ([W 77]) is a graph in which each vertex is associated to a certain horizontal level. More formally a hierarchic graph is denoted with H(V,E,L), where V is the set of vertices, E is the set of directed edges, and L is a function V—>1,2,...,k that associates a level to each vertex, in such a way that the following rules are satisfied:

rule 1) $L(s)=1$;
rule 2) if $(u,v)\in E$, then $L(u)<L(v)$.

The planarization step determines the topology of the drawing, taking into account aesthetics CROSS; it receives as input a hierarchic graph H(V,E,L) and produces a k-line embedding of H. A k-line embedding of a hierarchic graph specifies the ordering of vertices within each level. Such ordering can be represented with a function ORD that associates to each vertex its order number in its level.

Horizontal assignment considers aesthetics CENTER and BENDS, producing the final drawing by determining its horizontal metric: this step receives as input a k-line embedding, represented by a H(V,E,L) and a function ORD and produces the final coordinates of vertices and a description of the polygonals representing edges. An example of the inputs and outputs of the three steps is shown in figure 4.

4. The drawing algorithm for SD's

In this section we examine the three steps of the proposed algorithm, observing how aesthetics and constraints defined in section 2 are taken into account. We also outline several properties of the final drawing and the computational complexity of the different steps.

4.1 Vertical assignment

Step 1 produces a hierarchic graph H(V,E,L), starting from an acyclic digraph G(V,E) with only one source s, assigning a level L(v) to each vertex v of V. The algorithm assigns levels to vertices in such a way that rules 1 and 2 are enforced. It is a simple modification of the algorithm Minimum Completion Time for PERT diagrams (see for instance [E 79]) and can be summarized as follows:

Algorithm Vertical-assignment

(1) Count = $|E|$;
Define a label $d_{in}(v)$ that stores for each vertex
the number of ingoing edges;
Initialize a queue θ to empty queue and put s into θ;

(2) Extract a vertex v from θ and assign:

$$L(v) = \max_{(u,v)\in E} (L(u)) + 1;$$

For each w such that $(v,w)\in E$ do
begin
$d_{in}(w) = d_{in}(w) - 1$;
if $d_{in}(w) = 0$ then put w into θ;
end;
Count = Count - 1

(3) If Count = 0
then stop
else go to (2);

Notice that $d_{in}(v)$ is used to store the number of parents of v whose level has not been already computed and queue θ is used to store vertices v with $d_{in}(v) = 0$. The invariant of the algorithm is that L(v) is computed for a certain vertex v, only when it has been computed for all its parents. Using algorithm Vertical-assignment, rules 1 and 2 are clearly enforced and the height of the diagram is the minimum possible, to take into account aesthetic LENGTH. It is easy to state the following:

Property 1: Vertical assignment step is performed in $O(|E|)$ time.

In algorithm Vertical-assignment it is also possible to take into account ISO-LEVEL constraints. Suppose we want to constraint vertices u and v to stay on the same level. The first thing to say is that u cannot be an ancestor of v and vice-versa, otherwise we contradict aesthetics VERT; if this condition is satisfied it is possible to modify the level assignment in the following way:

suppose $L(u) < L(v)$
- $L(u) = L(v)$
- for each descendant w of u $L(w) = L(w) + L(v) - L(u)$.

V = {A,B,C,E,F,G,H,I,J,K,L,M,N}

E = {(A,B),(A,C),(B,E),(E,G),(G,N),(G,K),
(C,D),(D,G),(D,H),(H,L),(H,M),(H,J),
(C,F),(F,H),(F,I)}

graph

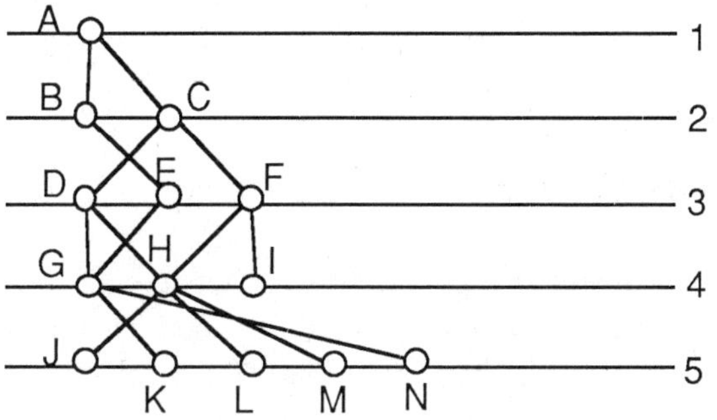

hierarchic graph

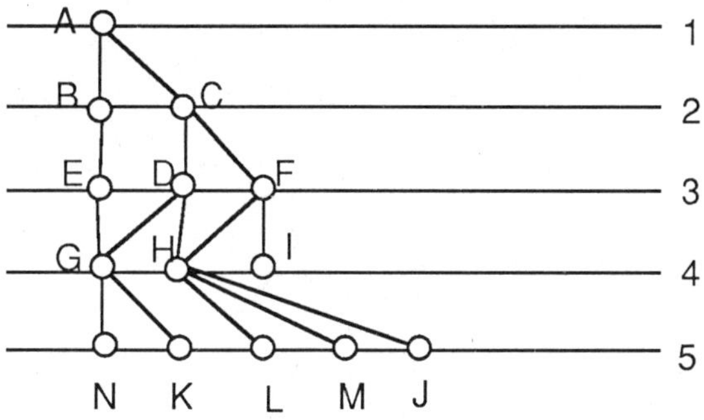

k-line embedding

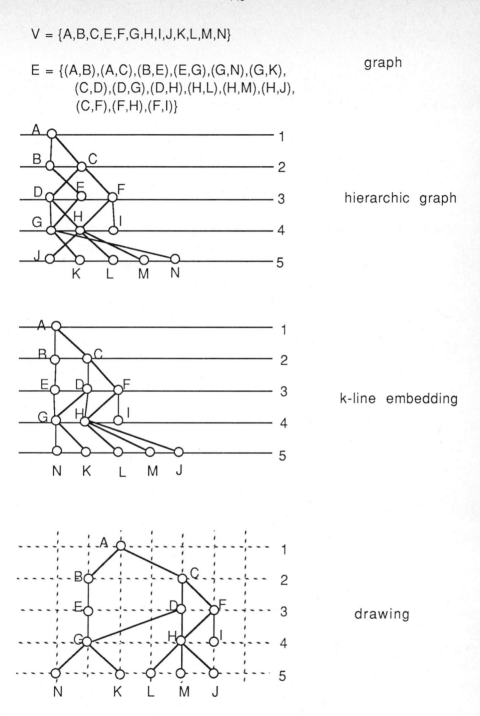

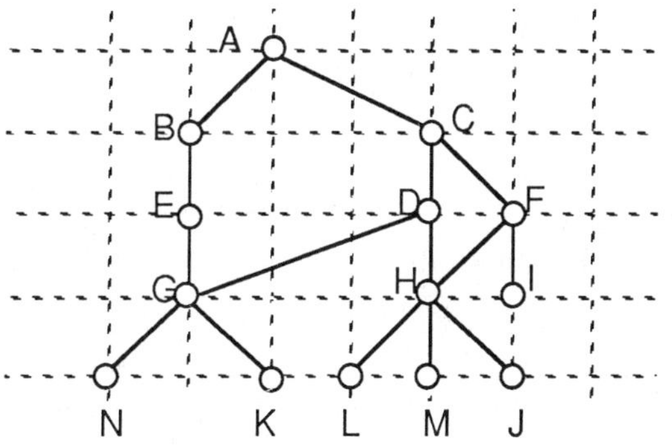

drawing

Figure 4: Examples of inputs and outputs of the algorithm steps.

4.2 Planarization

Step 2 produces the topology of thc final drawing, taking into account aesthetic CROSS. Notice that the problem here is quite different with respect to the typical planarity problem for undirected graphs, in fact here, due to aesthetic VERT, we must make two important assumptions:

- each vertex is constrained to stay on its level (assigned in step 1);
- polygonals representing edges must be monotonically decreasing in the vertical direction: this is done to enhance the verticality of the drawing.

With these assumptions the topology problem changes deeply, in fact see in figure 5 an example of a graph that is planar in the usual sense (i.e. it can be drawn without edge crossings), but that is not planar with the above assumptions: in figure 5(b) it is drawn without edge crossings using a polygonal that is not monotonically decreasing in the vertical direction; the drawing in figure 5(a) is not planar, but satisfies the two requirements.

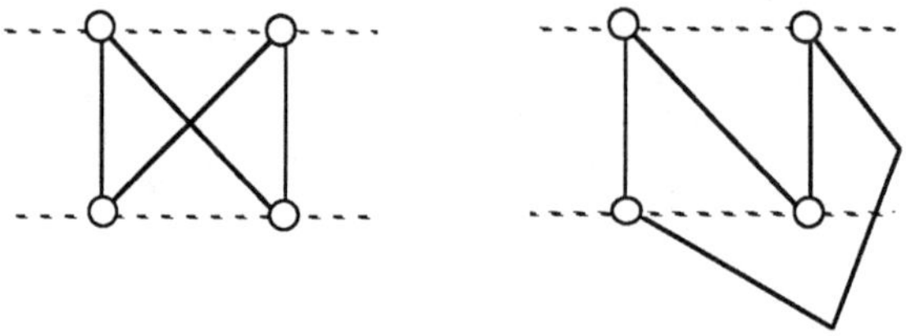

Figure 5: A graph that is planar in the usual sense but that is not planar if considered as hierarchic graph

In order to simplify the problem we replace edges connecting vertices belonging to non-consecutive levels with a sequence of edges, introducing a dummy vertex for each traversed level. In this way we can consider only edges (u,v) such that L(v) - L(u) = 1, we make also the assumption to represent each edge using a straight segment. Due to this assumption, an edge connecting two vertices of non-consecutive levels is represented here as a polygonal always decreasing in the vertical direction.

At this point the topology of the drawing is completely specified by the ordering of vertices (included dummy vertices) on each level, so the problem of minimizing the number of crossings is reduced to find an optimal ordering of vertices on the levels (optimal k-line embedding). This problem has been investigated in [EMW 86]; the paper shows how the minimization problem is NP-complete also if the number of levels is limited to 2. Besides this fact, in [BFN 85] it is shown that in this case an important sub-aesthetic can be "at least produce a drawing without edge crossings if it is possible". This is the aim of the proposed approach, that is based on a modification of the algorithm described in [DN 86].

We make two observations here that are crucial in the following algorithms:

1. If the hierarchic graph H is a directed tree rooted at s, it can be clearly drawn without edge crossings, i.e. it admits at least one planar k-line embedding;
2. If H admits a planar k-line embedding, for each directed spanning tree T of H, there exists at least one k-line planar embedding σ of T such that simply adding to σ the edges that are not on the spanning tree a planar embedding of H is obtained.

These two observations suggest an approach to the planarity problem based on the following algorithm:

1. Choose any directed spanning tree rooted at s of H; let E_T be the set of edges of T and E_o the set of edges that are not of T;
2. Determine the set S of all the possible k-line planar embeddings of T;
3. For each edge e of E_o
   - add the edge e to all the embeddings of S
   - remove from S all the embeddings that are not k-line planar after the addition of e

At the end of the execution, if H is planar then S contains at least one planar embedding, else it is empty. In order to take into account also the non-planar hierarchic graphs, the algorithm can be refined in the following way:

Algorithm Planarize

(1) Determine a directed spanning tree T of H, rooted at s;
    Let be E_T the set of edges of T and E_o the set of all the other edges;
(2) Determine the set S of all planar embeddings of T

(3) For each $e \in E_o$ do

begin

Determine the subset S' of S that remains planar

after the addition of e;

If S' is not empty

then S = S'

else put e into $E_{non\text{-}planar}$;

end;

(4) Find the embedding of S that better satisfy aesthetic CROSS

after the addition of the edges of $E_{non\text{-}planar}$;

Clearly the set of all the planar embeddings of T increases rapidly with the number of vertices (as a factorial in the worst case), so we have to use a data structure able to deal efficiently with such combinatorial explosion. The PQ-tree data structure (see [BL 76]), meets this criterion and allows compact representation of such embeddings. A precise description of steps 3 and 4 is out of the scope of this paper, for further details see [DN 86]. However the following properties can be proved for its behaviour:

Property 2: If the H-graph is k-line planar, algorithm Planarize produces a planar k-line embedding.

Property 3: Algorithm Planarize is performed in O(|E|) time, where E is the set of the edges of the hierarchic graph.

A proof of properties 2 and 3 can be found in [DN 86].

4.3 Horizontal assignment

At the end of step 2 we have found an ordering of vertices on each level, now we have to find the final abscissae of the vertices, taking into account aesthetics CENTER and BENDS. We have also to mantain the topology constraints, imposing that, for each couple u,v of vertices of the same level, if ORD(u)<ORD(v) then $x_u < x_v$. Here we use a heuristic technique, based on the Reingold and Tilford algorithm ([RT 81]). The approach can be formalized as follows:

Algorithm Horizontal-assignment

(1) Draw the directed spanning tree T, determined in step 2, using the algorithm

by Reingold and Tilford;
(2) Add to the drawing the edges that do not belong to T.

As noted in the introduction, the Reingold and Tilford algorithm is based on a divide and conquer strategy and can be summarized in the following way:

Algorithm Reingold-Tilford

If the tree is empty, or consists of a single vertex
    then trivially construct its drawing
    else
        Recursively draw all its subtrees;
        Place the drawings so obtained close one to the other, so that
            the minimum horizontal distance between them is D;
        Position the root at half way between the roots of the
            leftmost and rightmost subtrees

Notice that for the final drawing of tree T, the following properties are satisfied:

1. edges do not cross;
2. vertices in the same level are placed with a minimum distance D, one with respect to the other;
3. fathers are horizontally centered above their sons;
4. the width of the drawing is kept small
5. isomorphic subtrees have the same drawing
6. symmetric subtrees have mirror image drawings.

As a consequence, using the technique of algorithm Horizontal-assignment the following properties are satisfied:

Property 4: If the graph G is a tree, the horizontal placement is equal to the Reingold-Tilford placement.

Property 5: The overall number of bends is limited to $2|E_{long}|$, where E_{long} is the set of edges that span more than one level.

Property 4 is trivial; Property 5 can be proved observing that the only edges that can have bends are the edges that span more than one level and that there are at most two bends for each of them. Moreover we observe:

- SD's are typically sparse (the number of edges grows linearly in the number of vertices) and so they are usually "almost planar";
- SD's produced with the mentioned statistical data models have a few transitive edges.

So the following properties applies:

Property 6: If graph G is planar and has no transitive edges, then the number of bends of the final drawing is at most $2|V| - 6$.

In fact if G has no transitive edges, then $|E| \leq 2|V|-4$; $|V|-1$ edges belong to the spanning tree and therefore are drawn without bends; the algorithm causes at most two bends for each of the remaining edges.

5. Conclusions

We have shown an algorithm for automatic drawing of a wide class of statistical diagrams (the ones obtained with CSM, GRASS, SAM*, and SUBJECT models). The algorithm produces the final drawing through successive refinement steps in an overall $O(|E|)$ time complexity that makes it suitable also for on-line applications.

In order to identify the features of a readable drawing we have defined a set of aesthetics for statistical diagrams (coherently with [C et al. 87]), that are interpreted by the algorithm as optimization criteria.

The algorithm is under implementation at the Dipartimento di Informatica e Sistemistica dell'Universita' di Roma on an IBM PC and on a SUN workstation. It will be integrated into the tool for designing and accessing statistical databases described in [DT 86].

Acknowledgments

I am grateful to Carlo Batini for his encouragement during this research. I whish also to thank Enrico Nardelli and Roberto Tamassia for useful discussions and suggestions on this topic.

References

[BFN 85] C.Batini, L.Furlani, and E.Nardelli, "What Is a Good Diagram? A Pragmatic Approach," Proceedings of the 4th International Conference on the Entity- Relationship Approach, Chicago, 1985.

[BNT 86] C.Batini, E.Nardelli, and R.Tamassia, "A Layout Algorithm for Data Flow Diagrams," IEEE Transactions on Software Engineering, vol.SE-12, n.4, April 1986.

[BTT 84] C.Batini, M.Talamo, and R.Tamassia, "Computer Aided Layout of Entity-Relationship Diagrams," The Journal of Systems and Software, vol.4, 1984.

[BL 76] K.Booth and G.Lueker, "Testing for the Consecutive Ones Property, Interval Graphs, and Graph Planarity Using PQ-Tree Algorithms," Journal of Computer and System Sciences, vol.13, 1976.

[C et al. 87] D.Cable et al., "Interface Issues," in Report on the 3rd International Workshop on Statistical and Scientific Database Management, Luxembourg 1986, Statistical Software Newsletters, no.1 1987.

[C 80] M.J.Carpano, "Automatic Display of Hierarchized Graphs for Computer Aided Decision Analysis," IEEE Transactions on Systems, man, and Cybernetics, vol.SMC-10, no.11, 1980.

[CCO 86] R.Cubitt, B.Cooper, and G. Ozsoyoglu (editors), "Proc. of the 3rd International Workshop on Statistical and Scientific Database Management," Luxembourg, 1986.

[DFB 86] G.Di Battista, G.Ferranti, and C.Batini, "Design of Statistical Databases: a Methodology for the Conceptual Step," Technical Report of Dipartimento di Informatica e Sistemistica, Universita' di Roma RAP.12.86, 1986.

[DN 86] G.Di Battista and E.Nardelli, "An Algorithm for Testing Planarity of Hierarchical Graphs," in Lecture Notes in Computer Science vol.246, Proc. 12th International Workshop on Graphtheoretic Concepts in Computer Science, Bernried 1986.

[DT 86] G.Di Battista and R.Tamassia, "An Interactive Graphic System for Designing and Accessing Statistical Data Bases," Proc. 7th Symposium on Computational Statistics, Roma 1986.

[DT 88] G.Di Battista and R.Tamassia, "Algorithms for Plane Representations of Acyclic Digraphs," Theoretical Computer Science 1988 (to appear).

[E 79] S.Even, "Graph Algorithms," Computer Science Press, 1979.

[EMW 86] P.Eades, B.D.Mc Kay, and N.C.Wormald, "On an Edge Crossing Problem," Proceedings of the 9th Australian Computer Science Conference, Canberra 1986.

[OSV 82] T.W.Olle, H.G.Sol, and A.A.Verrijn-Stuart (eds.), "Information Systems Design Methodologies: A Comparative Review," Proc. of the IFIP WG 8.1 Working Conference on Comparative Review of Information Systems Design Methodologies, Noordwijkcrhout, Thc Netherlands, North IIolland 1982.

[RR 83] M.Rafanelli and F.Ricci, "Proposal for a Logical Model for Statistical Databases," Proc. of the 2nd Intl. Workshop on Statistical Database Management, Los Angeles 1983.

[RT 81] E.Reingold and J.Tilford, "Tidier Drawing of Trees," IEEE Transactions on Software Engineering, vol.SE-7, no.2, 1981.

[SC 80] A.Shoshani and P.Chan, "Subject: a Directory Driven System for Organizing and Accessing Large Statistical Databases," Proc. of the 2nd International Conference on Very Large Data Base (VLDB), 1980, pp.553-563.

[SW 85] A.Shoshani and H.K.T.Wong, "Statistical and Scientific Database Issues," IEEE Transactions on Software Engineering, vol. SE-11, N.10, October 1985.

[S 83] S.Y.W.Su, "SAM*: A Semantic Association Model for Corporate and Scientific-Statistical Databases," Information Sciences 29, 1983, pp. 151-199.

[STT 81] K.Sugiyama, S.Tagawa, and M.Toda, "Methods for Visual Understanding of Hierarchical System Structures," IEEE Transactions on Systems, Man, and Cybernetics, vol. SMC-11, no.2, 1981.

[T 87] R.Tamassia, "On Embedding a Graph in the Grid with the Minimum Number of Bends," SIAM Journal on Computing, vol.16, no.3, 1987.

[TDB 88] R.Tamassia, G.Di Battista, and C.Batini, "Automatic Graph Drawing and Readability of Diagrams," IEEE Trans. on Systems Man and Cybernetics, vol.SMC-18, no.1, 1988.

[TRC 79] H.Tardieu, A.Rochfeld, and R.Colletti, "Conception d'un Systeme d'Information: Construction de la Base de Donnees," Edition d' Organization, 1979.

[W 77] J.Warfield, "Crossing Theory and Hierarchy Mapping," IEEE Transactions on Systems, Man, and Cybernetics, vol.SMC-7, no.7, 1977.

THE CLASSIFICATION PROBLEM WITH SEMANTICALLY HETEROGENEOUS DATA

F. M. Malvestuto - C. Zuffada
(STUDI-DOC, ENEA: V. Regina Margherita 125, 00198 Roma - Italy)

ABSTRACT

Given a database fed by two alternative data sources using a common but not identical classification criterion, if we are able to state precisely the semantical connection between the two classification systems, we can derive new and more detailed summary data. Therefore, the question whether an aggregate information is derivable or not, is fundamental to a query-processing system. We state a necessary and sufficient condition which leads to a simple procedure for deciding the answerability of a summary query and evaluating it, if answerable. Surprisingly, the condition of derivability is independent of the database instance and is dependent only on the topological properties of the graph modelling the semantical connection of the classification systems adopted.

1. INTRODUCTION

A serious impediment to the integrated use of summary data produced by different data sources is the *semantic heterogeneity* of data. A common source of heterogeneity is the use of different *classification systems* [10]. Such a situation is frequent in dealing with geographically-based data (an example is the SEEDIS system [11]) as well as with data based on classification criteria such as "age", "industrial sector", "energy source" etc. [6, 19].
In order to benefit by a pool of data sources feeding a database, an *integration* of the supplied summary data sets is required. The integration problem is particularly felt by national and international organizations that are producers of comprehensive databases [3, 5, 6, 7, 9, 10, 11, 12, 16, 17, 18, 19].
A somewhat unpopular but often effective approach to avoiding problems of dealing with heterogeneous data is to impose standards. For example, the use of STANDARD METROPOLITAN STATISTICAL AREAS (SMSA) is effective in integrating U.S. data from multiple economic surveys. In cases where a set of standards can be imposed and maintained over a long period of time, the problems of heterogeneous data disappear. Unfortunately, such cases are rare.
The most obvious way to minimize the problems of connecting data categorized in two distinct ways is to provide complete data documentation (*metadata*). All of this information will be used to establish the semantic connections between the classification systems adopted.
Our goal is to provide a framework by which the question of source integration can be better understood as well as some methodology to combine source heterogeneous data.
The starting point is the representation of the semantic relationships between two

alternative classification systems using a labelled bipartite graph to be called a *correspondence graph.*
A first proof of the usefulness of correspondence graphs is given by the simplicity with which we shall be able to solve a classical problem of multi-source databases, namely, the "matching problem" [16, 17, 19] (or "comparison problem" [22]), where two data sources supply data on two different phenomena (i.e., two statistical variables, for example POPULATION and GROSS NATIONAL PRODUCT) using a common but not identical analysis criterion (for example, two different partitionings of the same geographic area).
As a second application of our graph-theoretical approach, we shall solve the "derivation problem" [9, 17], arising when we want to know whether a certain summary data is or not derivable from a database containing data related to a phenomenon that two distinct sources have measured using a common but not identical representation system. The derivation problem is not trivial in view of the fact that in most cases the two summary data sets have a "synergetic behavior", which leads to the exact evaluation of certain summary data having a higher detail level than stored data. For example, consider the data on the FINAL CONSUMPTION OF ELECTRICITY of Italy in 1985 broken down by Industral Sector. These data are available from two alternative data sources: Industry Department of Italy (abbr. IDI) [20] and Organization for Economic Co-operation and Development (abbr. OECD) [21]. The data are heterogeneous because the two sources adopt different units of measure and different classifications of industrial sectors (see Tables 1 and 2).

TABLE 1: FINAL CONSUMPTION OF ELECTRIC ENERGY (10^{12} Kcal) BY INDUSTRIAL SECTOR
Italy 1985 (source: IDI)

Industrial Sector	*Quantity*
Siderurgico	15.6
Metalli nonferrosi	4.9
Vetro e Ceramica	2.4
Materiali per costruzioni	5.7
Meccanico	11.8
Estrattivo	1.0
Agroalimentare	5.0
Carta e Grafica	4.8
Edile	0.8
Tessile e Abbigliamento	7.9
Chimico	11.5
Petrolchimico	5.2
Altro	3.0
Total	79.6

TABLE 2: FINAL CONSUMPTION OF ELECTRIC ENERGY (MToe) BY INDUSTRIAL SECTOR
Italy 1985 (source: OECD)

Industrial Sector	*Quantity*
Iron & Steel	1.56
Non-ferrous Metals	0.49
Non-metallic Minerals	0.81
Transport Equipment	0.25
Machinery	0.93
Mining & Quarrying	0.10
Food & Tobacco	0.50
Paper, Pulp & Printing	0.48
Construction	0.08
Textile & Leather	0.65
Chemical & Petrochemical	1.89
Wood & Wood Products	0.18
Industry (non-specified)	0.04
Total	7.96

The heterogeneity due to different units of measure is not a real problem because it is always possible (and easy) to convert data from one measure system to the other according to the simple formula of conversion: $1 \text{ MToe} \cong 10^{13} \text{ Kcal}$. On the contrary, the difference between the two classification systems for industrial sectors is not a trivial question. However, using a bilingual expert of energy information systems we have been able to determine which classes of one classification system correspond to which classes of the other.

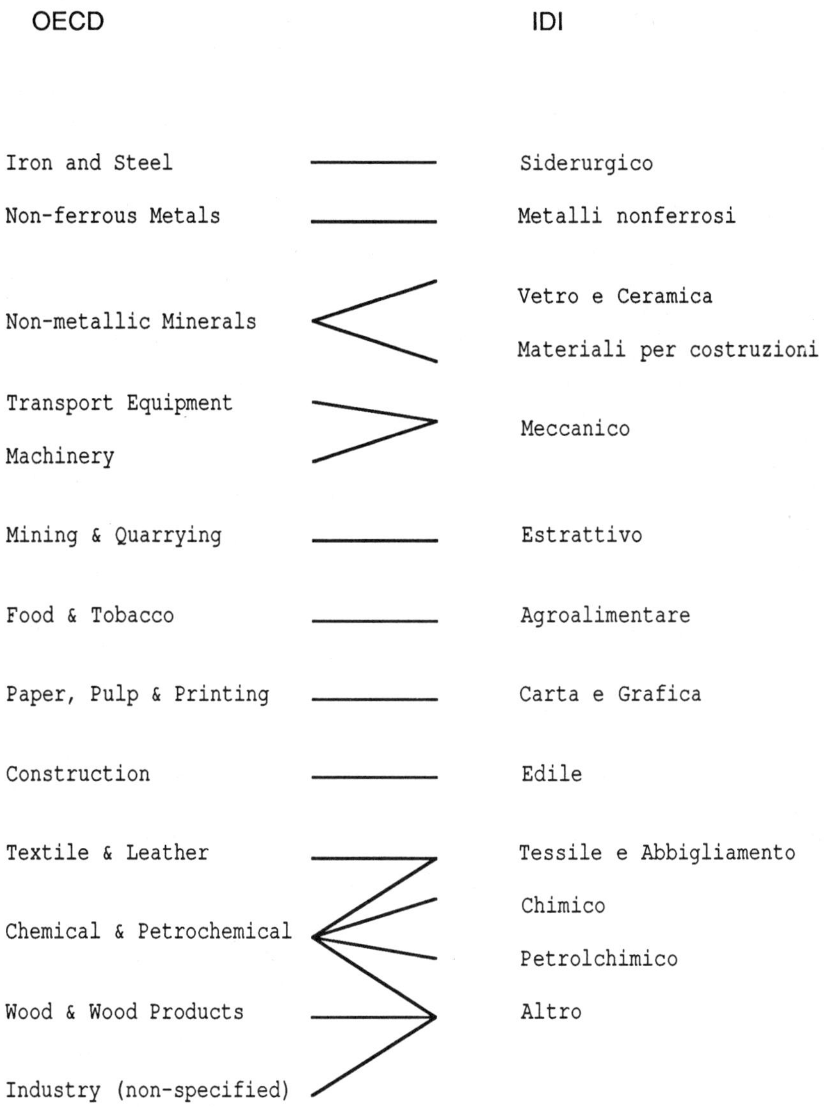

The above-mentioned information synergy between data sources entails that due to the knowledge of the correspondences between the two classification systems we can combine the data reported in Table 1 and Table 2, and determine a more detailed distribution in an exact way.
For example, the distribution reported in the following Table 3 is derivable from Tables 1 and 2 and is more detailed than both, in that it uses a finer classification [9].

TABLE 3: FINAL CONSUMPTION OF ELECTRIC ENERGY (Mtoe) BY INDUSTRIAL SECTOR Italy 1985 (sources: IDI, OECD)

Sector	*Quantity*
Iron & Steel	1.56
Non-ferrous Metals	0.49
Glass & Pottery	0.24
Construction-ware	0.57
Transport Equipment	0.25
Machinery	0.93
Mining & Quarrying	0.10
Food & Tobacco	0.50
Paper, Pulp & Printing	0.48
Construction	0.08
Textile & Leather	0.65
Clothing Chemical	0.14
Chemical	1.15
Petrochemical	0.52
Other Chemical	0.08
Wood & Wood Products	0.18
Industry (non-specified)	0.04
Total	7.96

Although it is well-known that one can in most cases derive new and more detailed data, no necessary and sufficient condition for the derivability of an arbitrary summary data was previously known. The major result of this paper is that such a condition can be obtained by exploiting the topological properties of correspondence graphs and is independent of the database instance at issue.

The paper is organized as follows. Section 2 gives the basic definitions and introduces some terminology. Section 3 analyses the algebraic structure of the set of all possible classifications of a given universe of objects. There we shall introduce the notion of a *correspondence graph* to model the semantical connection between two homogeneous classification systems. In Section 4 we give a first application of correspondence graphs to solve the "matching problem". In Section 5 we shall introduce the formal definitions of *derivability* for a summary data and of *measurability* for an aggregate of objects. There, we shall attempt to give an axiomatic characterization of measurable aggregates.
Section 6 contains the main result of the paper, that is, two theorems which state a necessary and sufficient condition for an arbitrary aggregate to be measurable. Section 7 deals with some computational details of the test for derivability. In Section 8 some suggestions for future research are given.

2. BASIC DEFINITIONS AND TERMINOLOGY

Suppose we are given a collection of *data* related to a certain *universe* Ω of *objects* (individuals, things, ...). Such data are the result of observation or measurement of some "important" properties (referred to as *attributes*) of the elements of the universe. The result is a relational data table (or simply a *relation*) $\mathbf{r}$, which has as many columns as the attributes and has as many rows as the objects. A subset of the universe, henceforth referred to as an *aggregate*, is then represented by a subrelation of $\mathbf{r}$.

As we limit our considerations to statistical users, we are interested in designing a "database abstract", we call a *statistical distribution*, which provides a synthetic representation of the phenomenon under examination. Generally speaking, a statistical distribution is obtained by grouping the individual data according to some aggregation criterion. Data aggregation can be thought of as the result of two conceptually distinct operations: (a) the *classification* of the underlying universe, (b) the computation of certain statistical functions over the aggregates resulting from the classifications.

Classification Variables

A classification can be specified by assigning a collection of so-called *classification variables* (sometimes called *category attributes*). A classification variable can be either an attribute of the objects or a new variable functionally dependent on the set of attributes.

Two classifications of a given universe will be called *homogeneous* if they are based on the same set of attributes.

Usually, a classification system consists of a certain number of aggregates, called *classes* and *superclasses.* A *class* is an aggregate specified by assigning a value for each classification variable, so that the set of all classes forms a *partition* P of the underlying universe. A *superclass* is a meaningful aggregate resulting from the union of a certain number of classes. From a mathematical point of view, classes and superclasses can be regarded as *P-aggregates*, these being defined to be the elements of the set field **P** associated with the partition P of the universe.

Statistical Variables

After grouping the objects of a universe according to some classification system, we have to introduce an *aggregate function* , f , in order to compute some "important" (numeric) characteristic, x , over the classes and the superclasses resulting from the classification of the universe. The quantity x is called a *statistical variable* (sometimes, a *summary attribute*).

From the mathematical point of view, f is a real-valued additive function defined on the field **P** : if A and B are two P-aggregates, then

$$f(A) + f(B) = f(A \cup B) + f(A \cap B) \ .$$

Due to the additivity property, all the information about the statistical variable x can be condensed into the distribution $x = f_P(p)$ which specifies the values taken by the aggregate function on the classes of the partition P . Once we know this distribution, the value $x = f(A)$ corresponding to the P-aggregate A amounts to

$$f(A) = \Sigma_{p \in A} \ f_P(p) \ .$$

3. ALGEBRA OF PARTITIONS

Consider a database storing two statistical distributions provided by two distinct data sources that use a common but not identical classification criterion on a given universe Ω . In order to compare the two statistics, we must be able to state precisely the connection between the semantical structures underlying the two classification systems. This leads us to introduce the algebra of partitions as the natural framework where any two partitions corresponding to two different classification systems can be compared.

The family $\wp$ of the partitions of the universe Ω has the algebraic structure of a lattice based on the relationship of refinement: we say that partition P is *finer* than partition Q (or that Q is *coarser* than P), denoted $P \geq Q$, if each class of P is a subset of some class of Q or, equivalently, if each class of Q is a P-aggregate, i.e., a union of classes from P. It is easy to check that this defines a partial order on $\wp$.

The *unit* (**1**) of the lattice is given by the *point partition*, whose classes are the singleton aggregates of the universe.

The *zero* (**0**) of the lattice is given by the *trivial partition*, which consists of only one class, i.e., the universal class.

The lattice operation *join* of two partitions P and Q , denoted by (P, Q) , is defined as the coarsest partition among those which refine both P and Q .

The lattice operation *meet* of two partitions P and Q , denoted by [P, Q] , is defined as the finest partition among those which are coarser than both P and Q .

With regard to the algebraic notions of "greatest lower bound" (glb) (or "infimum") and "least upper bound" (lub) (or "supremum"), we have

$$\text{glb}(P, Q) = [P, Q] \qquad \text{and} \qquad \text{lub}(P, Q) = (P, Q) .$$

The structure of the algebra of partitions is shown in the figure below, where the level of detail decreases in the direction going from the point partition, **1** , to the trivial partition, **0** .

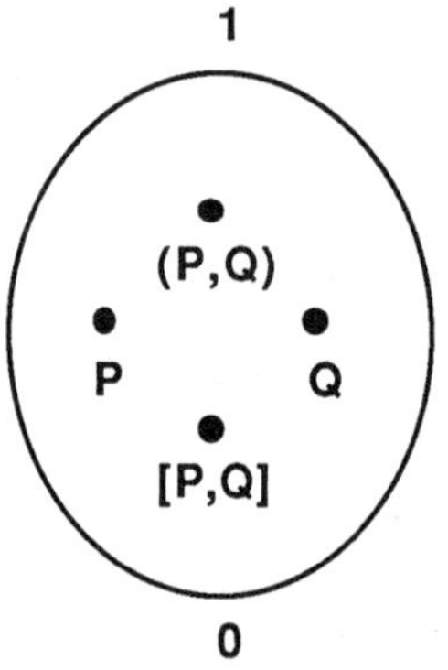

In order to represent the semantic connection between two given partitions P and Q , we shall make use of a bipartite graph G(P, Q) to be called the *correspondence graph.* An edge (p, q) in this graph is formed by a pair of "overlapping" classes p and q , i.e., casses having at least one object in common.

An example of a correspondence graph was given in the previous section.

As another example, consider two geographical partitions of the twenty Italian Regions: Abruzzi (1), Basilicata (2), Calabria (3), Campania (4), Emilia Romagna (5), Friuli Venezia Giulia (6), Lazio (7), Liguria (8), Lombardia (9), Marche (10), Molise (11), Piemonte (12), Puglia (13), Sardegna (14), Sicilia (15), Toscana (16), Trentino Alto Adige (17), Umbria (18), Valle d'Aosta (19), Veneto (20).

The first partitioning P groups the Regions by GEOGRAPHICAL_AREA: North, Centre, South, Isles.
The second partitioning Q groups the Regions by ALTIMETRIC_ZONE: Mountain, Hill, Plain.
The correspondence graph G(P, Q) looks like as follows:

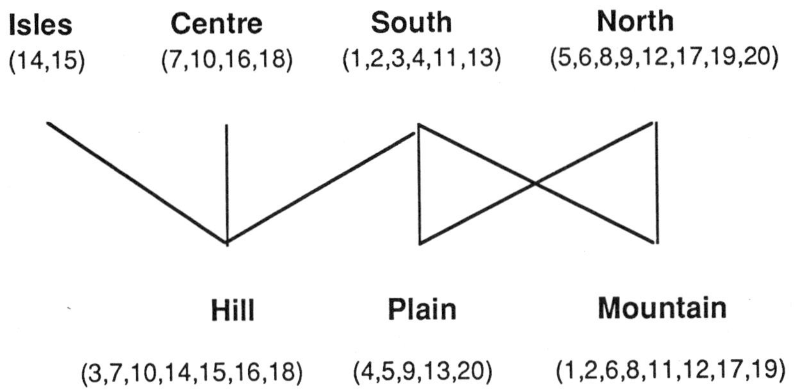

The lattice operations join and meet find a simple representation in a correspondence graph, G(P, Q) . The classes of the join-partition (P, Q) and of the meet-partition [P, Q] correspond respectively to the edges and to the connected components of G(P, Q) .
In our case, the join-partition (P, Q) is made up of seven classes, which we may call: Isles, Centre, Hilly South, Plain South, Mountainous South, Plain North and Mountainous North. The meet-partition [P, Q] coincides with the trivial partition **0** .
The notion of correspondence graph is useful in analysis of semantic connections between homogeneous classifications.
Two partitions P and Q are called *independent* if G(P, Q) is complete, that is, if all pairs (p, q) are overlapping. Otherwise, they are called *dependent.*
Two partitions P and Q are *tree dependent* if G(P, Q) is *acyclic*, that is, if its connected components are *trees* (a tree is a graph with *n* nodes and *n -1* edges).
A trivial example of tree dependent partitions occurs when partition P is finer than partition Q .

Acyclic correspondence graphs possess nice properties [9].

4. THE MATCHING PROBLEM

The matching problem arises when two summary data related to distinct statistical variables are compared, for example in order to compute their correlation.
Consider the following example taken from Sato [18]. We are given two summary data, as shown in the table below.

LAND TRANPORT INDUSTRY (P)	TOTAL_WAGES
Railroad Passenger Transport (p1)	68
Railroad Freight Transport (p2)	23
Other Passenger Land Transport (p3)	586
Local Trucking (p4)	493
Long-Distance Trucking (p5)	295

LAND TRANPORT INDUSTRY (Q)	PROFITS
Railroad Transport (q1)	4
Other Mass Passenger Transport (q2)	26
Other passenger Land Transport (q3)	53
Trucking (q4)	92

Now suppose that a user wants to compare the profits with the wages by industry. In this case we have to find a common comparison basis for wages and profits. The finest partition which makes such a comparison feasible is the meet of P and Q , [P, Q] . As we know, this classification is formed by the connected components of the following correspondence graph.

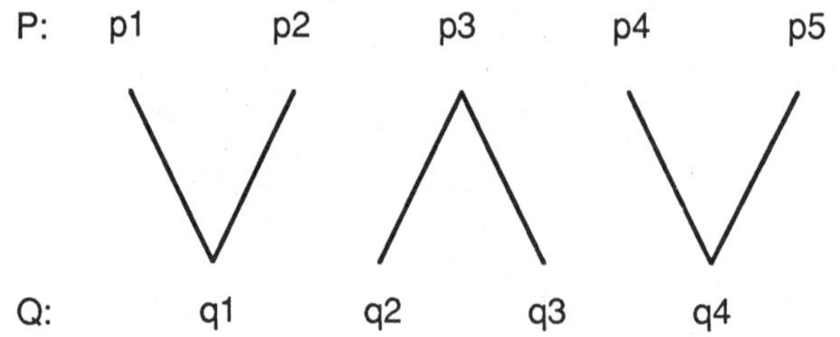

Here we find three connected components which correspond to q1 (Railroad Transport), p3 (Other Passenger Land Transport), q4 (Trucking) . So, we have

LAND TRANSPORT INDUSTRY [P, Q]	TOTAL_WAGES	PROFITS
Railroad Transport (q1)	91	4
Other Passenger Land Transport (p3)	586	79
Trucking (q4)	788	92

5. THE DERIVATION PROBLEM

Consider a database fed by two alternative data sources. This is to say, the data stored in our database refer to two distributions, $x = f_P(p)$ and $x = f_Q(q)$, which share the same measurement variable, x , and are based on two homogeneous classifications, whose partitions are P and Q . When a user interrogates such a database, the database management system is to be able to decide whether the information required by the user is available. If it is available, then and only then can the query be satisfied. The notion of "available information" encompasses two types of information: the information that is explicitly represented (i.e., stored data) and the information that is implicitly represented (i.e., derivable data). We focus on this latter type of information.
Of course, the application of any set function to a set of stored data yields something which may be properly called a derived data. However, here we focus on the derivability of summary data of the type $x = f(A)$, A being an arbitrary aggregate. In other terms, we are looking for a necessary and sufficient condition for a summary data to be uniquely determined by the stored data. If this is the case, then the corresponding aggregate will be called *measurable.*
We begin by assuming that the database $\{x = f_P(p)\ ,\ x = f_Q(q)\}$ be consistent in that there exists at least one joint distribution $x = f(p, q)$ defined on the join-partition (P, Q) which satisfies the "marginal constraints":

$$\begin{aligned} \Sigma_{q \in Q(p)}\ f(p, q) &= f_P(p) \\ \Sigma_{p \in P(q)}\ f(p, q) &= f_Q(q) \end{aligned} \qquad (1)$$

where Q(p) denotes the set of Q-nodes linked to the node p and P(q) denotes the set of P-nodes linked to the node q . The consistency assumption implies that

$$\Sigma_p\ f_P(p) = \Sigma_q\ f_Q(q)\ . \qquad (2)$$

Usually, there is a more or less large set $\mathfrak{F}$ of joint distributions $x = f(p, q)$. Now, consider an arbitrary aggregate A defined as the union of some of the classes of the join-partition. Aggregate A is said to be *measurable* if all the joint distributions in $\mathfrak{F}$ assign the same value to A . If this is the case, we also say that the summary data f(A) is *derivable* from the database. In other words, A is measurable if the equation

$$\Sigma_{(p,q) \in A}\ f(p, q) = f(A)$$

is *linearly dependent* on the equation system (1).
In order to find a formal expression for all aggregates, let us make a change of notation. After indexing the edges of $G(P, Q)$ by $k = 1, ..., s$, we shall represent any aggregate by its characteristic vector $\gamma_A = [\gamma_A(1), ..., \gamma_A(s)]$ where $\gamma(k)$ is equal to 1 if the k.th edge is in A and equal to 0 otherwise. Let α_i be the aggregate (or, equivalently, its characteristic vector) made up of the edges incident to the node p_i and β_j be the aggregate (or, equivalently, its characteristic vector) made up of the edges incident to the node q_j. Furthermore, let u_i, v_j and $x(A)$ respectively denote $f_P(p)$, $f_Q(q)$ and $f(A)$. Then the equation system (1) can be written as

$$\begin{aligned} \Sigma_k \ \alpha_i(k)\ x(k) &= u_i \\ \Sigma_k \ \beta_j(k)\ x(k) &= v_j \end{aligned} \qquad (3)$$

We know that if A is measurable, then the equation

$$\Sigma_k \ \gamma_A(k)\ x(k) = x(A)$$

is linearly dependent on the equation system (3). This means that the rank of the matrix

$$\begin{bmatrix} \alpha_1(1) & \alpha_1(2) & \cdots\cdots & \alpha_1(s) \\ \cdots\cdots & \cdots\cdots & \cdots\cdots & \cdots\cdots \\ \alpha_m(1) & \alpha_m(2) & \cdots\cdots & \alpha_m(s) \\ \beta_1(1) & \beta_1(2) & \cdots\cdots & \beta_1(s) \\ \cdots\cdots & \cdots\cdots & \cdots\cdots & \cdots\cdots \\ \beta_n(1) & \beta_n(2) & \cdots\cdots & \beta_n(s) \end{bmatrix}$$

does not change if we add the row

$$\gamma_A(1) \qquad \gamma_A(2) \qquad \cdots\cdots \qquad \gamma_A(s)$$

A consequence is that if A is measurable, then its characteristic vector γ_A can be written as a linear combination of the α_i's and β_j's, that is,

$$\gamma_A = \Sigma_i \ \rho_i \alpha_i + \Sigma_j \ \sigma_j \beta_j$$

where $(\rho_1, ..., \rho_m, \sigma_1, ..., \sigma_n) \neq (0, ..., 0, 0, ..., 0)$, and its summary value $x(A)$ is

$$x(A) = \Sigma_i \ \rho_i u_i + \Sigma_j \ \sigma_j v_j \ .$$

We conjecture that the values taken by the coefficients ρ_i's and σ_j's can be only $0, \pm 1$ and that there are only four possible forms of representation for γ_A :

(i) A is a P-aggregate:

$$\gamma_A = \Sigma_{i \in I} \ \alpha_i$$

(ii) A is a Q-aggregate:

$$\gamma_A = \Sigma_{j \in J} \ \beta_j$$

(iii) A is the union of a P-aggregate and a Q-aggregate, having an empty intersection:

$$\gamma_A = \Sigma_{i \in I} \ \alpha_i + \Sigma_{j \in J} \ \beta_j$$

(iv) A is the difference of a P-aggregate and one of its subsets which is also a Q-aggregate:

$$\gamma_A = \Sigma_{i \in I} \ \alpha_i - \Sigma_{j \in J} \ \beta_j$$

Consequently, there are only four possible types of derivable summary data:

(i) $x(A) = \Sigma_{i \in I} \ u_i$

(ii) $x(A) = \Sigma_{j \in J} \ v_j$

(iii) $x(A) = \Sigma_{i \in I} \ u_i + \Sigma_{j \in J} \ v_j$

(iv) $x(A) = \Sigma_{i \in I} \ u_i - \Sigma_{j \in J} \ v_j$

If our conjecture is true, the family $\aleph$ of measurable aggregates can be axiomatically characterized in a sound and complete way by the following formal system:

P-Axiom:	$\forall i \in M \ \ \alpha_i \in \aleph$
Q-Axiom:	$\forall j \in N \ \ \beta_j \in \aleph$
Union Rule:	$\forall A, B \in \aleph$ **if** $A \cap B = \varnothing$ **then** $A \cup B \in \aleph$
Difference Rule:	$\forall A, B \in \aleph$ **if** $A \supseteq B$ **then** $A - B \in \aleph$

6. TESTING DERIVABILITY

In the preceding section we attempted to give a formal theory of measurability for aggregates. In this section we present a test for deciding the measurability of a given aggregate, A . We shall see that in the case that the underlying correspondence graph is acyclic, the test for measurability is a trivial question, since it is sufficient to check that the aggregate A is a (P, Q)-aggregate. On the contrary, if the underlying correspondence graph is cyclic, then the answer is not immediate and we shall state a necessary and sufficient condition based on the graph-theoretical notion of "circulation" on a digraph [2].

Acyclic case
We say that two distributions, $x(p_i) = u_i$ and $x(q_j) = v_j$, are *joinable* if each class of the join-partition (P, Q) is measurable, that is, if there is one and only one joint distribution x(i, j) .

RESULT 1 [9]. Let P and Q be two homogeneous partitions of the same universe and let G(P, Q) be their correspondence graph. A necessary and sufficient condition for two arbitrary consistent distributions, defined respectively over P and Q , to be joinable is that P and Q be tree dependent.

A simple computation procedure of *the* joint distribution x(i, j) is given in [9] and it was used to compute the entries of Table 3 in Section 2.
An important consequence of this result is that the database is equivalent to a single distribution, i.e., the joint distribution. So, we can decide whether the summary data x(A) is derivable or not simply by checking if A is a (P, Q)-aggregate.

Cyclic case
Suppose now that the correspondence graph is not acyclic. As an example, we may consider two sources that have aggregated the census data of the Italian people according to the two geographic partitionings P and Q , shown in Section 3.

TABLE 4: DISTRIBUTION OF THE ITALIAN PEOPLE BY GEOGRAPHICAL AREA
Italy 1981 (1000 individuals)

Geographical Area	*Count*
Isles (p1)	6501
Centre (p2)	10803
South (p3)	13552
North (p4)	25701

TABLE 5: DISTRIBUTION OF THE ITALIAN PEOPLE BY ALTIMETRIC ZONE
Italy 1985 (1000 individuals)

Altimetric Zone	*Count*
Hill (q1)	19365
Plain (q2)	26530
Mountain (q3)	10662

Consider the correspondence graph G(P, Q) . By a *block* (or "biconnected component") of G(P, Q) we mean a maximal connected subgraph such that the removal of a node (and of its incident edges) is not enough to disconnect it [15]. A block is called *degenerate* if it consists of only one edge; otherwise it is called *nondegenerate.*
In our case, for example, the correspondence graph G(P, Q) contains four blocks:

Isles	*(degenerate block)*
Centre	*(degenerate block)*
Hilly South	*(degenerate block)*
Plain-Mountain	*(nondegenerate block)*

An important property of blocks is the following.

RESULT 2 [9]. Each block of the correspondence graph G(P, Q) is measurable.

By applying the computation procedure given in [9] to our case, we find:

COUNT (Isles)		= f(p1)	=	6501
COUNT (Centre)		= f(p2)	=	10803
COUNT (Hilly South) =	COUNT(Hill) - COUNT(Isles) - COUNT(Centre)	= f(q1) - f(p1) - f(p2)	=	2061
COUNT(Plain-Mountain) =	COUNT(Plain) + COUNT(Mountain)	= f(q2) + f(q3)	=	37192

The information gain following from the evaluability of the blocks is sometimes remarkable. For example, in the case under examination, assume that the regional population data are "sensitive" data not to be revealed to users. Knowledge of the correspondence graph G(P, Q) allows a user to infer the population (2,061,000 individuals) of Region 3 (Calabria), which is the only region of the degenerate block Hilly South. That is, the security of our statistical database is "compromised" [4, 13].

Generally speaking, we may view an arbitrary aggregate A as

$$A = A_{deg} \cup A_{nondeg}$$

where A_{deg} is an aggregate resulting from the union of a certain number of degenerate blocks, and A_{nondeg} is an aggregate resulting from the union of a certain number of edges belonging to nondegenerate blocks.
By virtue of Result 2, it is clear that A_{deg} is always measurable, so that if $A_{nondeg} = \emptyset$, then the whole aggregate A is measurable. Therefore, we can limit our considerations to the case where $A_{deg} = \emptyset$ and $A_{nondeg} \neq \emptyset$. Furthermore, without loss of generality, we shall assume that the edges of G(P, Q) included in A_{nondeg} come from the same nondegenerate block.
Take the following example of a nondegenerate block, B(P, Q) .

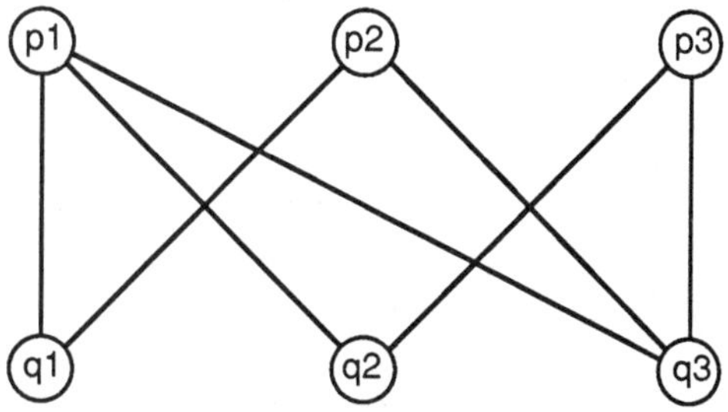

Let $M = \{i = 1,..., m\}$ and $N = \{j = 1,..., n\}$ be the index sets of the P-node set and the Q-node set, respectively. Let s be the number of edges in B(P, Q) . Consider the equation system (1) over B(P, Q) , which we rewrite as follows:

$$\sum_{j \in N(i)} x(i, j) = u_i$$
$$\sum_{i \in M(j)} x(i, j) = v_j \qquad (4)$$

where N(i) is the subset of N indexing Q(p) and M(j) is the subset of M indexing P(q) .
In general, such a system has s unknowns, x(i, j) , and m + n - 1 independent equations in view of the constraint (2). Therefore, it admits ∞^r distinct solutions (i.e., joint distributions) , where r = s - (m+n-1) . For instance, the system (4) has r = 2 degrees of freedom. Consequently, the solutions are dependent on two parameters, say λ_1 and λ_2 , and can be built up explicitly in the following way. We begin by arbitrarily choosing an edge to which we associate the undetermined value λ_1 . Next, we will try to express the values of the other edges as functions of λ_1 . When we can't proceed further, then we shall introduce the parameter λ_2 .

STEP 1. Put $x(1, 1) = \lambda_1$. Hence the functional expressions for the edges (2, 1) e (2, 3) :

$$x(2, 1) = v_1 - \lambda_1$$
$$x(2, 3) = u_2 - v_1 + \lambda_1$$

STEP 2. Put $x(1, 2) = \lambda_2$. Hence the functional expressions for the remaining edges:

$$x(1, 3) = u_1 - (\lambda_1 + \lambda_2)$$
$$x(3, 2) = v_2 - \lambda_2$$
$$x(3, 3) = u_3 - v_2 + \lambda_2$$

To summarize, we have the following parametric expression of a joint distribution over the block B(P, Q) :

(i, j)	x(i, j)
(1, 1)	λ_1
(1, 2)	λ_2
(1, 3)	$u_1 - (\lambda_1 + \lambda_2)$
(2, 1)	$v_1 - \lambda_1$
(2, 3)	$u_2 - v_1 + \lambda_1$
(3, 2)	$v_2 - \lambda_2$
(3, 3)	$u_3 - v_2 + \lambda_2$

At this point, let us decompose $x(i, j)$ as follows

$$x(i, j) = x_0(i, j) + c(i, j)$$

where

(i, j)	$x_0(i, j)$	$c(i, j)$
$(1, 1)$	0	λ_1
$(1, 2)$	0	λ_2
$(1, 3)$	u_1	$-(\lambda_1 + \lambda_2)$
$(2, 1)$	v_1	$-\lambda_1$
$(2, 3)$	$u_2 - v_1$	λ_1
$(3, 2)$	v_2	$-\lambda_2$
$(3, 3)$	$u_3 - v_2$	λ_2

The distribution $x_0(i, j)$ represents a particular solution of the system (4). For an arbitrary (P, Q)-aggregate A we put

$$x_0(A) = \Sigma\, x_0(i, j)$$

the summation extended over all the edges (i, j) in A.
The distribution $c(i, j)$ represents a solution of the homogeneous system associated with the system (4):

$$\begin{aligned} \Sigma_{j \in N(i)}\ x(i, j) &= 0 \\ \Sigma_{i \in M(j)}\ x(i, j) &= 0 \end{aligned} \tag{5}$$

If we interpret $B(P, Q)$ as a "flow network", the system (5) gives the conditions of "flow conservation" at the nodes of the network. According to this interpretation, a solution $c(i, j)$ is said to be a *circulation* [2, page 212]. If A is an arbitrary aggregate of edges from $B(P, Q)$, we put:

$$c(A) = \Sigma\, c(i, j)\ .$$

THEOREM 1. Let $x(i) = u_i$ and $x(j) = v_j$ be two consistent distributions. Assume that the correspondence graph is a nondegenerate block, $B(P, Q)$. A necessary and sufficient condition for an arbitrary aggregate A to be measurable is that for every circulation $c(i, j)$ on $B(P, Q)$ one has $c(A) = 0$.
Proof. The theorem is an immediate consequence of the decomposition

$$x(A) = x_0(A) + c(A)\ .$$

In order to deduce from Theorem 1 a procedure for testing measurability, we must analyse in depth the mathematical structure of the set $\mathbb{C}$ of all the possible circulations we can define on B(P, Q) . It is well-known that $\mathbb{C}$ is a vector space, whose dimension r is given by the degrees of freedom of the system (4) and, therefore, $r = s - (m+n-1)$ [1, 2] . A representation basis for $\mathbb{C}$, $\{c_1(i, j), \ldots, c_r(i, j)\}$, can be obtained by taking the coefficients of the parameters $\lambda_1, \ldots, \lambda_r$.
For example, for the case under examination we have

$$c(i, j) = \lambda_1 c_1(i, j) + \lambda_2 c_2(i, j)$$

where

(i, j)	$c_1(i, j)$	$c_2(i, j)$
(1, 1)	1	0
(1, 2)	0	1
(1, 3)	- 1	- 1
(2, 1)	-1	0
(2, 3)	1	0
(3, 2)	0	-1
(3, 3)	0	1

Once we have constructed a representation basis for the circulations in $\mathbb{C}$, we can finally test the measurability of an arbitrary aggregate using the following theorem.

THEOREM 2. Let $x(i) = u_i$ and $x(j) = v_j$ be two consistent distributions. Assume that the correspondence graph is a nondegenerate block, B(P, Q) . Let $\{c_1(i, j), \ldots, c_r(i, j)\}$ be a representation basis of $\mathbb{C}$. A necessary and sufficient condition for an arbitrary aggregate A to be measurable is that $c_h(A) = 0$ for $h = 1, \ldots, r$.
Proof. The theorem is an immediate consequence of the decomposition

$$x(A) = x_o(A) + \Sigma_{h=1, \ldots, r} \lambda_h c_h(A) \quad .$$

It should be noticed that, in view of Theorem 2, the measurability of an aggregate is decidable a priori, that is, without knowing the values of the base distributions $x(i) = u_i$ and $x(j) = v_j$. All the information we need is the knowledge of how the correspondence graph is made.
In the next paragraph, we shall present an automatic procedure for getting a representation basis for $\mathbb{C}$.

7. SEARCH FOR A BASIS OF $\mathbb{C}$

In this section, we present an algorithmic procedure for the research of a representation basis of $\mathbb{C}$. To this end, we have to give an *orientation* to B(P, Q) in order to convert it into a digraph. Let us choose to orient B(P, Q) from P-nodes to Q-nodes, so that the P-node of an edge is its "initial" node and the Q-node is its "terminal"node.
Then, a *cycle* is defined to be a chain of edges $\mathbf{c} = (e_1, \ldots, e_q)$ such that:

(1) each edge e_t $(1 < t < q)$ has one node in common with the preceding edge e_{t-1} and one node in common with the subsequent edge e_{t+1} ;
(2) the chain does not use the same edge twice;
(3) the initial node and the terminal node are the same.

A *circuit* (or an *elementary* cycle) is a cycle, in which, in addition

(4) no node is encountered more than once (except the initial node which is also the terminal node).

With cycle $\mathbf{c}$ we shall associate an integer-valued circulation in the following way. Let us denote by $\mathbf{c}^+$ the set of all edges in $\mathbf{c}$ that are in the direction in that the cycle is traversed, and denote by $\mathbf{c}^-$ the set of the other edges in $\mathbf{c}$. The circulation $c(i, j)$ associated with $\mathbf{c}$ is defined as follows:

$$c(i,j) = \begin{cases} 0 & \text{if } (i,j) \notin \mathbf{c} \\ +1 & \text{if } (i,j) \in \mathbf{c}^+ \\ -1 & \text{if } (i,j) \in \mathbf{c}^- \end{cases}$$

Henceforth, a cycle and its circulation will be used interchangeably.

The cycles $\mathbf{c}_1, \ldots, \mathbf{c}_r$ are said to be *dependent* if there exists a circulation equation of the form

$$\lambda_1 c_1(i,j) + \ldots + \lambda_r c_r(i,j) = 0 \qquad (i = 1, \ldots, m; j = 1, \ldots, n)$$

where $\lambda_1 , \ldots , \lambda_r$ are real numbers, not all zero. If the cycles are not dependent, they are said to be *independent.*

A *cycle basis* is defined to be a set $\{\mathbf{c}_1, \ldots, \mathbf{c}_r\}$ of independent circuits, such that the circulation associated with any cycle $\mathbf{c}$ can be written as

$$c(i,j) = \Sigma_{h=1,\ldots,r} \ \lambda_h c_h(i,j)$$

where $\lambda_1 , \ldots , \lambda_r$ are real numbers. A cycle basis can serve as a basis of the cycle space $\mathbb{C}$. For this reason, the dimension r of $\mathbb{C}$ is called the "cyclomatic number" of B(P, Q) .
Now, let τ be a spanning tree of B(P, Q) . The search of a spanning tree is a classical

graph-theoretical problem solvable by using the "Greedy" Algorithm [8]. For instance, the following tree spans the block B(P, Q) considered in the previous section.

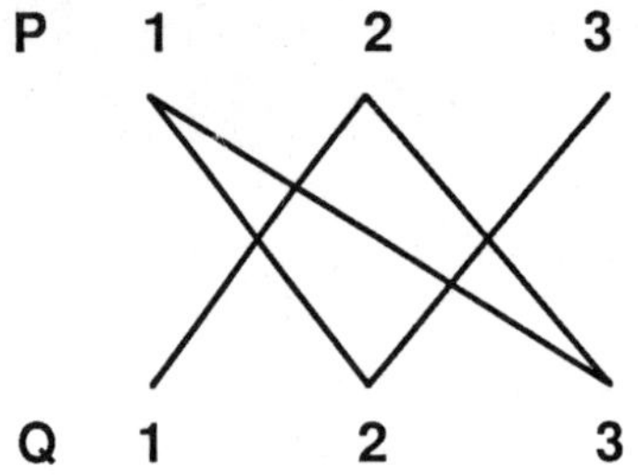

In general, if edge $(i, j) \notin \tau$ then $\tau \cup (i, j)$ contains exactly one circuit. As B(P, Q) contains s edges and a spanning tree τ contains $m + n - 1$ edges, we can construct

$$r = s - (m + n - 1)$$

circuits, which form a cycle basis.
For example, in the case under examination we find two circuits $\mathbf{c}_1$ and $\mathbf{c}_2$ obtained by adding to τ respectively the edges (1, 1) and (3, 3) as starting edges:

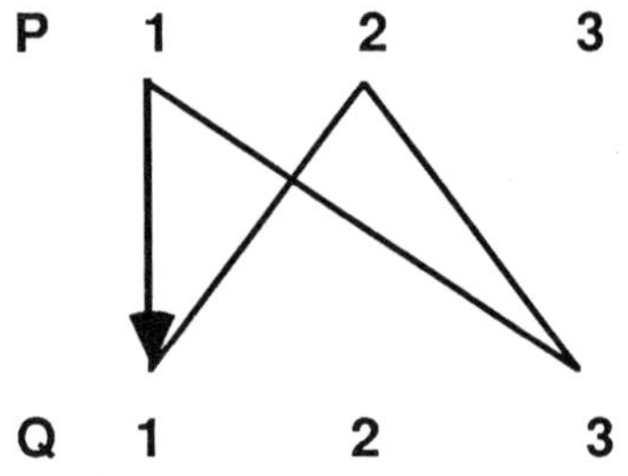

circuit $\mathbf{c}_1$

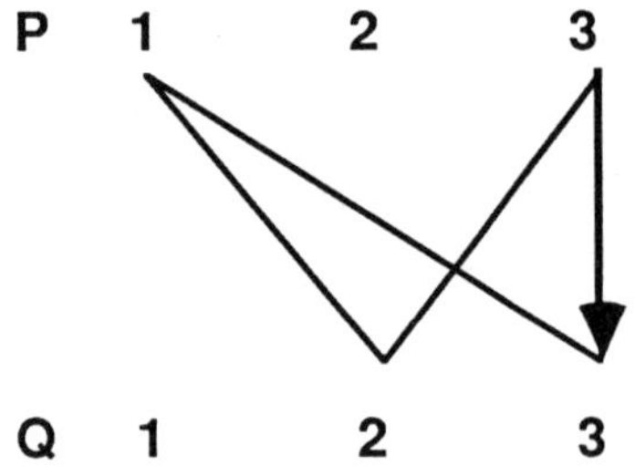

circuit $\mathbf{c}_2$

The circulations $c_1(i, j)$ and $c_2(i, j)$ associated with the circuits $\mathbf{c}_1$ and $\mathbf{c}_2$ are exactly those used as a basis of $\mathbb{C}$ in the preceding section.

8. CONCLUSIONS

We have presented a way to decide the measurability of an aggregate or, equivalently, the derivability of a summary data from a two-source statistical database, once the semantic connection between the two classification systems adopted have been stated precisely by means of a correspondence graph. The proposed algorithm, based on the condition of "zero-circulation", answers the question of measurability without accessing the data stored in the statistical database, by exploiting the topological properties of the underlying correspondence graph.

We expect that future research on summary data derivation will have to answer at least the following four open questions:

(1) Is the formal system presented in Section 5 complete?

(2) How to test consistency and how to manage inconsistency?

(3) When the aggregate function is restrained to take only nonnegative values (e.g., frequency distributions, probability distributions etc.), then the conditions stated by the above Theorems 1 and 2 turn out to be sufficient but not necessary. How must we change them in order to have an effective test for measurability?

(3) How to extend the above results to the case of multi-source databases?

REFERENCES

1. C. Berge, *Graphs and hypergraphs* . NORTH HOLLAND, 1973
2. J.A. Bondy and U.S.R. Murty, *Graph theory with applications* . 1976
3. C. Chen and P. Hernon, *Numeric databases* , Ablex Publishing Corporation, 1984
4. D. E. Denning,P. J. Denning and M. D. Schwartz,The Tracker: A Threat to Statistical Database Security, *ACM Trans. on Datab. Syst.* 4: 1 (1979) 76-96
5. S. Heiler and A. T. Maness, "Connecting Heterogeneous Systems and Data Sources", Working Group Notes: 2 Int. Workshop on Statistical & Scientific Database Management, in *Database Engineering* , 7: 1 (1984) 23-29
6. R. Johnson, "Modelling Summary Data", *Proc. ACM SIGMOD 1981 Conf on "Data Management"* , 93-97
7. R. Johnson, "A Data Model for Integrating Statistical Interpretations",*TR UCLR-86765* (1981)
8. E. L. Lawler, *Combinatorial Optimization: Networks and Matroids*, RINEHART & WINSTON, New York, 1976.
9. F. M. Malvestuto, "The derivation problem for summary data", *Proc ACM SIGMOD 1988 Conf on "Data Management"* , 82-89
10. F. M. Malvestuto, M. Rafanelli, C. Zuffada, *Many-source databases: some problems and solutions.* IASI-CNR Tech. Rep. 218 (June 1988)
11. J. L. McCarthy et al., "The SEEDIS Project: A summary Overview of the Social, Economic, Environmental, Demographic Information System", Lawrence Berkeley Laboratory document PUB-424 , April 1982
12. D. Merrill, "Problems in Spatial Data Analysis", *Proc. VII SAS User Group Int. Conf.*, San Francisco 1982
13. Z. Michalewicz, Compromisability of a Statistical Database, *Information Systems* 6: 4 (1983) 301-304
14. Z. M. Ozsoyoglu and G. Ozsoyoglu, "An Extension of Relational Algebra for Summary Tables", *Proc. 2 Int. Workshop on Statistical & Scientific Database Management* 1983, 202-211
15. E. M. Reingold, J. Nievergelt and N. Deo, *Combinatorial Algorithms: Theory and Practice.* PRENTICE-HALL, 1977
16. R. Ruggles and N. Ruggles, "The Role of Microdata in the National Economic and Social Accounts", *Review of Income and Wealth* , June 1975, 203-216
17. H. Sato, "Handling Summary Information in a Database: Derivability", *Proc. ACM SIGMOD 1981* , 98-107
18. H. Sato, "Fundamental Concepts of Social/Regional Summary Data and Inferences in their Databases", *Doctoral Thesis, The Faculty of Engineering, Tokyo University* (1982)
19. A. Shoshani, "Statistical Databases: Characteristics, Problems and Some Solutions", *Proc. VIII Int. Conf. on VERY LARGE DATA BASES* (1982)
20. Ministero dell'Industria,*Bilancio Energetico Nazionale.* Roma, 1986
21. OECD, *Energy Balance of OECD countries.* Paris, 1987
22. UNITED NATIONS, "Towards a System of Social and Demographic Statistics", ST/EST/ STAT/SER.F/18, New York, 1975

DATABASE ISSUES FOR A VETERINARY MEDICAL EXPERT SYSTEM

by

Mary McLeish †

Matthew Cecile

and

Alex Lopez-Suarez

Department of Computing and Information Science

University of Guelph

Guelph, Ontario

Canada N1G 2W1

Abstract

A large project to build an expert system for veterinary medicine was recently begun at the University of Guelph in collaboration with the Ontario Veterinary College. The hospital database system has been collecting on-line medical data for almost ten years. The prototype being developed is comparing several knowledge acquisition techniques from data which limit the number of rules used for diagnosis. Along with this approach, a fuzzy relational system involving all test results is being implemented. This paper shows how a combination of INGRES and the 'C' programming language can be used as a knowledge base to support these different methodologies.

Introduction

The prototype diagnostic system being developed concerns the diagnosis of surgical versus medical colic in horses [10, 11]. This is a significant problem in veterinary medicine, having led to many studies, diagnostic charts etc., to aid owners and veterinarians in the difficult process of recognizing serious cases. Horses suspected of requiring surgery must be shipped at a significant cost to the veterinary hospital, where further tests are conducted and a final decision is made. We are endeavoring to provide a computerized diagnostic tool in the hospital, which would also be available for practicing veterinarians on a remote access basis, to assist with this difficult task.

The animal hospital at the University of Guelph has a computer system called VMIMS (Veterinary Medical Information Management System). This system has been custom programmed in Sharp APL and handles usual admission and billing procedures. However, the

† This project was supported by the NSERC operating grant of Dr. Mary McLeish, #A4515

system also stores a considerable amount of medical information on each patient, including bacteriology, clinical pathology, parasitology, radiology, and other patient information such as age, sex, breed, presenting complaint, treatment procedures, diagnoses and outcome. In radiology and other diagnostic areas, computer terminals have replaced typewriters for preparing medical summaries. In clinical pathology, much of the data is electronically generated by the lab equipment. Terminals throughout the hospital provide daily printouts, including round lists, admit-discharge reports and census reports. The database currently holds about 48,000 records with 30,000 unique cases, requiring 500 megabytes of disk storage. The current project is being designed to make use of the vast amounts of data stored in VMIMS to aid the diagnostic process.

In Section 1, we discuss the downloading and restructuring of the data into INGRES from the VMIMS mainframe environment. This was done partly to alleviate problems of costly access time and a slow and costly information retrieval package. It also allowed the data to be easily fed into a variety knowledge acquisition routines.

Indeed, the diagnostic problem involves approximately 50 parameters (tests). A variety of techniques exist for inferring rules from data ranging from classical statistical techniques of discriminant analysis and logistic regression [17, 19], machine learning techniques involving event-covering [6], maximum entropy [21] and clustering [23]. These techniques find significant variables and, depending on the methodology, may omit factors if they do not exhibit certain types of behaviour (such as linearity). We have implemented and run all the methodologies just mentioned [c.f. 18,19].

Many other methods exist for handling uncertainty in facts and rules for medical expert systems [4]. Much controversy exists here, especially between Bayesian and non-Bayesian approaches. Bayesian methodologies incur difficulties with the assessment of priors and the requirement for information about dependencies (usually conditional independence assumptions are made). One such approach due to Zadeh [28] is particularly useful for representing uncertain information by linguistic variables. His methodology has been implemented by a group in Europe (Adlassnig *et al.* [1, 2].). This approach makes use of fuzzy relations involving diseases and symptoms (tests performed). Several fuzzy operations and other calculations need to be performed on data with their associated characteristic functions.

There have been a number of papers devoted to the implementation of fuzzy databases [3, 5, 29]. Recently a paper by Raju and Majumdar [22] discusses a formal fuzzy relational database system (but not the implementational problems). In Section 2.1 we review some basic notions from fuzzy sets and in 2.2 we present the particular forms necessary for the application at hand. Section 3 presents the methodology we have adopted for implementing a fuzzy relational system. Examples are provided of the fuzzy operations required. The types of membership functions used are explained and how they are handled within INGRES and C demonstrated. The advantages of using this particular combination of languages for the medical diagnostic system is explained. Section 4 provides the summary and future directions for the project.

Section 1

The current database system for the OVC hospital (VMIMS) is implemented on IBM 4381 equipment and operates under DOS and MVS operating systems. The Sharp APL package MABRA is available for information retrieval purposes. For the colic study, 3 sources of data are used, only one coming from VMIMS. The very initial symptom data in paragraph form is stored on hard copy. An independent colic study was conducted by a clinician (P. Pascoe).

Initially, the basic purpose of the knowledge management component of the project was to act as an interface between the three sources of data and the requirements of the decision support system. In several of the earlier medical expert systems such as the MYCIN project [4], the database component was set up expressly to meet the needs of the expert system. A contrasting approach was taken in the Rx project [27], where the initial data were obtained from existing databases and a collection of transpositions were then performed on the data attributes to allow for the efficient extraction of knowledge.

In order to provide a common database to the decision support system, it was decided to design a relational database. Since in a relational database structure data is stored simply as a collection of 'flat' files, there will be sufficient flexibility to reorganize the base relations to meet the requirements of the artificial intelligence, statistical and mathematical algorithms. The Database Management System Ingres [8] is being used to implement the relational database. The decision to use a relational database structure and Ingres provides the following additional benefits:

- Several of the advantages provided by standard Data Base Management Systems (DBMS) such as secondary storage management, security, backup, and recovery are readily available in Ingres.
- The data independence provided by the use of a DBMS [7], will give us the ability to reorganize the database without having to rewrite the Expert System component to accommodate the changes.
- The use of Ingres will allow us to build extensions to the relational database system required by the decision system components.

The last item is a particularly strong advantage since several recent research efforts in this area have already shown the benefits of this approach. Kung *et al* [15] propose extensions to support the selection and execution of algorithms for shortest path search problems. Held and Carlis [13] have developed a new operator that extends the relational data model to do pattern matching with very complex stored patterns. Snodgrass and Ilsoo [24] are developing a set of relational operators to support queries that involve time (temporal databases). Stonebraker *et al* describe a method that allows the implementation of expert rules in a relational database system [25, 26]. Grundy [12] reports possible extensions to DBMS to support statistical cluster analysis.

The Conversion Process

In our project we have three sources of data, each with very different characteristics. The clinical data and the symptoms data will require a one-time only conversion to the new knowledge-based system. The conversion software for these phases is written in JCL, APL and Fortran on the IBM equipment. The relational database is implemented on Microvax equipment. On the Microvax equipment the software is written in C and the Equel query language provided by Ingres. Figure 1 shows the general structure of this stage of the project.

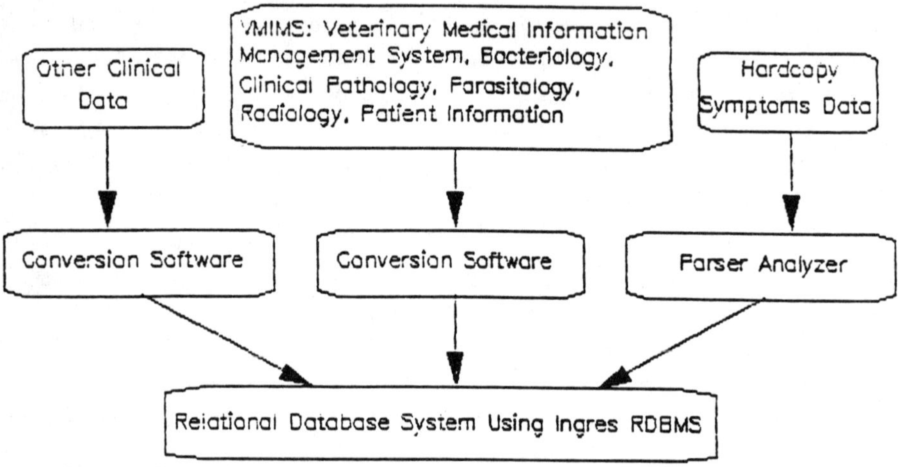

Figure 1. Sources of data for the Knowledge-Based Veterinary Medical Diagnostic System

The VMIMS database will provide the primary source of data during the operation of the prototype system. The logical structure of VMIMS is hierarchical. Figure 2. shows the main hierarchical structures available in VMIMS that are relevant to this project. During the conversion process, use is being made of the Sharp APL package MABRA. This phase converts the hierarchical structures into nested array structures common in APL. These arrays are then transferred into the Microvax equipment and converted (using C and Equel) into a format suitable for creating the base relations in Ingres.

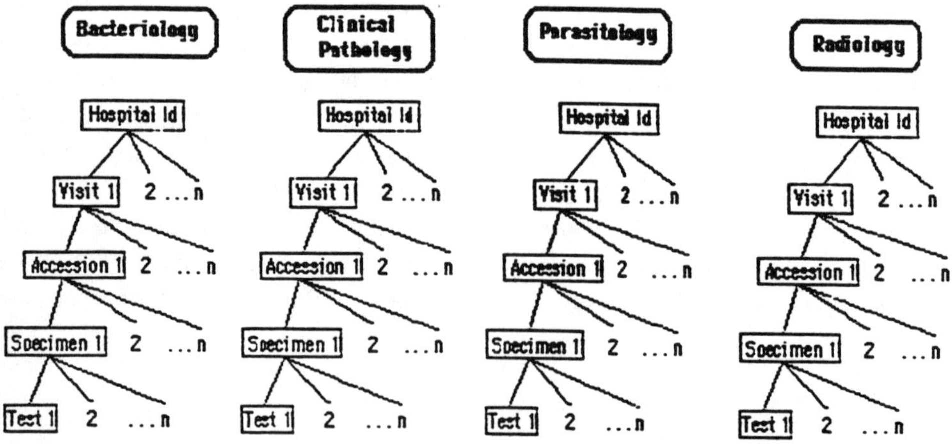

Figure 2. Logical structure of VMIMS database.

In order to retain the associations between each branch in the VMIMS database and the corresponding tuple in a relation, additional fields need to be included in the base relations. For example, Table (a) below shows the fields added to each row in the bacteriology relation that corresponds to a branch in the VMIMS hierarchy.

VMIMS BRANCH	BASE RELATION
hosptNum	hospNum
visitNum	visitNum
	visitCounter
accessionNum	accessionNum
	accessionCounter
specimenCd	specimenCd
specDescrip	specDescrip
specInterp	specInterp
	specCounter
testReq	testReq
testInterp	testInterp
	testCounter
organism	organism
orginterp	orginterp

* The order of the items is not the actual order in the VMIMS database

Table (a). Example of a VMIMS branch and its corresponding relation tuple.

The relational structure of the knowledge base has several additional advantages over the hierarchical structure of VMIMS. For example, in some pattern analysis methods it is required to perform horizontal access to VMIMS data across patients to gather information about specific bacteriological specimens. A similar process is required to collect information about some parasitological specimens. The resulting relations are then used to apply different pattern recognition techniques in order to attempt to identify possible cause-effect relationships. With the relational database, this data collection is performed through simple select and join operations.

Figure 3. below shows the interaction between the different components of the entire system which connect onto the DBMS component of Figure 1.

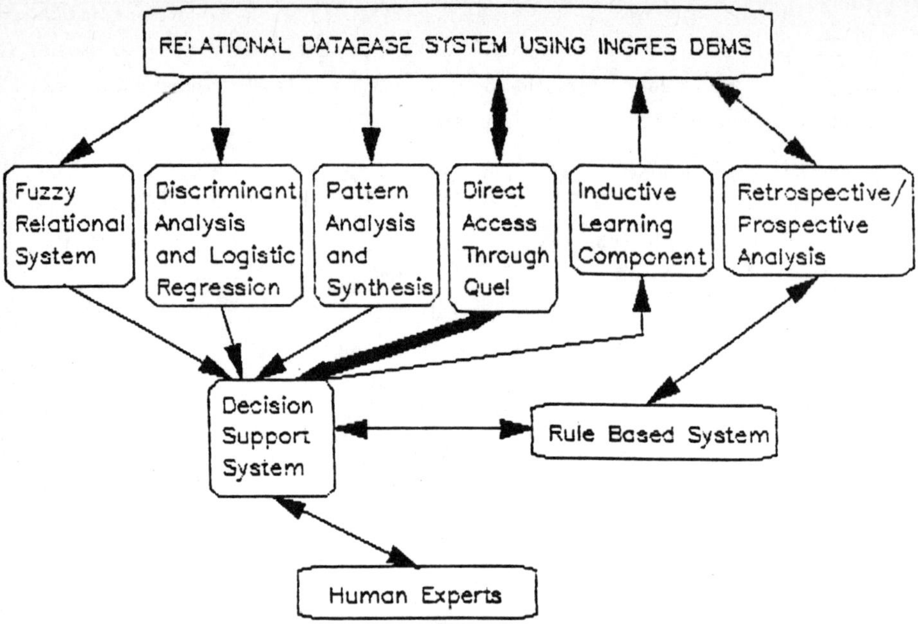

Figure 3. System Components

Sections 2 and 3 will focus on the use of INGRES for the fuzzy relational system.

Section 2.1

The most difficult component to link to the system is the complicated fuzzy relational system. The methods used will be described in detail in Section 3. We review here some basic concepts from the theory of fuzzy sets.

Following [9,14], we define a fuzzy subset A of a Universe X to be a set of pairs $\{x,\ \mu_A(\chi),\ \chi \in X\}$ where $\mu_A(\chi)$ is a real valued function taking values in [0, 1]. This function is sometimes called the grade of membership of χ in A. Classical union and intersection can be extended to fuzzy sets according to:

$$\mu_{A \cap B}(\chi) = \min\,(\mu_A(\chi), \mu_B(\chi))$$

$$\mu_{A \cup B}(\chi) = \max\,(\mu_A(\chi), \mu_B(\chi))$$

and the complement, $\overline{A}$, of A is defined by $\forall\ \chi \in X,\ \mu_{\overline{A}}(\chi) = 1 - \mu_A(\chi)$.

Other operations, such as products, bounded difference, concatenation etc. [cf. 9, 14] can be defined, but for our application, the next most significant operation arises from the extension principle.

Suppose we have an n-ary function of which is a mapping from the cartesian product $X_1 \cdots X_N$ to a set Y such that $y = f(\chi_1, \ldots \chi_N)$. Then there is a fuzzy set F induced on Y from the fuzzy subsets A_i of X_i where

$$\mu_F(Y) = \sup_{\substack{\chi_1 \cdots \chi_N \\ y = f(\chi_1 \cdots \chi_N)}} \min [\mu_{A_1}(\chi_1) \cdots \mu_{A_N}(\chi_N)]$$

Related to this concept is the idea of an n-ary fuzzy relation which is a fuzzy subset of $X_1 \times \cdots X_N$.

In the binary case, the composition of R and S on $X \times Y$ and $Y \times Z$ is given by:

$$\mu_{R \bigcirc S}(\chi, z) = \sup_{y \in Y} \min [\mu_R(\chi, y), \mu_S(y, z)] \; \forall \chi \in X, \; z \in Z$$

There are many properties of fuzzy relations and fuzzy sets that may be proven and these can be found in the material referenced already. However, for the application at hand, the types of expressions shown here are the ones that need to be computed.

Section 2.2

The CADIAG system [1, 2] uses fuzzy set theory to formalize medical relationships and fuzzy logic to model the diagnostic process. Our system imitates many of the features of this system. In order to indicate the type of calculations that will need to be performed a few of the features of this system are presented.

Symptoms S_i take values μ_{S_i} *in* $[0,1] \cup \{\nu\}$ where μ_{S_i} is the membership function and ν implies symptom S_i has not yet been examined. A binary fuzzy relationship $R_{PS} \subset \Pi \times \Sigma$ is then established defined by $\mu_R(P_q, S_i) = \mu_{S_i}$ for patient P_q where $P_q \in \Pi = \{P_1, \ldots P_r\}$, $S_i \in \Sigma = \{S_1, \ldots S_M\}$.

There is also a disease-patient relation defined similarly on $\Pi \times \Delta = \{D_1, \ldots D_N\}$. Other combinations are utilized when indicated: (i.e.) symptom, symptom relations, various intermediate combinations, etc. The fuzzy logical connectives are those of fuzzy set theory with a slight variation to account for the addition of ν:

$$\text{e.g. } \chi_1 \cup \chi_2 = \begin{bmatrix} \max(\chi_1, \chi_2) \text{ if } \chi_1, \chi_2 \in [0,1] \\ \chi_1 \text{ if } \chi_1 \in [0,1], \chi_2 = \nu \\ \chi_2 \text{ if } \chi_1 = \nu \text{ and } \chi_2 \in [0,1] \\ \nu \text{ if } \chi_1 = \chi_2 = \nu \end{bmatrix}$$

Notions of frequency of occurrence (o) and strength of confirmation (c) are introduced. These are themselves fuzzy values (or linguistic variables). The relations R_{SD} may be further described as R^c_{SD} or R^o_{SD} depending on the manner in which the membership function was defined. Composition for (hypothesis and confirmation) becomes:

$$R_{PD} = R_{PS} \bigcirc R_{SD}^{c},\ \mu_{R_{PD}}(P_q, D_j) = \max_{S_i} \min \left[\mu_{R_{PS}}(P_q, S_i),\ \mu_{SD}^{c}(S_i, D_j) \right]$$

A number of other compositions are used. Other notions of excluded diagnoses and possible diagnoses are introduced. Disease-disease relationships allow the inference of further diagnoses (confirmed or excluded).

Section 3 explains how a system has been designed to extend our existing INGRES implementation to a fuzzy relational system.

3.0 Fuzzy Relations and Their Manipulation using Ingres

This section discusses a prototype diagnostic system which has been implemented using fuzzy relations and embedded QUEL (EQUEL) in Ingres. It will be shown in this section how Ingres has made possible a quick implementation of a fuzzy relational scheme. Section 3.1 describes the structure of the fuzzy relations in our database and provides the EQUEL code used for the evaluation of membership functions. Section 3.2 discusses the issues related to the fuzzy composition operator and provides the EQUEL code used for this operation. Lastly section 3.3 provides an example of a typical iteration of the diagnostic process and shows its effect on the relations in the database.

3.1 Database Structure and Implementation

Several features of Ingres have greatly facilitated the implementation of a fuzzy relational system. Among these, the ability to embed QUEL commands into C programs has allowed us to perform many operations that would otherwise be very difficult or impossible. For example, Ingres tuples may be retrieved and stored in C data structures where complicated numerical operations may be performed on them. Other Ingres functions include aggregates like "max" and "min" that can find the maximum or minimum of an attribute over a selected set of tuples. These aggregates and embedded QUEL facilities of Ingres provide us with the means to quickly and simply implement a fuzzy relational diagnostic system.

A simple and efficient database structure is essential to facilitate fast, easy coding of our inference engine, learning mechanism, and fuzzy operations. Tables 1 to 4 show examples of the basic type of relations that exist in our database. The membership relation (table 1) is derived by the system designer in cooperation with the field expert. The reader should note that although a simple coordinate representation is adopted here, our final system may use another representation such as B-Splines [17]. The symptom group relation (table 2) and the symptom group-disease relation (table 3) are derived through knowledge aquisition methods (conditional probabilities, event-covering, etc.) and/or learning techniques. The patient-symptom group relation (table 4) is updated through the evaluation of the membership functions whenever a test is performed upon a patient and reflects the extent to which a patient has a symptom group. The patient-disease relation (table 5) is modified through the use of the composition operator described in the next section. The database also includes information pertaining to prior patient's test results and diagnoses, as previously described.

Test	Sympt	X1	X2	Y1	Y2
1	1	0	70	1.0	1.0
1	1	70	80	1.0	0.0
1	1	80	150	0.0	0.0
1	2	0	60	0.0	0.0
1	2	60	85	0.0	1.0
1	2	85	100	1.0	1.0
1	2	100	110	1.0	0.0
1	2	110	150	0.0	0.0
2	3	0	105	0.0	0.0
2	3	105	115	0.0	1.0
2	3	115	150	1.0	1.0

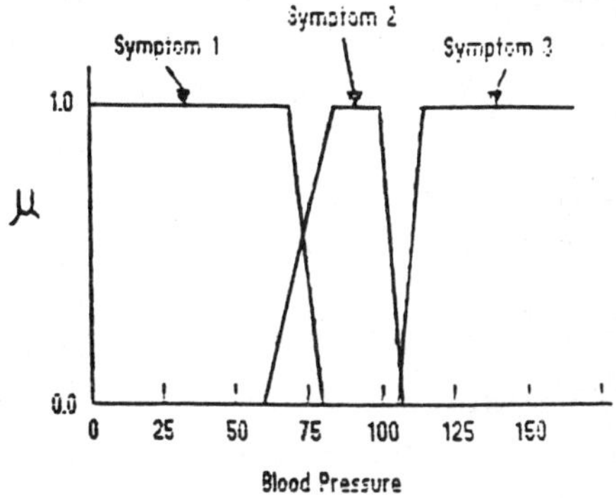

Table 1. Membership Relation and Associated Function
(X1, X2, Y1, Y2 correspond to the coordinate pairs (x1,y1) & (x2,y2))

Symptom Group	Test	Sympt	Strength
1	1	1	1.0
2	1	2	0.2
2	2	1	0.67
		.	
		.	
		.	

Table 2. Symptom Group Relation

Patient Number	Symptom Group	Strength
1	1	1.0
1	2	0.2
1	3	0.67
	.	
	.	
	.	

Table 4. Patient-Symptom Group Relation

Symptom Group	Disease	Strength
1	2	1.0
1	3	0.67
2	1	0.9
2	2	0.4
	.	
	.	
	.	

Table 3. Symptom Group-Disease Relation

Patient Number	Disease Number	Strength
1	1	0.2
1	2	1.0
1	3	0.67
	.	
	.	
	.	

Table 5. Patient-Disease Relation

When a test result is reported to the diagnostic system all of the relevant membership functions can be discovered from the symptom group relation through a simple retrieval on the symptom group relation. For each of these membership functions we need retrieve only those portions that contain our observed value (i.e. are on that particular line segment) and then interpolate. Next the symptom group relation is updated to reflect the information aquired from this test. It can now be determined if the truth level of any of the symptom groups have changed - do we know any more about the symptom patterns of this patient. If we have learned something new about this patient's symptom groups we then update the patient-symptom group relation to reflect this. The EQUEL code required to perform these steps is presented in the appendix and is a slightly simplified version of actual code.

3.2 Implementation of the Composition Operator

As described previously, we can determine the extent to which patients have symptom groups through the evaluation of membership functions. Through knowledge aquisition and learning methods we can determine the extent to which different symptom groups imply diseases. Our ultimate goal, however, is to determine the strength of the relationships between patients and diseases. Thus we have,

P -> SG

and SG -> D

we require P -> D

To do this we require the fuzzy join operator: the composition P $\bigcirc$ D which is defined as:

$$P_i \bigcirc D_k = \max_{SG_j} min\,[R_{P_i SG_j}, R_{SG_j D_k}] \qquad \textbf{(3.1)}$$

where:

$R_{SG_j D_k}$ is a value in [0,1] which measures the strength of association between symptom group SG_j that implies disease D_k

$R_{P_i SG_j}$ is the strength in [0,1] of the patient having symptom SG_j

The composition operator thus provides us with our basic inference mechanism, our modus ponens.

Table 5 shows a sample patient-disease relation. Given the patient-symptom group relation (table 3) and the symptom group-disease relation (table 4) it can be seen that we can form the patient-disease relation by joining these two relations by the symptom group: $R_{PSG} \bowtie_{SG} R_{SGD}$. The strength factor of this new relation can be determined using the formalism given in (3.1). Specifically the strength is the maximum over all symptom groups of the minimum of R_{PSG} and R_{SGD}.

In implementing these extensions to Ingres it has been our design philosophy to use the power of Ingres whenever possible. There are, however, cases when Ingres simply does not provide the facilities needed. In these instances "C" is used to manipulate the data objects. When implementing the composition operator the choice of "C" vs Ingres becomes difficult. Using mainly Ingres we could perform the following steps:

(1) Join R_{PSG} and R_{SGD} and store the result in a temporary relation. While doing this store the minimum of the two strengths in a "C" array.

(2) Create a new relation using the first temporary relation and the "C" array.

(3) Find the max over all S_j implying D_k and update the patient-disease relation.

Alternatively, relying on "C" we could do the following:

(1) Join R_{PSG} and R_{SGD} and store the result in a "C" data structure computing the min of the two strengths while retrieving.

(2) Find the max over all S_j implying D_k using a "C" subroutine and update the patient-disease relation.

Step 2 of the "C" implementation could also store the results of step 1 in an Ingres relation and use the built in max aggregate but seems to be a high computational price to pay to avoid creating a "C" max function.

One would expect the "C" implementation to be less expensive computationally and take longer to implement. The "C" implementation is more appealing because it does not require intermediate relations to be formed. As Ingres does not provide the ability to assign the minimum of two values of a row to an attribute of a new relation it is necessary to do this in "C" (Step 2 of Ingres implementation). For these reasons we found it necessary to deviate from our design philosophy in implementing this operation which was possible using almost all Ingres but was very inefficient and awkward.

3.3 A Sample Inference Cycle

Our sample inference begins with the reporting of the results of two tests: test 1 (blood pressure) = 65 and test 2 (temperature) = 97.0. Retrieving the symptom group relation (table 8) informs us that in test 1 symptoms 1 and 2 are important for diagnoses and in test 2 symptom 2 is important. Using this information we retrieve the tuples in membership relation (table 7) that are important (marked with a * in the table).

Having retrieved the membership values for the symptoms we then update the symptom group relation strength as shown in table 8 and determine if the patient-symptom strengths require updating. Once the patient-symptom relation has been updated as shown in table 10 we now perform the composition operation to determine the patient-disease relation.

Given the symptom group-disease relation shown in table 6 and the patient-symptom group relation shown in table 10 we now perform $R_{PSG} \underset{SG}{\bowtie} R_{SGD}$ and evaluate the min of the two strengths. This provides us with the data structure shown in table 9. Finally using a "C" max subroutine over all symptom groups implying a disease creates the tuples shown in table 11.

From the evaluation of two tests we have confirmed disease 2 and set the possibilities of disease 1 and 3 to 0.2 and 0.67 respectively. A typical diagnostic session will consist of many iterations through this inference cycle.

Symptom Group	Disease	Strength
1	2	1.0
1	3	0.67
2	1	0.9
2	2	0.4
	.	
	.	

Table 6. Symptom Group-Disease Relation

Test	Sympt	X1	X2	Y1	Y2	
1	1	0	70	1.0	1.0	*
1	1	70	80	1.0	0.0	
1	1	80	150	0.0	0.0	
1	2	0	60	0.0	0.0	
1	2	60	85	0.0	1.0	*
1	2	85	100	1.0	1.0	
1	2	100	110	1.0	0.0	
1	2	110	150	0.0	0.0	
2	1	0	96	1.0	1.0	
2	1	96	97.5	1.0	0.0	*
2	1	97.5	120	0.0	0.0	
		.	.			
		.	.			

Table 7. Membership Function Relation

Symptom Group	Test	Sympt	Strength
1	1	1	1.0
2	1	2	0.2
2	2	1	0.67
		.	
		.	

Table 8. Symptom Group Relation

Symptom Group	Patient	Disease	Strength
1	1	2	1.0
1	1	3	0.67
2	1	1	0.2
2	1	2	0.2
	.	.	
	.	.	

Table 9. "C" Data Structure Used For Intermediate Step

Patient Number	Symptom Group	Strength
1	1	1.0
1	2	0.2
1	3	0.67
	.	
	.	
	.	

Table 10. Patient-Symptom Group Relation

Patient Number	Disease Number	Strength
1	1	0.2
1	2	1.0
1	3	0.67

Table 11. Patient-Disease Relation

4. Conclusions

This paper has presented some of the database problems we have encountered in developing an expert system which uses large sources of data and complicated analysis techniques. More attention has been given recently to using data and statistical techniques in conjunction with 'expert' opinions in medical expert systems. This requires a database of information suitable for scientific and statistical computations. We have shown an approach which solves many of the necessary problems.

Further ongoing work concerns the exploration of alternate uncertainty management techniques. The methodology used by Adlassnig et al [1] is being compared to a scheme which uses fuzzy sets as a representation for evidence but evidence combination rules which are probabilistic in nature. Use is made of the notion of the probability of a fuzzy event as proposed by Yager [28]. Evidence combination is performed using two methods. The first technique is called the corroboration of evidence formula and is due to I.J. Good: C(H:E) $= \sum_i W(H{:}E_i)$ where $W(H{:}E_i)$ is the weight of evidence towards the hypothesis H and is measured by $\ln\left(\frac{P(H/E_i)}{P(\bar{H}/E_i)}\right)$. The second method uses a primarily intuitive formulation of the form: $P(H/E) = \frac{\sum_i W(H{:}E_i) * P(H{:}E_i)}{\sum_i W(H{:}E_i)}$. The details of this work are described in [18]. Results so far indicate similar final diagnostic behaviour from Good's formulation and the weighted average form. These results have also been compared to logistic regression and Bayesian inductive inference. In a particular case, where surgical lesion was misdiagnosed and the horse died, logistic regression provided an answer close to p = 0.5 (unconclusive) whereas the other three methods strongly indicated the need for surgery.

In the above formulations just described, instead of working with individual pieces of evidence, a method has been formulated to look for significant groups of evidence (symptom groups). This process is combinatorially explosive. However, the OVC hospital has recently converted the VMIMS system to a Sequent parallel processing machine using the Oracle RDBMS. We have consequently converted our applications to take advantage of the parallel power of this machine in order to alleviate the computational problems encountered in the search for significant groups of symptoms.

Acknowledgement

The authors wish to thank especially Dr.'s Peter Pascoe and T. Stirtzinger from the Ontario Veterinary College for their help with collection and interpretation of the data.

Appendix

The following is the EQUEL code required for membership evaluation and updating of the symptom group and patient-symptom relations.

```
                                  /* newtest is the test number taken           */
                                  /* testval is the actual value of the test     */
                                  /* First discover relevant tests ...           */
  x = 0;
## range of s is s_group
## retrieve unique (s.testnum = newtest)
## {
       rel[x].symptom = s.symptom_num;
       ++x;
## }
  numtuples = x - 1;
  for (x = 0; x <= numtuples; ++x)
  {
                                  /* Retrieve the relevant membership portions  */
                                  /* and interpolate                            */
##     range of m is memberships
##     retrieve (m.testnum = newtest, m.symptomnum = rel[x].symptom)
##     where ((m.xcoord1 <= testval)and(m.xcoord2 > testval))
##     {
           rel[x].strength = ((testval/(m.xcoord1-m.xcoord2)*
                        (m.ycoord2 - m.ycoord1)) + m.ycoord1;
##     }
                                  /* Now update the symptom group relation and */
                                  /* keep track of which groups have changed    */
    y = 0;
##     replace (s.strength = rel[x].strength)
##     where ((s.testnum = newtest)and(s.symptomnum = rel[x].symptom))
##     {
           updated[y].group = s.group;
           ++y;
##     }
  }

                                  /* For all those groups that changed reevaluate */
                                  /* the symptom group                             */
  numupdated = y - 1;
  for (y = 0; y <= numupdated; ++y)
  {
##     retrieve (s.group = updated[y].group)
##     where (s.strength = min(s.strength))
##     {
           updated[y].strength = s.strength;
##     }

                                  /* Update the patient-symptom group relation    */
##     range of p is rel_ps
##     replace (p.strength = updated[y].strength)
##     where (p.sgroup = updated[y].group)
##     {
##     }
  }
```

References

1. Adlassnig, K. P., "Fuzzy Set Theory in Medical Diagnosis", IEEE Transactions on Systems, Man and Cybernetics, vol 16, 1986, pp. 260-265.

2. Adlassnig, K. P., "A Survey on Medical Diagnosis and Fuzzy Subsets", Approximate Reasoning in Decision Analysis, North Holland Press, 1982, pp. 203-217.

3. Anvari, M. and Rose, G. F., "Fuzzy Relational Databases", Proceedings of the First International Conference on Fuzzy Information Processing", Kuaui, Hawaii, 1984, B-6-3.

4. Buchanan, B. and Shortliffe, E., "Rule-Based Expert Systems: The MYCIN Experiments of the Stanford Heuristic Programming Project", Addison-Wesley, 1986.

5. Buckles, B. P. and Petry, F. E., "A Fuzzy Representation for Relational Databases", Fuzzy Sets and Systems 7, (1982), pp. 213-226.

6. Chiu, D. K. Y. and Wong, A. K. C., "Synthesizing Knowledge: A Cluster Analysis Approach Using Event-Covering", IEEE Transactions on Systems, Man and Cybernetics, vol. SMC-16, no. 2, March/April 1986, pp. 251-259.

7. Date, C. J., "An Introduction to Database Systems", Addison-Wesley, 1987.

8. Date, C. J., "A Guide to Ingres", Addison-Wesley, 1987.

9. Dubois, D. and Prade, H., "Fuzzy Sets and Systems: Theory and Applications", Academic Press, 1980.

10. Ducharme, N., Pascoe, P. J., Ducharme, G. and Lumsden, T., "A Computer-Derived Protocol to Aid in Deciding Medical or Surgical Treatment of Horses with Abdominal Pain", private communication.

11. Ducharme, N., Ducharme G., Pascoe, P. J. and Horney, F. D., "Positive Predictive Value of Clinical Explanation in Selecting Medical or Surgical Treatment of Horses with Abdominal Pain", Proc. Eq. Colic. Res., pp. 200-230, 1986.

12. Grundy, "Cluster-Analysis - A Series of Database Views", Proceedings of the Third International Workshop on Statistical and Scientific Database Management, July, 1986, pp. 208-211.

13. Held, J. and Carlis, J., "Match: A New High-Level Relational Operator for Pattern Matching", Comm. ACM, 30, 1(January 1987), pp. 62-75.

14. Kandel, A., "Fuzzy Mathematical Techniques with Applications", Addison-Wesley, 1986.

15. Kung, R., Hanson, E., Lonnidis, Y., Sellis, T., Shapiro, L. and Stonebraker, M.,"Heuristic Search in Database Systems", In *Expert Database Systems*, edited by Kerschberg, L., Addison-Wesley, 1987.

16. Kung, R., "A Database Management System Base on an Object-Oriented Model", In *Expert Database Systems*, edited by Kerschberg, L., Addison-Wesley, 1987.

17. Lewis J. W., "B-Fuzzy Sets", submitted to *Uncertainty in Artificial Intelligence*, edited by Tod Levitt, North Holland Press.

18. McLeish, M., Cecile, M., "Induction and Uncertainty Management Techniques Applied to Veterinary Medical Diagnoses", submitted to AAAI Uncertainty Management Workshop, August 1988.

19. McLeish, M., "Exploring Knowledge Aquisition Tools for a Veterinary Medical Expert System", First International Conference on A.I. and Expert Systems, IEA/AIE-88.

20. Nau, D. and Reggia, J., "Relationship Between Deductive and Abductive Inference in Knowledge-Based Diagnostic Problem Solving", In *Expert Database Systems*, edited by Kerschberg, L., Addison-Wesley, 1987.

21. Quinlan, U. Ross, "Learning Efficient Classification Procedures and Their Application to Chess End Games", 1983, In *Machine Learning: An Artificial Intelligence Approach*, edited by Ryszard Michalski, Tioga, pp. 463-482.

22. Raju, K. and Majumdar A., "Fuzzy Functional Dependencies and Lossless Join Decomposition of Fuzzy Relational Database Systems", accepted for publication in ACMTODS.

23. Rendell, L., "A General Framework for Induction and a Study of Selective Induction", Machine Learning, Kluwar Pub., 1986, vol. 1, pp. 177-226.

24. Snodgrass, R. and Ilsoo, A., "Temporal Databases", IEEE Computer 19(9), September 1986, pp. 35-42.

25. Stonebraker, M., Woodfill, J. and Anderson, E., "Implementation of Rules in Relational Database Systems", Database Engineering, 6,4(December 1983).

26. Stonebraker, M. (Editor), "The Ingres Papers: Anatomy of a Relational Database System", Addison-Wesley, 1986.

27. Weiderhold, G., Blum, R. L. and Walker, M., "An Integration of Knowledge and Data Representation", in *On Knowledge Base Management Systems*, edited by Brodie, M. L. and Mylopoulos, J., Springer-Verlag, 1986.

28. Yager, R.R., "A Note on Probabilities of Fuzzy Events", Information Sciences 18, 1979, pp. 113 - 129

29. Zadeh, L. A., "Fuzzy Logic and Approximate Reasoning", Synthese 30, (1975), pp. 407-428.

30. Zvieli, A. and Chen, P. P., "Entity-relationship Modeling and Fuzzy Databases" , Proceedings of the Second International Conference on Data Engineering, Los Angeles, California, 1986, pp. 320-327.

Ranges and Trackers in Statistical Databases

Zbigniew Michalewicz* Keh-Wei Chen†

Abstract

The goal of statistical databases is to provide statistics about groups of individuals while protecting their privacy. Sometimes, by correlating enough statistics, sensitive data about individual can be inferred. The problem of protecting against such indirect disclosures of confidential data is called the inference problem and a protecting mechanism - an inference control. A good inference control mechanism should be effective (it should provide security to a reasonable extent) and feasible (a practical way exists to enforce it). At the same time it should retain the richness of the information revealed to the users. During the last few years several techniques were developed for controlling inferences. One of the earliest inference controls for statistical databases restricts the responses computed over too small or too large query-sets. However, this technique is easily subverted. In the previous paper (see [5]) we proposed a new query-set size inference control which is based on the idea of multiranges and has better performance then the original one. In this paper we go further investigating the consequences of non-uniform distribution of ranges, for which queries are unanswerable.

1 Introduction

The goal of statistical databases is to provide statistics about groups of individuals while protecting their privacy. Sometimes, by correlating enough statistics, sensitive data about individual can be inferred. When this happens, the personal records are compromised—we say, the database is *compromisable.* The problem of protecting against such indirect disclosures of confidential data is called the *inference problem.* During the last few years several techniques were developed for controlling inferences. One of the earliest inference controls for statistical databases (see [1], [4] and [7]) restricts the responses computed over too small or too large query-sets; later (see [3]) it was classified as one of the cell restriction techniques. This technique is easily subverted—the most powerful tools to do it

*Department of Computer Science The University of North Carolina at Charlotte, Charlotte, NC 28223; on leave from Victoria University of Wellington, Wellington, New Zealand

†Department of Mathematics, The University of North Carolina at Charlotte, Charlotte, NC 28223

are called *trackers* (we will define them later in the text). However, query-set size controls are trivial to implement. Moreover, they can be valuable when combined with other protection techniques (see [3]), so they are worth some deeper examination. In this paper we will generalize the query-set size control technique to include multi-ranges with non-uniform distribution and indicate the usefulness of our new approach.

A *statistical database* consists of a collection X of some number n of *records*, each containing a fixed number of confidential *fields*. Some of the fields are considered to be *category fields* and some to be *data fields* (the set of category fields need not be disjoint from the set of data fields). It is assumed that for any category field there is a given finite set of possible values that may occur in this field for each of the records. Data fields are usually numerical, *i.e.* it is meaningful to sum them up.

A *statistical query* has the form $COUNT(C)$, where C is an arbitrary expression built up from category-values (specifying a particular value for given category fields) by means of operators $AND(\cdot)$, $OR(+)$, and $NOT(\sim)$. The set of those records which satisfy the conditions expressed by C is called the *query set* X_C. The query-set size inference control is based on the following definition of the response to the query $COUNT(C)$:

$$COUNT(C) = \begin{cases} |X_C| & if \;\; k \leq |X_C| \leq n-k \\ \# & otherwise \end{cases}$$

where $|X_C|$ is the size (cardinality) of X_C; k is a certain integer, fixed for a given database, $0 \leq k \leq n/2$; and # denotes the fact that the query is unanswerable, *i.e.* the database refuses to disclose $|X_C|$ for the query (the intuition being that the disclosure of $|X_C|$ for the query set X_C may lead to the disclosure of some individual record characteristic, which should not occur in a statistical database). Usually the set of allowable queries in statistical database also includes other queries, such as averages, sums and other statistics, as:

$$SUM(C;j) = \begin{cases} \sum_{i \in X_C} v_{ij} & if \;\; k \leq |X_C| \leq n-k \\ \# & otherwise \end{cases}$$

where j is a data field and v_{ij} is the value in field j of record i. Generally, we will deal with arbitrary queries $q(C)$ satisfying the condition

$$X_{C_1} \cap X_{C_2} = \emptyset \implies q(C_1 + C_2) = q(C_1) + q(C_2),$$

or equivalently,

$$q(C_1 + C_2) = q(C_1) + q(C_2) - q(C_1 \cdot C_2)$$

(this condition is clearly satisfied for COUNT and SUM).

We say, a statistical query $q(C)$ is ***answerable*** if $k \leq |X_C| \leq n - k$.

Before we concentrate on the compromisability problem for statistical databases for query-set size controls, let us introduce some useful notation. Suppose that there are m category fields, and that i-th category field can take n_i values $v_{i1}, \ldots, v_{in_i}$. We may form

$$N = \prod_{i=1}^{m} n_i$$

elementary conjunctions, each of them being a conjunction of a different combination of category-values. Let $C_1, C_2, \ldots, C_s$ be all the elementary conjunctions with a non empty (elementary) query set. Let $P = (p_1, p_2, \ldots, p_s)$ be a sequence such that $p_i = |X_{C_i}|$. Let $g = \max_{1 \leq i \leq s} P$. Additionally, let us denote $A = \{i : p_i \geq k\}$. Clearly A is the set of indices i such that the query C_i is answerable. In the case when $|A| = 1$, let $A = \{i\}$ and define $B = \{j : j \neq i \text{ and } p_i + p_j > n - k\}$. Clearly B is the set of indices $j \neq i$ such that the query $q(C_i + C_j)$ is unanswerable.

2 The compromisability problem

What is crucial in compromising a statistical database is to have a method to get around the condition $k \leq |X_C| \leq n - k$, and to be able to calculate $|X_C|$ (and $q(C)$) for any C. In this section we summarize briefly research made in this direction.

A *general tracker* (see [1]) is a formula T such that $2k \leq COUNT(T) \leq n - 2k$. Now, if the query $q(C)$ is unanswerable (because either $|X_C| < k$ or $|X_C| > n - k$) then the following formula is used for calculating the value $q(C)$:

$$q(C) = \begin{cases} q(C + T) + q(C + \sim T) - Q & if\ |X_C| < k \\ 2Q - q(\sim C + T) - q(\sim C + \sim T) & if\ |X_C| > n - k \end{cases}$$

where $Q = q(T) + q(\sim T)$.

It is easily seen that all subqueries, namely $q(C + T)$, $q(C + \sim T)$ (when $|X_C| < k$), $q(\sim C + T)$, $q(\sim C + \sim T)$ (when $|X_C| > n - k$), $q(T)$, and $q(\sim T)$ are answerable, so at least one of the equations is calculable.

The general tracker is not guaranteed to work when $k > n/4$, that is, when more then half the range of query set sizes is disallowed. However, this does not imply that the database is secure because there may exist a formula T (for a given C) such that one of the above equations is calculable, or there may exist a double tracker (see [1]) using which we may compromise the database when $n/4 < k \leq n/3$.

During the last few years a significant effort was made to find sufficient conditions for compromisability of a statistical database for such query-set size controls:

- $s \geq 2k+1$ and $k \leq n/4$ is a sufficient condition for existence of a general tracker (and consequently for compromisability), given by Denning *et al* [1];
- $s \geq 2k+1$ and $k \leq n/3$ is a sufficient condition for existence of a double tracker (and consequently for compromisability), given by Denning *et al* [1];
- $k \leq n/2 - g$ is a sufficient condition for compromisability, given by Schlörer [7];
- $|A| = 1$ and $|B| > 1$ or $|A| = 0$ and $k > n/3$ are the only conditions when the database may be non-compromisable, given by Michalewicz [4].

However, only [2] gives an algorithm to construct a tracker and consequently to compromise the database. So the trackers are the most serious threat to statistical database security.

In [5] we proposed a new query-set size inference control which deals with trackers much better. We summary the main results of [5] in the next section.

3 Multiranges, multitrackers and compromisability

Let us modify the formula for the response to a query $q(C)$ in the following way:

$$q(C) = \begin{cases} q(C) & if\ p \leq |X_C| \leq \frac{n-p}{2}\ or\ \frac{n+p}{2} \leq |X_C| \leq n-p \\ \# & otherwise \end{cases}$$

This means that we also restrict the middle range of the interval $\langle 0, n \rangle$ (note that we informally use $q(C)$ to denote both the response to the query, and the query itself). We will describe a method to get around this condition.

A *3-range tracker* is either a formula T_1 such that $2p \leq COUNT(T_1) \leq \frac{n-3p}{2}$ or a formula T_2 such that $\frac{n+3p}{2} \leq COUNT(T_2) \leq n - 2p$. If the query $q(C)$ is unanswerable, then the size of the query set $|X_C|$ is in one of the three restricted ranges: $\langle 0, p)$, $(\frac{n-p}{2}, \frac{n+p}{2})$, or $(n-p, n\rangle$.

3-RANGE TRACKER COMPROMISE. The value of any unanswerable query q(C) can be computed using any 3-range tracker T_m as follows. First calculate

$$Q = q(T_m) + q(\sim T_m)$$

If $COUNT(C) \in \langle 0, p)$, then the queries on the right-hand side of this equation are answerable:

$$q(C) = q(C + T_m) - q(\sim C \cdot T_m)$$

If $COUNT(C) \in (\frac{n-p}{2}, \frac{n+p}{2})$, then the queries on the right-hand side of at least one equation are answerable:

$$q(C) = \begin{cases} q(C+T_m) - q(\sim C \cdot T_m) \\ \qquad \textit{if } q(C+T_m) \textit{ and } q(\sim C \cdot T_m) \textit{ are answerable} \\ Q - q(\sim C + T_m) + q(C \cdot T_m) \\ \qquad \textit{if } q(\sim C + T_m) \textit{ and } q(C \cdot T_m) \textit{ are answerable} \\ q(C+T_m) - q(T_m) + q(C \cdot T_m) \\ \qquad \textit{if } q(C+T_m) \textit{ and } q(C \cdot T_m) \textit{ are answerable} \\ Q - q(\sim C + T_m) + q(T_m) - q(\sim C \cdot T_m) \\ \qquad \textit{if } q(\sim C + T_m) \textit{ and } q(\sim C \cdot T_m) \textit{ are answerable} \end{cases}$$

Otherwise $COUNT(C) \in (n-p, n\rangle$ and the queries on the right-hand side of this equation are answerable:

$$q(C) = Q - q(\sim C + T_m) + q(C \cdot T_m)$$

In all cases m is equal 1 or 2.

For proof the reader is referred to [5].

Note that the necessary condition for the existence of a 3-range tracker is $p \leq n/7$. To compare this result with the previous one ($k \leq n/4$) let us assume that $3p = 2k$, *i.e.* the total length of the restricted ranges of the query set sizes are equal. In that case, $p = \frac{2k}{3}$, and consequently, $k \leq \frac{3n}{14} < \frac{n}{4}$. So the introduction of a ***middle restricted range*** decreased the upper limit for the value of the parameter k for which a tracker may exist. In other words, it should be ***harder*** to find a tracker in the modified query-set size control.

We may make a similar argument in the case when the number of restricted ranges is 5. Again, the length of each closed range is p, the gaps between the closed ranges are of equal length. Then a

5-range tracker is a formula T_m, such that $T_m \in \langle \frac{(n-p)\cdot(m-1)}{4} + 2p, \frac{(n-p)\cdot m}{4} - p \rangle$ for some $1 \leq m \leq 3$.

An interesting result is that the necessary condition for existence of 5-range tracker is $p \leq n/13$. Again, comparing it with the original and the previous cases ($5p = 2k$) we get $k \leq \frac{5n}{26} < \frac{3n}{14} < \frac{n}{4}$ (*i.e.* $k \leq .192 \cdot n < .214 \cdot n < .250 \cdot n$).

Let us generalize the above approach. Let us assume we have $(2q+1)$ restricted ranges (the odd number is kept only to simplify calculations), all of the same length p with $2q$ gaps between them, all of equal length.

So the *i-th* restricted range I_i is given as: $I_i = (\frac{(n-p)\cdot(i-1)}{2q}, \frac{(n-p)\cdot(i-1)}{2q} + p)$, where $i = 1, 2, \ldots, 2q+1$ (note also that the first range should be closed on the left side and the last one on the right side). Let us denote $I = \bigcup_{i=1}^{2q+1} I_i$.

The formula for the response to the query $q(C)$ is:

$$q(C) = \begin{cases} q(C) & if\ |X_C| \notin I \\ \# & otherwise \end{cases}$$

(again, we informally use $q(C)$ to denote both the response to the query, and the query itself).

We will describe a method to get around this condition.

A *(2q+1)-range tracker* (or simply *multitracker*) is a formula T_j such that $COUNT(T_j) \in \langle \frac{(n-p)\cdot(j-1)}{2q} + 2p, \frac{(n-p)\cdot j}{2q} - p \rangle$ for some $(j = 1, 2, \ldots, 2q)$. The idea behind this definition is that for any multitracker T_m: $\min_{x \in I}\{|x - COUNT(T_m)|\} > p$.

MULTITRACKER COMPROMISE. The value of any unanswerable query q(C) can be computed as follows using any multitracker T_m. First calculate

$$Q = q(T_m) + q(\sim T_m)$$

Then the queries on at least one right-hand side of the equation are answerable:

$$q(C) = \begin{cases} q(C + T_m) - q(\sim C \cdot T_m) \\ \quad if\ q(C+T_m)\ and\ q(\sim C \cdot T_m)\ are\ answerable \\ Q - q(\sim C + T_m) + q(C \cdot T_m) \\ \quad if\ q(\sim C + T_m)\ and\ q(C \cdot T_m)\ are\ answerable \\ q(C + T_m) - q(T_m) + q(C \cdot T_m) \\ \quad if\ q(C + T_m)\ and\ q(C \cdot T_m)\ are\ answerable \\ Q - q(\sim C + T_m) + q(T_m) - q(\sim C \cdot T_m) \\ \quad if\ q(\sim C + T_m)\ and\ q(\sim C \cdot T_m)\ are\ answerable \end{cases}$$

For proof the reader is referred to [5].

It is easily seen that a necessary condition for an existence of a multitracker is $p \leq \frac{n}{6q+1}$. Again, let us compare this result with the original one, when only 2 ranges are restricted. Assume the total lengths of the restricted ranges in both cases are equal: *i.e.* $p \cdot (2q+1) = 2k$. Then we get our result:

$$k \leq \frac{n \cdot (2q+1)}{2 \cdot (6q+1)}$$

For large q the upper limit for a parameter k is close to $n/6$, so indeed the inference control based on multiple ranges changes the necessary condition for an existence of a multitracker from $k \leq n/4$ (2 ranges, case of general tracker) into $k \leq v$, where v is close to $n/6$ and $2k$ is the total length of all restricted ranges.

4 Non-uniform distribution of multiranges

The main weakness of the multiranges approach is that some 'safe' queries (for which the query-set size is around $n/2$) are restricted, decreasing the richness of the information revealed to the users. It is intuitively clear that a new approach which is based on some non-uniform distribution of ranges (*i.e.* restricted ranges are longer and closer to each other at the ends of the range $\langle 0, n \rangle$ than towards the middle of this range) should give much better performance (at least as the usefulness of the system is concern).

In the next subsections we will consider two (non-uniform) distribution of restricted ranges based on arithmetical and geometrical progresses and we will investigate their properties.

4.1 Arithmetical distribution

Let us assume we have $(2q+1)$ restricted ranges I_i, $i = 1, 2, \ldots, 2q+1$ (the odd number is kept only to simplify calculations). The 'middle' range I_{q+1} is given as: $I_{q+1} = (\frac{n-a}{2}, \frac{n+a}{2})$, *i.e.* the length of this range is a, which is a parameter of the method (it need not be an integer). The two ranges 'next' to the 'middle' range, namely I_q and I_{q+2} have lengths $a + r$ each ($r > 0$ is another parameter of the method), the ranges I_{q-1} and I_{q+3} have lengths $a + 2r$ each, and so on. The last pair of ranges, I_1 and I_{2q+1} have lengths $a + qr$ each. In the same time, the lengths of 'gaps' between restricted ranges vary as well. The length of the gap between I_1 and I_2 (as well as the length of the gap between I_{2q} and I_{2q+1}) is equal to b (next parameter of the method), the length of the gap between I_2 and I_3 (and between I_{2q-1} and I_{2q}), is equal to $b+r$, and so on. The largest length of the gap is between restricted ranges I_q and I_{q+1}, and between I_{q+1} and I_{q+2}, and is equal to $b+(q-1)r$.

In general, the *i-th* restricted range I_i is given as:

$$I_i = \begin{cases} (x_i, x_i + a + (q-i+1)r) & if\ i = 1, 2, \ldots, q+1 \\ (n - x_{2q+2-i} - a - (i-q-1)r, n - x_{2q+2-i}) & if\ i = q+2, \ldots, 2q+1 \end{cases}$$

where $x_i = (i-1)a + (i-1)b + \frac{r}{2}(q(q+1) + (i-2)(i-1) - (q-i+1)(q-i+2))$ for $i = 1, 2, \ldots, q+1$. Note that the first range should be closed on the left side and the last one on the right side.

Note also that in fact we have only two parameters (apart of q) of the method, since the sum of all lengths of restricted ranges and lengths of gaps between them is equal to n, *i.e.* $(2a+qr)(q+1) - a + (2b+(q-1)r)q = n$.

Let us denote $I = \bigcup_{i=1}^{2q+1} I_i$.

The formula for the response to the query $q(C)$ is again:

$$q(C) = \begin{cases} q(C) & if\ |X_C| \notin I \\ \# & otherwise \end{cases}$$

(again, we informally use $q(C)$ to denote both the response to the query, and the query itself).

There are two advantages of this approach in comparison with ranges with uniform distribution. Firstly, as we already discussed it, a 'safe' query is more likely to fall into unrestricted range, *i.e.* into gap between restricted ranges. Secondly, it is not clear, whether there is any method for finding a tracker in that case. The previous approach which determines trackers if the length of a gap between restricted ranges is sufficient, does not work in this case. For example, let us consider a formula T_m such that $COUNT(T_m)$ is between restricted ranges I_i and I_{i+1} ($i = 1, 2, \ldots, q$) and the distance between the right point of the range I_i and $COUNT(T_m)$ (as well as the distance between $COUNT(T_m)$ and the left point of the range I_{i+1}) is larger than the length of the largest range, *i.e.* I_1. In other words, the value of $COUNT(T_m)$ is between $x_i + (q-i+1)r + a + qr$ and $x_{i+1} - a - qr$ ($i = 1, 2, \ldots, q$). Because of non-uniform distribution of restricted ranges, in general we can not guarantee any longer that at least one of the queries $q(C \cdot T_m)$ and $q(\sim C \cdot T_m)$, or one of the queries $q(C + T_m)$ and $q(\sim C + T_m)$ is answerable, which is the basis for proving that the formula T_m can be used to compromise the database, *i.e.* the formula T_m is a tracker.

However, in some special cases it is possible to find a tracker. Let us consider the case where $q = 1$, *i.e.* we have 3 restricted ranges. These are: $I_1 = \langle 0, a + r)$, $I_2 = (\frac{n-a}{2}, \frac{n+a}{2})$, and $I_3 = (n - a - r, n\rangle$. If we find a formula T_m such that $COUNT(T_m)$ is in the range $\langle 2a + 2r, \frac{n-3a}{2} - r\rangle$ or in the range $\langle \frac{n+3a}{2} + r, n - 2a - 2r\rangle$, then we can prove that at least one of the queries $q(C \cdot T_m)$ and $q(\sim C \cdot T_m)$, and at least one of the queries $q(C + T_m)$ and $q(\sim C + T_m)$ are answerable—the conclusion is that the formula T_m can serve as a multitracker. However, even in this case, the results are better than those of the case of uniform distribution of restricted ranges. Here the necessary condition for existence of a such tracker is $a \leq \frac{n-6r}{7}$. To compare this result with the previous one we assume that the total length of restricted ranges is $2k$, *i.e.* $3a + 2r = 2k$, which gives us $a = \frac{2k-2r}{3}$, and consequently, $k \leq \frac{3n-4r}{14} < \frac{3}{14}n$, which is true for all $r > 0$. So even in the case when a tracker exists, the introduction of non-uniform distribution decreased the upper limit for the value of the parameter k for which a tracker can exist, *i.e.* it should be harder to find a tracker in this case.

In a general case when the number of restricted ranges is $2q + 1$, the necessary condition for the existence of a tracker is $b + (q-1)r \geq 2a + 2qr$, *i.e.* the length of the largest gap is not smaller than the double length of the largest range. Replacing b the inequality becomes: $\frac{n+a-(2a+qr)(q+1)}{2q} + \frac{(q-1)r}{2} \geq 2a + 2qr$, which gives: $a \leq \frac{n-2qr(2q+1)}{6q+1}$. Again, assuming that the total length of all restricted ranges is equal to $2k$, *i.e.* $a = \frac{2k-qr(q+1)}{2q+1}$, we get our final result: $k \leq \frac{(2q+1)n-qr(2q^2+q+1)}{2(6q+1)} < \frac{2q+1}{2(6q+1)}n$ for all $q > 0$ and $r > 0$.

4.2 Geometrical distribution

Let us assume again, we have $(2q+1)$ restricted ranges I_i, $i = 1, 2, \ldots, 2q+1$ (the odd number is kept only to simplify calculations). The 'middle' range I_{q+1} is given as: $I_{q+1} = (\frac{n-a}{2}, \frac{n+a}{2})$, *i.e.* the length of this range is a, which is a parameter of the method. The two ranges 'next' to the 'middle' range, namely I_q and I_{q+2} have lengths σa each ($\sigma > 1$ is another parameter of the method), the ranges I_{q-1} and I_{q+3} have lengths $\sigma^2 a$ each, and so on. The last pair of ranges, I_1 and I_{2q+1} have lengths $\sigma^q a$ each. In the same time, the lengths of 'gaps' between restricted ranges vary as well. The length of the gap between I_1 and I_2 (as well as the length of the gap between I_{2q} and I_{2q+1}) is equal to b (a parameter of the method), the length of the gap between I_2 and I_3, and between I_{2q-1} and I_{2q}, is equal to σb, and so on. The largest length of the gap is between restricted ranges I_q and I_{q+1}, and between I_{q+1} and I_{q+2}, and is equal to $\sigma^{q-1}b$.

In general, the *i-th* restricted range I_i is given as:

$$I_i = \begin{cases} (x_i, x_i + \sigma^{q-i+1}a) & if\ i = 1,2,\ldots,q+1 \\ (n - x_{2q+2-i} - \sigma^{i-q-1}a, n - x_{2q+2-i}) & if\ i = q+2,\ldots,2q+1 \end{cases}$$

where $x_i = \frac{\sigma^{i-1}-1}{\sigma-1}(a\sigma^{q-i+2} + b)$, for $i = 1,2,\ldots,q+1$. Note that the first range should be closed on the left side and the last one on the right side.

Note also that in fact we have only two parameters (apart from q) of the method, since the sum of all lengths of restricted ranges and lengths of gaps between them is equal to n, *i.e.* $2b\frac{\sigma^q-1}{\sigma-1} + 2a(\frac{\sigma^{q+1}-1}{\sigma-1} - \frac{1}{2}) = n$.

Again, let us consider the case when $q = 1$, *i.e.* there are three restricted ranges. These are: $I_1 = \langle 0, a\sigma)$, $I_2 = (\frac{n-a}{2}, \frac{n+a}{2})$, and $I_3 = (n - a\sigma, n\rangle$. If we find a formula T_m such that $COUNT(T_m)$ is in the range $\langle 2a\sigma, \frac{n-a}{2} - a\sigma\rangle$ or in the range $\langle \frac{n+a}{2} + a\sigma, n - 2a\sigma\rangle$, then we can prove that at least one of the queries $q(C \cdot T_m)$ and $q(\sim C \cdot T_m)$, and at least one of the queries $q(C + T_m)$ and $q(\sim C + T_m)$ are answerable—the conclusion is that the formula T_m can serve as a multitracker. In this case the necessary condition for existence of a such tracker is $2a\sigma \leq \frac{n-a}{2} - a\sigma$, which is $a \leq \frac{n}{6\sigma+1}$. To compare this result with the previous results we assume that the total length of restricted ranges is $2k$, *i.e.* $(2\sigma+1)a = 2k$, which gives us $a = \frac{2k}{2\sigma+1}$, and consequently, $k \leq \frac{2\sigma+1}{12\sigma+2}n$, which, for any $\sigma > 1$ gives $k < \frac{3}{14}n$ (better result as for uniform distribution), for $\sigma = 2$ gives $k \leq \frac{5}{26}n$ (the same result as for non-uniform distribution considered in the previous subsection), and for $\sigma > 2$, the results are better—again we decreased the upper limit for the value of parameter k for which a tracker may exist $(\frac{2\sigma+1}{12\sigma+2} < \frac{5}{26}$ for $\sigma > 2)$.

In a general case when the number of restricted ranges is $2q + 1$, the necessary condition for the existence of a tracker is $\sigma^{q-1}b \geq 2\sigma^q a$, *i.e.* the length of the largest gap is not smaller than the double length of the largest range. Replacing b the inequality becomes: $a \leq \frac{(\sigma-1)n}{6\sigma^{q+1}-5\sigma-1}$. Again, assuming that the total length of all restricted ranges is equal to $2k$, *i.e.* $a = \frac{2k(\sigma-1)}{2\sigma^{q+1}-\sigma-1}$, we get our final result: $k \leq \frac{2\sigma^{q+1}-\sigma-1}{2(6\sigma^{q+1}-5\sigma-1)}n$, which again is better than in the case of uniform distribution (if $\sigma^{q+1} > q(\sigma-1) + \sigma$, which is true for all $\sigma \geq 2$ and all $q \geq 1$).

5 Usability and security of statistical databases

As we mentioned in Introduction, the goal of statistical databases is to provide statistics about groups of individuals while protecting their privacy. In other words, we should balance between **usability** (provision of statistics) and **security** (protecting privacy) in statistical databases. These two concepts are essential in evaluating any inference control mechanism and they work against each other: it is intuitively clear that stronger security measures decrease usability of statistical database. In particular, a database which refuses to answer any query (null usability) has perfect security.

Before we proceed further we try to define in a formal way these two fundamental concepts. Having these definitions we will measure the "goodness" of different inference control mechanisms based on the idea of multiranges (with uniform and nonuniform distribution). This evaluation will serve as a basis for optimal selection of ranges (see [6]).

Let us assume that Q is a set of statistical queries $q(C)$ (we denote by X_C the query set of $q(C)$. Let us denote $f_Q : \langle 0,n\rangle \longrightarrow \langle 0,\infty)$, a function which, for any integer j from the range $\langle 0,n\rangle$, returns the number of queries $q(C)$ from the set Q such that $|X_C| = j$. We will call the function f_Q the query distribution function and assume that it satisfies $f_Q(i) = f_Q(n-i)$ for all $1 \leq i \leq n$.

Again, we assume that we have a number of restricted ranges I_i, $i = 1, 2, \ldots, m$, and we denote $I = \bigcup_{i=1}^{m} I_i$.

The formula for the response to the query $q(C)$ is again:

$$q(C) = \begin{cases} q(C) & if\ |X_C| \notin I \\ \# & otherwise \end{cases}$$

(again, we informally use $q(C)$ to denote both the response to the query, and the query itself).

The *usability* U of a statistical database is a function of f_Q and I and is defined as follows:

$$U(f_Q, I) = \frac{\sum_{j \notin I} f_Q(j)}{\sum_{j=0}^{n} f_Q(j)}$$

It is clear that $U(f_Q, I)$ gives the fraction of answerable queries from the set of queries Q. In other words it measures the richness of the information revealed to the users.

We do not make any assumptions on the set of queries Q (for example, it can be the set of all statistical queries contained in some applications, against which we build some inference control mechanism). Because of this we will write f instead of f_Q understanding, that the set of queries is fixed (in fact we need not know anything about this distribution of values for query-sets and we can *assume* the normal distribution, see [6]).

Let us denote by $|I_i|$ the length of the *i-th* restricted range given by $\langle x_i, y_i\rangle$. Let $\mathcal{M}$ be the maximum length of all restricted ranges, *i.e.* $\mathcal{M} = \max_{1\leq i\leq m}\{y_i - x_i\}$, and $\mathcal{G}$ be the maximum length of all gaps between restricted ranges, *i.e.* $\mathcal{G} = \max_{1\leq i\leq m-1}\{x_{i+1} - y_i\}$.

From the Section 3 follows that the necessary condition for existence of a multitracker in the case of uniform restricted ranges is $2\mathcal{M} \leq \mathcal{G}$. For non-uniform distribution of restricted ranges this need not be a necessary condition, however (see Section 4.1) in some cases excludes the possibility for existance of a multitracker. Thus (to provide some level of security to a statistical database) we impose an additional constraint: $\mathcal{G} < 2\mathcal{M}$. Note that this condition will not provide a database an absolute security, but (in some cases) it will prevent users from finding a multitracker, which is the most serious threat to security of a statistical database.

Let us introduce some additional notation. Let $k = \theta n$ $(0 < \theta < \frac{1}{2})$ and $F(x) = \frac{\sum_{i\leq x} f(i)}{\sum_{i=1}^{n} f(i)}$. $F(x)$ is a cumulative distribution function (in statistical sense). Now we can express the usability $U(f, I)$ of a statistical database in terms of the function$F(x)$:

$$U(f,I) = 1 - \sum_{i=1}^{m}[F(y_i) - F(x_i)]$$

In [6] we made an attempt to maximize usability of a statistical database while $\mathcal{G} < 2\mathcal{M}$; then the significance of non-uniform distribution of the closed ranges is clearly visible.

6 Conclusions

A *good* inference control mechanism should be effective (it should provide security to a reasonable extent) and feasible (a practical way exists to enforce it). At the same time it should retain the richness of the information revealed to the users.

If we compare a traditional approach (two restricted ranges, general trackers) with a modified one (multiranges with uniform or non-uniform distribution), we observe that:

- The inference control mechanism based on multiple ranges is more effective than a traditional one, since there are some values of the parameter k (here $2k$ is the total length of all restricted ranges) for which there is no multitracker, even though a general tracker may exist.

- Both methods are equally feasible,
- The method based on multiranges with non-uniform distribution can reveal richer information to the users.
- If the user does not know the exact values of parameters (say, q, σ, and r in the last method), it should be relatively harder to construct a multitracker than a general tracker.

It would be interesting to check the usefulness of this approach when combined with other protection techniques and compare it against traditional combination (two ranges combined with other protection techniques).

There are also some interesting problems which remain open:

- It should be clear that we can introduce more parameters, *i.e.* r_1, r_2, and σ_1, σ_2 instead of the same r and σ in arithmetical and geometrical distributions, to make the inference control more effective. On the other hand we can consider the distribution of values $COUNT(C)$ for all expressions C and find an *optimal* distribution of the lengths of restricted ranges and the gaps between them.
- The general formulation of the above problem can be stated as follows: In the definition of the answerability of a statistical query

$$q(C) = \begin{cases} q(C) & if \ |X_C| \in \mathcal{B} \\ \# & otherwise \end{cases}$$

  find the *optimal* set $\mathcal{B}$, where $\mathcal{B}$ is an arbitrary, but fixed subset of $\{0, 1, \ldots, n\}$.
- Find a procedure to produce a multitracker.
- Find a sufficient condition for existence of a multitracker (and consequently for the compromisability) and compare it with sufficient conditions for the original query-set size inference control given in the Section 2.

References:

1. Denning, D.E., Denning, P.J. and Schwartz, M.D., *The Tracker: A Threat to Statistical Database Security*, ACMToDS, Vol.4, No.1, March 1979, pp.76-96,

2. Denning, D.E. and Schlörer, J., *A Fast Procedure for Finding a Tracker in a Statistical Database*, ACMToDS, Vol.5, No.1, March 1980, pp.88-102,

3. Denning, D.E. and Schlörer, J., *Inference Controls for Statistical Databases*, Computer, Vol.16, No.7, July 1983, pp.69-85,

4. Michalewicz, Z., *Compromisability of a Statistical Database*, Inf. Syst., Vol.6, No.4, Dec. 1981, pp.301-304,

5. Michalewicz, Z. and Yeo, A., *Multiranges and multitrackers in Statistical Databases*, Fundamenta Informaticae, Vol. X, No.4, Dec. 1987,

6. Michalewicz, Z. and Chen, K-W., *Usability and Security of Statistical Databases*, submitted for publication,

7. Schlörer, J., *Disclosure from Statistical Databases: Quantitative Aspects of Trackers*, ACMToDS, Vol.5, No.4, Dec. 1980, pp.467-492.

A Data Model, Knowledge Base, and Natural Language Processing for Sharing a Large Statistical Database

Hideto SATO
The Institute of Social and Economic Research
Osaka University
6-1 Mihogaoka, Ibaraki, Osaka 567, Japan

Abstract

Most existing statistical databases are mere collections of statistical files gathered for specific purposes. Consequently, as they grow in size, users are faced with difficulties in identifying and finding the data they need.

In order to obtain data descriptions independent of specific purposes, this paper proposes an object-oriented data design, which distinguishes between data conceptually obtainable and data actually stored in a database, and specifies relationships among classifications and categories independent of particular data files.

This is followed by a discussion of the representation of knowledge about data and classifications on a knowledge base, giving clear definitions of hierarchies and relationships among statistical data concepts.

Finally, a natural language query system using the knowledge base is demonstrated, which proves the advantage of the proposed statistical data concepts.

1. Introduction

Statistical data are obtained for specific purposes, which guide the selection and categorization of data objects. Statistical files have complicated structures which reflect their purpose.

So far, many statistical data models were proposed for describing this complicated structure as it exists (e.g. SUBJECT [Chan 81] and [Ozsoyoglu 83]). Applying such a model to a large statistical database (DB), however, is questionable, because it neglects the logical design of the entire DB.

We proposed a statistical data model under which the semantic structure of statistical data is uniquely described, and applied it to a logical redesign of a statistical DB in the National Land Agency [Sato 86]. Thereafter, in order to employ design results to locate and retrieve data, we constructed a frame system knowledge base which stores information about data structures and about relationships among classifications. Using this knowledge base we then developed an experimental version of a natural language query system.

Section 2 explains the problems of existing statistical DBs and the essential factors of our statistical data model, restating the subject of our previous work [Sato 86]. Sections 3 and 4 are devoted to the representation of knowledge about data and classifications in a knowledge base. Section 5 introduces a browser of the knowledge base. Sections 6 and 7 discuss natural language query processing. Finally, Section 8 summarizes our approach. For details of our approach, refer to [Sato 88].

2. Problems in Sharing Large Statistical Databases and their Solution

As existing statistical DBs grow larger, such problems arise that even data administrators hardly understand the whole contents of their DBs, and users have trouble finding data they need and joining related data ([NLA 86] and [Cubitt 83]).

Statistical data are filed in these DBs from survey to survey. Fig.2-1 shows a simplified structure of such files. This type of file has the following features.

(1) Similar data are distributed among many files; e.g. the same "population" data are recorded in both Fig.2-1 files.

(2) The logical structure of a file depends on a specific purpose or for recording convenience, which is peculiar to the file; e.g. regional categories are used for categorizing the rows in (a) but for the columns in (b).

(3) Data about many types of objects are mixed in a file; e.g. data about persons and households are mixed in (a).

Fig.2-1 Existing Statistical Files

(a) Population Census

year (every 5th year)	prefecture/ city	population by sex			number of household by tenure of dwelling		
		total	male	female	total	owned houses	rented houses
1980	TOKYO-prefecture	11618	5856	5762	4013	2646	1367
1980	TOKYO-city	8351	4236	4115	3106	2138	968
.	.	.	.	.	.	.	.
.	.	.	.	.	.	.	.
.	.	.	.	.	.	.	.

(b) Vital Statistics

year	Population			
	TOKYO	KYOTO	OSAKA	. .
1980	11609	2527	8473	. .
1981	11621	2552	8489	. .
.	.	.	.	.
.	.	.	.	.

The above-mentioned problems are mainly due to these features of statistical files, which are no more than reflections of the fact that the DBs consist of files of differing purposes.

In case of business data, it is usual for a DB to be logically designed by arranging data by every type of object, creating data files neutral against any specific purposes. A similar approach can be adopted to statistical data in Fig.2-1, obtaining files such as shown in Fig.2-2.

Fig.2-2 Statistical Object File

statistical object	category attribute			summary attribute
PERSONS	time	region	sex	population
Population Census	1980	TOKYO-prefecture	total	11618
	1980	TOKYO-prefecture	male	5856
	1980	TOKYO-prefecture	female	5762
	1980	TOKYO-city	total	8351
	.	.	.	.
	.	.	.	.
Vital Statistics	1980	TOKYO-prefecture	total	11609
	1981	TOKYO-prefecture	total	11621
	.	.	.	.
	.	.	.	.
	.	.	.	.

We call this type of file a statistical object file, because it is a collection of records about objects of statistics that have the same properties. The term representing the objects becomes the name of the object file. For example, "PERSONS" in Fig.2-2 is the name of the file and at the same time it represents the objects described in the file. The properties categorizing the objects such as "time," "region" and "sex" are called category attributes, and the properties representing statistical summaries such as "population" are called summary attributes according to [Shoshani 82]. A DB consisting of only such statistical object files is easy to understand because the same kind of data are gathered in a single, standardized file structure.

This is, however, not sufficient for a statistical DB. For example, assume a request of "Retrieve male population at TOKYO-city at 1981" to the file in Fig. 2-2. The DBMS will merely answer "No such data exists." Note that an expert of statistics will not give such an answer but reply "Population by sex by city is observed only at 5 year intervals. Male population at TOKYO-city at 1980 or total population in TOKYO-prefecture at 1981 is available." The difference in the answers is due to the following two characteristics of statistical data which is distinct from normal business data.

First, in case of a statistical DB, the closed world assumptions [Reiter 78] cannot be established; i.e. it is impossible for statistics to extract data corresponding to every fact in the object world of the DB. Accordingly, a user needs to know which part of data of interest is available. But this kind of information is lost in the simple data arrangement above.

Secondly, an object of statistical data is not an atomic object but a group of atomic ones grouped by a certain categorization. Hence, it is possible that the very data requested do not exist while alternative candidates exist. For example, no population data exists for TOKYO-city but for TOKYO-prefecture (the latter covers the former).

The first point requires discrimination between data conceptually obtainable in the object world (a conceptual file) and data actually stored in the DB (a DB file) and to describe correspondence between them [Sato 86]. Fig. 2-3 presents such a description concerning the file in Fig. 2-2. In this figure, "PERSONS" represents a conceptual file and both "PERSONS-FOR-CENSUS" and "PERSONS-FOR-SURVEY" represent DB files.

Fig.2-3 Correspondence of Conceptual File to DB Files

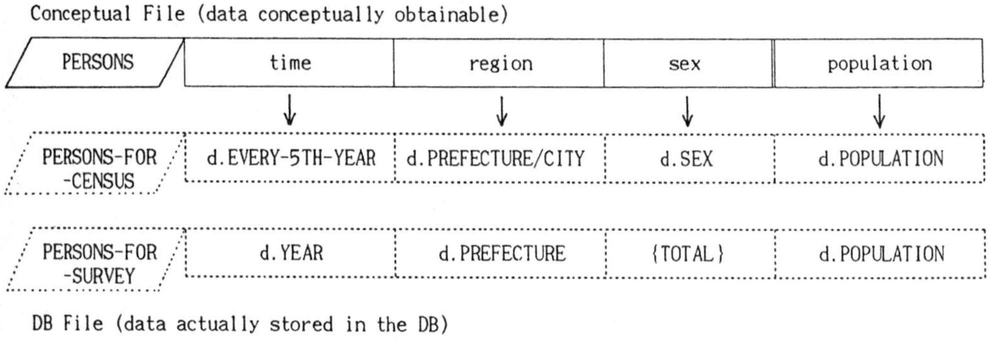

In this description, "d.YEAR" in the scheme of "PERSONS-FOR-SURVEY" represents the domain of the attribute "time," i.e. a set of values allowed for the attribute. This is also true for "d.PREFECTURE" and so forth. A domain of a category attribute is a set of categories and it is the same as a classification in the statistics field. This terminology will be used hereafter. In addition, {TOTAL} also indicates a classification of the attribute "sex" and means that a single value "TOTAL" can be taken for the attribute "sex" in the DB file; i.e. the objects of the file are not classified by sex.

By looking at a description of correspondence between a conceptual file and DB files, users can understand which parts of data in the conceptual file are available in the DB.

The second point, the problem of group objects, is concerned only with classifications used in grouping objects, so that it is solved by providing a description of relationships between categories in the classifications. This matter is discussed in Section 4.

3. Knowledge Representation of Statistical Data

3.1 Statistical Data Dictionary and Frames

We developed a statistical data dictionary which stores metadata according to our data model mentioned in the previous section. We used a frame system, one form of knowledge base systems, for managing the dictionary, but not an existing DBMS or a DD/DS. The main reasons are as follows.

1) Hierarchies of objects and categories are easy to describe.
2) Navigating relationships among objects is fast.
3) Demons (procedural knowledge) and their inheritance can be used for holding integrity constraints on hierarchies and on relationships.
4) Easy to combine with natural language query processing.

3.2 Statistical Object File and Knowledge Representation

In a statistical object file, a record represents attribute information about each object; e.g. "males in TOKYO-prefecture at 1980" is an object for the second record in Fig.2-2. On the other hand, in a frame system, an instance frame represents each object accompanying attribute information as its slot values, and a prototype frame represents abstraction of instance frames which have the same slots (attributes). Accordingly, data in Fig.2-2 can be represented as a collection of frames connected by <u>is-a links</u> (relationships of instances to a type) as seen in Fig.3-1.

Fig.3-1 Frame Representation of Statistical Data

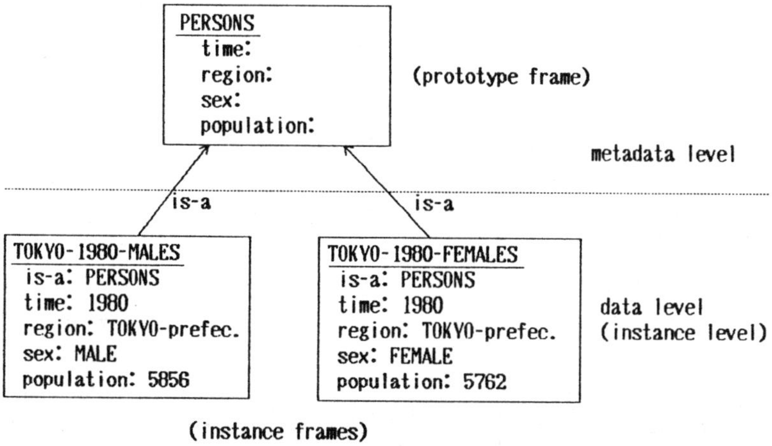

This representation is, however, unsuitable for a query on a metadata level such as "What data are categorized by sex ?" This is because the name "sex" in this representation is not a frame name but a slot name, and only a frame name can be a retrieval key in a frame system. For this reason, we regarded category attributes and summary attributes themselves as objects (meta-objects) to be described, and represented them by frames as seen in Fig.3-2.

Fig.3-2 Frame Representation of Statistical Metadata

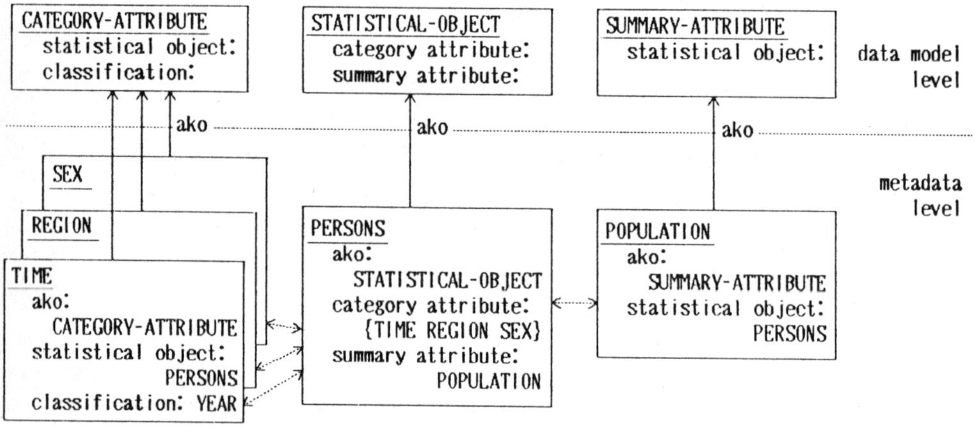

Then, we described the relationship between a statistical object and its attribute as a link between frames representing them. Fig.3-2 shows that the frame "PERSONS" represents a type of statistical object that has "TIME," "REGION" and "SEX" as its category attributes and "POPULATION" as its summary attribute.

The representation in Fig.3-1 corresponds to DB data, and that in Fig.3-2 corresponds to data-dictionary data.

3.3 Representation of a Conceptual File and a DB File

A statistical object in a metadata level such as "PERSONS" in Fig.3-2 represents a single statistical file as seen in Fig.2-2. As mentioned in Section 2 for a statistical file, correspondence between the conceptual file and the underlying DB files must be considered.

In order to describe this correspondence, we divide the metadata level into two sub-levels, the conceptual level and the DB level, as seen in Fig.3-3. Then we describe the above correspondence as an ako link (a-kind-of link, i.e. a relationship of a subtype to a supertype) between a statistical object in the conceptual level and that in the DB level. As seen in the figure, statistical objects in the DB level, "PERSONS-FOR-CENSUS" and "PERSONS-FOR-SURVEY", represent DB files underlying the conceptual file represented by the conceptual level object, "PERSONS."

In addition, statistical objects in the conceptual level can be arranged into a generalization hierarchy according to the commonality of their attributes [Smith 77]. This generalization hierarchy is described in use of ako links in the conceptual level. In Fig.3-3, "PERSONS" and "EMPLOYEES" are subtypes of "GENERIC-PERSONS."

Fig.3-3 Conceptual/DB Levels and Generalization Hierarchy

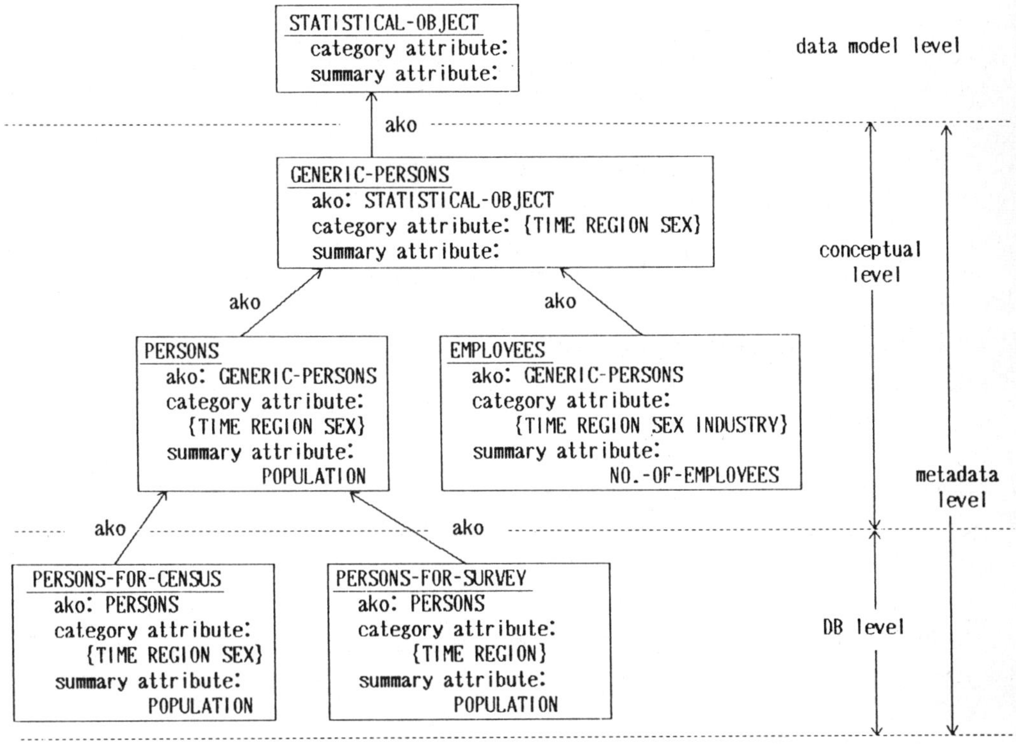

4. Representation of Classification Knowledge

4.1 Relation of a Statistical File to Classifications

As a statistical object is categorized by certain classifications, it may relate to other statistical objects via the classifications. Therefore, it is necessary when describing relationships between statistical files to specify relations of the files to the classifications and to identify relationships among the classifications and among the categories constituting the classifications.

So far, many researchers of statistical DBs described the above knowledge about classifications as a part of the description of individual statistical files and solely considered inclusion relationships among categories belonging to each classification.

In this approach, however, it is difficult to describe relationships among statistical files related to classifications including partly different categories. For example, it is impossible to describe which part is common between population data categorized by prefecture and that by prefecture/city. Besides, the inclusion relationships among categories is not enough to represent knowledge about classifications. It is, of course, possible to describe the fact that "Tokyo-city" is a part of "Tokyo-prefecture", but it is impossible to describe such general information that a "city" category is a finer category than a "prefecture" category.

To improve on these situations, we describe knowledge about classifications separately from the description of individual statistical files, and express a relation of a file to a classification as a link between frames representing them. In Fig.3-2, a value in the "classification" slot in a "CATEGORY-ATTRIBUTE" frame represents this link.

Next, we precisely describe relationships among classifications and categories in the frame form. This matter is explained in the following sub-section.

4.2 Relationships among Classifications and Categories

A classification is a set of categories and categories have inclusion relationships among them. For this reason, there are many kinds of relationships between classifications, between categories and between a classification and a category. Let us take two categories (category instances), "MACHINERY" and "MANUFACTURING" for instance, and think of category types and classifications related to them. The following relationships can be considered among them.

<u>Relationship between category instances</u>

① MACHINERY is a part of MANUFACTURING.

<u>Relationship between a category instance and a category type</u>

② MACHINERY is an INDUSTRIAL-CATEGORY.

③ MANUFACTURING is an INDUSTRIAL-CATEGORY.

④ MACHINERY is an INTERMEDIATE-INDUSTRIAL-CATEGORY.

⑤ MANUFACTURING is a MAJOR-INDUSTRIAL-CATEGORY.

<u>Relationship between category types</u>

⑥ An INTERMEDIATE-INDUSTRIAL-CATEGORY is an INDUSTRIAL-CATEGORY.

⑦ A MAJOR-INDUSTRIAL-CATEGORY is an INDUSTRIAL-CATEGORY.

⑧ An INTERMEDIATE-INDUSTRIAL-CATEGORY is finer than a MAJOR-INDUSTRIAL-CATEGORY.

<u>Relationship between a category instance and a classification instance</u>

⑨ MANUFACTURING is one of the categories in the INDUSTRIAL-CLASSIFICATION-IN-LABOUR-FORCE-SURVEY.

⑩ MACHINERY is one of the categories in the STANDARD-INDUSTRIAL-CLASSIFICATION.

⑪ MANUFACTURING is one of the categories in the STANDARD-INDUSTRIAL-CLASSIFICATION.

<u>Relationship between classification instances</u>

⑫ The INDUSTRIAL-CLASSIFICATION-IN-LABOUR-FORCE-SURVEY is a part of the STANDARD-INDUSTRIAL-CLASSIFICATION.

<u>Relationship between a classification instance and a classification type</u>

⑬ The INDUSTRIAL-CLASSIFICATION-IN-LABOUR-FORCE-SURVEY is an INDUSTRIAL-CLASSIFICATION.

⑭ The STANDARD-INDUSTRIAL-CLASSIFICATION is an INDUSTRIAL-CLASSIFICATION.

Fig.4-1 Knowledge Representation of Classification

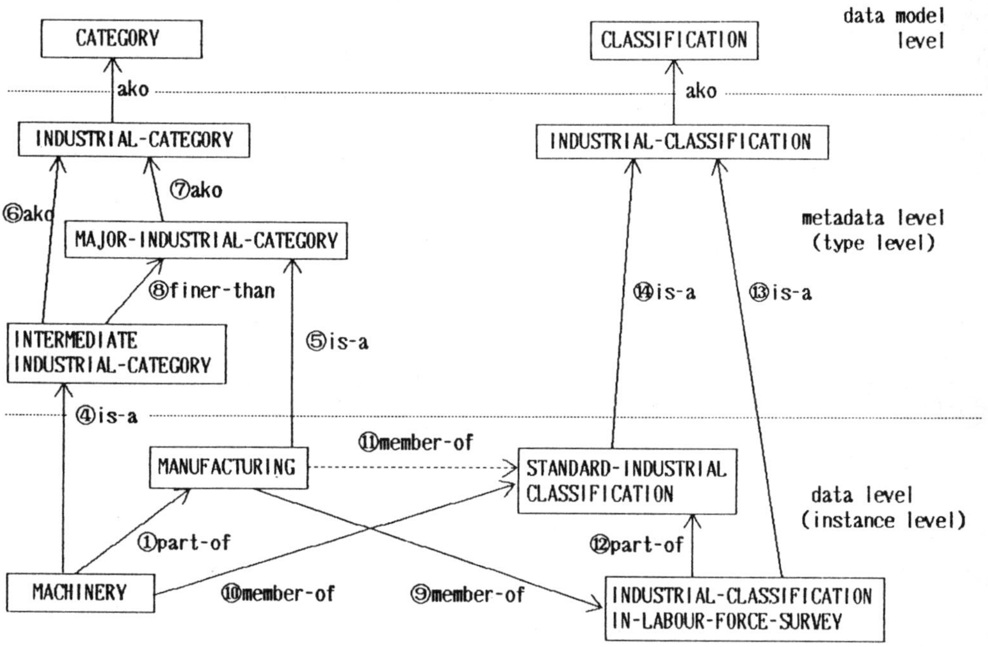

These relationships are implicitly/explicitly used when users of statistics speak of statistical data. Therefore, it is important to represent and to manage them for the sake of sharing statistical data.

We use links of frames to represent the relationships and introduce three types of link, <u>part-of</u> (part-whole relationship), <u>member-of</u> (member-set relationship) and <u>finer-than</u> (finer-coarser relationship), in addition to previously introduced is-a and ako links.

Fig.4-1 presents representation of the above sample relationships among classifications and categories, using the above five kinds of linkages. The number attached to an arrow in the figure corresponds to the number of the sample relationship. The type of link is also identified for each arrow, of which roles are summarized as follows.

A part-of link represents a part-whole relationship between two instance frames and is used for describing an inclusion relationship between categories (①) or between classifications (⑫). A finer-than link represents a relationship between two prototype frames whose instances have part-whole relationships (⑧). A **member-of** link represents a relationship between a member instance and a set instance and is used for describing relationships between constituent categories and a classification (⑨,⑩).

Incidentally, the arrow, ⑪"MANUFACTURING —member-of→ STANDARD-INDUSTRIAL-CLASSIFICATION" drawn by a dotted line in the figure, indicates an inference from ⑨ and ⑫ by the rule,

A —member-of→ B & B —part-of→ C ⟹ A —member-of→ C.

Similarly, ②"MACHINERY —is-a→ INDUSTRIAL-CATEGORY" and ③"MANUFACTURING —is-a→ INDUSTRIAL-CATEGORY" are inferable by the rule,

A —is-a→ B & B —ako→ C ⟹ A —is-a→ C.

4.3 Category Hierarchy and Classification Hierarchy

In the representation of classification knowledge as explained above, the following five types of hierarchical relationship are embedded.

1) <u>Category hierarchy</u>

A hierarchy of category instances, such as "WHOLE-INDUSTRY ← SECONDARY-INDUSTRY ← MANUFACTURING ← MACHINERY," formed by chains of part-of links (① in Fig.4-1).

2) <u>Finer-coarser hierarchy of category types</u>

A hierarchy of category types, such as "MAJOR-INDUSTRIAL-CATEGORY ← INTERMEDIATE-INDUSTRIAL-CATEGORY ← MINOR-INDUSTRIAL-CATEGORY," formed by chains of finer-than links (⑧).

3) <u>Generalization hierarchy of category types</u>

An ordinary generalization hierarchy of category types, such as "CATEGORY ← LIVING-THINGS-CATEGORY ← ANIMALS-CATEGORY ← MAMMALS-CATEGORY ← DOGS-CATEGORY," formed by chains of ako links (⑥, ⑦).

4) <u>Classification hierarchy</u>

A hierarchy of classification instances according to the ranges covered by the classifications, such as "CLASSIFICATION-COVERING-WHOLE-INDUSTRY ← CLASSIFICATION-COVERING-SECONDARY-INDUSTRY ← CLASSIFICATION-COVERING-MANUFACTURING," formed by chains of part-of links (⑫).

5) <u>Generalization hierarchy of classification types</u>

An ordinary generalization hierarchy of classification types, such as "CLASSIFICATION ← LIVING-THINGS-CLASSIFICATION ← ANIMALS-CLASSIFICATION ← MAMMALS-CLASSIFICATION ← DOGS-CLASSIFICATION," formed by chains of ako links.

5. Statistical Data Dictionary and Its Usage

We applied the knowledge representation of statistical data and classifications explained in Section 3 and 4 to describe statistical files collected at the National Land Agency, and constructed a statistical data dictionary storing the descriptions. In this section, we introduce this data dictionary and its browser.

5.1 Statistical Data Dictionary Configuration

Our statistical data dictionary consists of frames representing five data model concepts, including (1)STATISTICAL-OBJECT, (2)CATEGORY-ATTRIBUTE, (3)SUMMARY-ATTRIBUTE in Fig.3-2, (4)CLASSIFICATION and (5)CATEGORY in Fig.4-1. It also includes frames representing their inferior concepts.

5.2 A Browser for a Statistical Data Dictionary

We developed a browser for browsing the statistical data dictionary. It has hierarchically organized menu screens reflecting the hierarchical structure of the dictionary frames. It can answer 22 kinds of questions/orders using links among frames as explained in Section 3 and 4.

The following are partial examples of such questions/orders.

1) Which files include data about the category "MALE" ?
2) What categories constitute the classification "MAJOR-INDUSTRIAL-CLASSIFICATION" ?
3) Which files include data categorized by the classification "CITIES" ?
4) Which files include data by residential place (i.e. data categorized by classifications for the category attribute "RESIDENTIAL-PLACE") ?
5) How is the statistical object "EMPLOYEES" categorized ?
6) Which files include data about the summary attribute "INCOMES" ?

5.3 Utilizing the Correspondence between a Conceptual File and DB Files

Our statistical data dictionary contains information about correspondence between conceptual files and DB files as mentioned in Section 3.3. The browser can generate a correspondence table which expresses this information. Fig.5-1 (a) is a sample screen displaying such a table. This table shows correspondence of a conceptual file "Employees" with DB files underlying it. The items in the left hand column represent the attributes included in the conceptual file. The head items are the names of DB files. Each column provides information about attributes included in the DB file.

Fig.5-1 Sample Screens concerning Conceptual/DB Files

Correspondence Table between Conceptual-DB Files of Employees

Attribute/DB File	¥A	¥B	¥C	¥D	¥E	¥F	¥G	¥H
time	A1	A1	A2	A2	A3	A3	A4	A4
region	B1	B1	B1	B1	B1	B1	B2	B2
age group	C1	-	C1	-	C1	-	-	-
employment status	-	D1	-	D1	-	D1	-	-
industry	E1	-	E1	-	E1	-	E1	-
occupation	-	-	-	-	-	-	-	F1
number of employees	0	0	0	0	0	0	0	0

SBX > (C D1)

(a)

⇩

Display of Categories

1. employees and exectives (C-7901-1)
2. self-employed with employees (C-7901-2)
3. self-employed without employees (C-7901-3)
4. family workers (C-7901-4)

(b)

Let us focus on the column "¥F." It shows that "number of employees" data are recorded by time (A3), by region (B1) and by employment status (D1) in the DB file named "¥F." In this figure, A3, B1 and D1 represent classifications adopted for respective category attributes. Users are able to learn the categories constituting each of the classifications by inputting a command. In the sample, an order "List categories constituting D1" (a command (C D1)) is given, and the categories are displayed as (b).

As seen above, users can discover the mutual differences and similarities of DB files underlying a conceptual file.

6. Statistical Data Query and Semantic Grammar

Although the browser explained in the previous section can guide a user to data, it is not friendly enough for a casual user because he must understand data model concepts prior to using the browser. In our experience, many users had difficulty understanding the difference between a category attribute, a classification and a category. For this reason as well as others, we began to develop a natural language query system for the statistical DB.

This section explains data query sentences and a semantic grammar for analyzing them.

6.1 Purpose of Analyzing Data Query Sentences

The most important type of sentence used in DB queries is a data query sentence, which expresses the structure of data to be retrieved. A sample is seen in Fig.6-1 (a). (b) in the figure shows the scheme of data requested by the sentence (a). The notation of the scheme is close to the correspondence scheme between a conceptual file and DB files shown in Fig.2-3. The scheme (b) represents "population" data categorized by "sex," restricted to "Tokyo and Kyoto" in "region," and also to "1980" in time.

The main purpose of natural language query processing is to analyze a data query sentence like (a) and to generate a corresponding requested data scheme like (b).

Fig.6-1 Data Query Sentence and Requested Data Scheme

(a) Data Query Sentence : "I need population data by sex in Tokyo and Kyoto at 1980."

(b) Requested Data Scheme

PERSONS	sex	region	time	population
REQUESTED-DATA	d.SEX	{Tokyo Kyoto}	{1980}	d.POPULATION

6.2 Semantic Grammar and Parsing the Query Sentence

As most sentences demanding statistical data have certain patterns, we can formulate a semantic grammar based on our data model concepts. Fig.6-2 shows the result of parsing the sentence in Fig.6-1 (a) by using such a grammar. (Provided that our original grammar is of Japanese sentences.)

As seen in the figure, the non-terminal symbols of the grammar correspond to our data model concepts and the result of the parsing is directly interpreted into the information about the requested data scheme in Fig.6-1 (b). In fact, the "SUMMARY-ATTRIBUTE-PHRASE" corresponds to the summary attribute "population" in (b), and the three "CATEGORY-ATTRIBUTE-PHRASEs" to "d.SEX", {Tokyo Kyoto} and {1980}, respectively.

Fig.6-2 Parsing Tree for Statistical Data Query Sentence

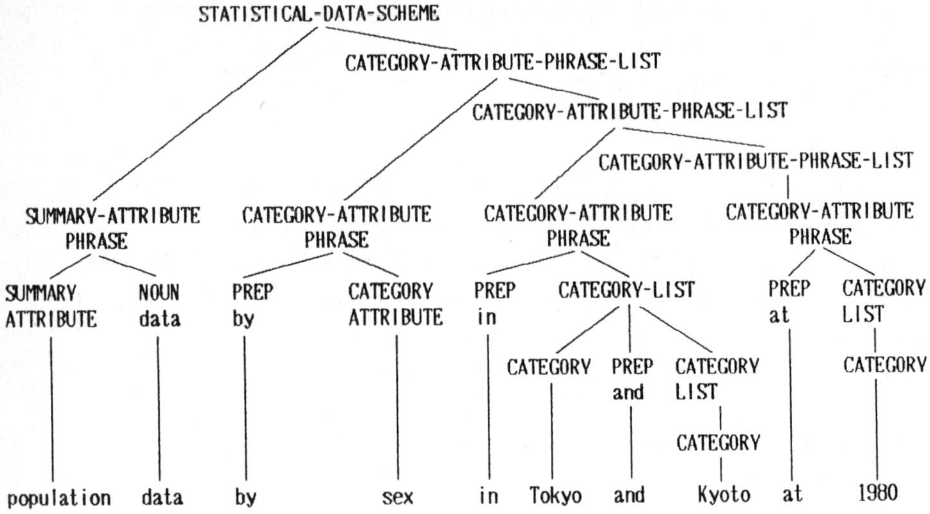

Therefore, recognizing "Tokyo" and "Kyoto" as "regional" categories, "1980" as a "time" category, and "population" as a summary attribute of "PERSONS", the scheme information of (b) can be generated.

7. The Practice of Natural Language Query Processing

We developed an experimental version of a natural language query processing system for statistical data, named LIDS86 (Land Information Dictionary System developed in 1986), which uses the statistical data dictionary explained in Section 5.1.

The characteristics of this system are (1) use of the semantic grammar based on the statistical data model, and (2) capability of presenting neighboring data by using correspondence between a conceptual file and DB files, even if right-fit data does not exist in the DB. This system is written in UTI-LISP and Prolog/KR, both of which were developed at Tokyo University.

Fig.7-1 shows the architecture of LIDS86. The knowledge bases in it are explained in Section 7.1 and the other components and their functions are introduced in the explanation of sample query processing in Section 7.2.

Fig.7-1 Architecture of LIDS86

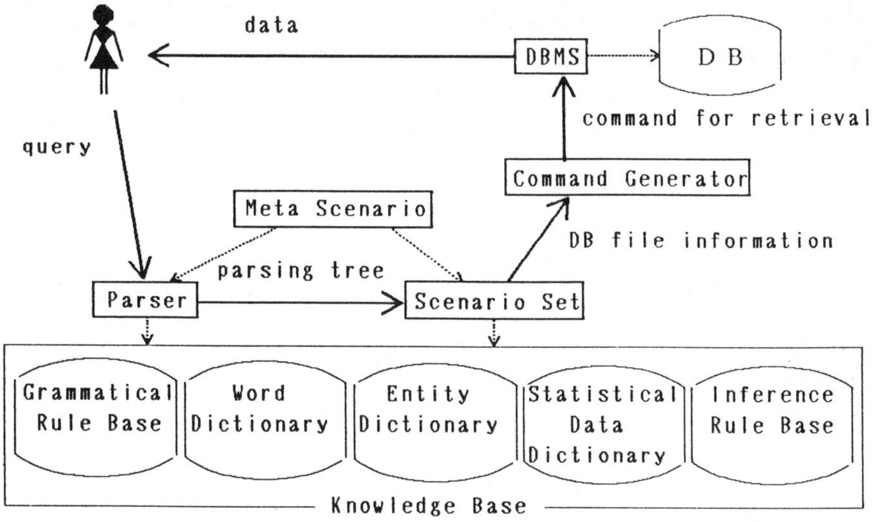

7.1 Knowledge Bases in LIDS86

LIDS86 has the following three kinds of knowledge bases.

1) <u>Grammatical rule base</u>: This is a collection of the grammatical rules of the semantic grammar mentioned in Section 6.

2) <u>Inference rule base</u>: This is a collection of production rules used in semantic analyses of a natural language sentence and in processing retrievals of statistical data.

3) <u>Dictionaries</u>: LIDS86 has the following three dictionaries.

A <u>word dictionary</u> consisting of word frames, each of which represents a single word and has links to corresponding entity frames. An <u>entity dictionary</u> consisting of entity frames, each of which represents an entity (an object in the object world) and has links to corresponding statistical concept frames. Entity frames also have semantic links like those among classifications and categories in Fig.4.1. A <u>statistical data dictionary</u> consisting of statistical concept frames as enumerated in Section 5.1.

The distinction between a word frame and an entity frame is due to dealing with synonyms and homonyms. In addition, entity frames serve to describe conventional knowledge separately from statistical data knowledge described in statistical concept frames.

7.2 A Sample Query and Its Processing

Fig.7-2 shows a sample dialogue between a user and LIDS86. Using this sample, we demonstrate a processing process performed by LIDS86.

Fig.7-2 Sample Dialogues with LIDS86

(translation from Japanese)

```
LIDS: Welcome to LIDS86. What do you want ?
USER: I need number of employees by sex in Tokyo at 1975.
LIDS: Do you mean Tokyo-prefecture by "Tokyo" ?
USER: No.
LIDS: You mean Tokyo-city, don't you ?
USER: Yes.
LIDS: 《data about employees》
      We have data such as number of employees by TIME-POINTS-10 by CITY-AND
      -PREFECTURE by MAJOR-INDUSTRIAL-CLASSIFICATION-A. But it is not
      categorized by sex. Is this OK ?
USER: Do you have another one ?
LIDS: 《data about employees》
      We have data such as number of employees by TIME-POINTS-10 by CITY-AND
      -PREFECTURE by MAJOR-OCCUPATION-CLASSIFICATION. But it is not
      categorized by sex. Is this OK ?
USER: What is MAJOR-INDUSTRIAL-CLASSIFICATION-A ?
LIDS: MAJOR-INDUSTRIAL-CLASSIFICATION-A consists of the following
      categories.
      * WHOLE INDUSTRY
      * AGRICULTURE
      * FORESTRY
         .
         .
USER: The first data, please.
LIDS: 《data about employees》
      We have data such as number of employees by TIME-POINTS-10 by CITY-AND
      -PREFECTURE by MAJOR-INDUSTRIAL-CLASSIFICATION-A. But it is not
      categorized by sex. Is this OK ?
USER: Yes.

    GET INTO  T164644
              (time, region, industry, number-of-employees)
               JITEN,TIIKI,SANGYO,SHUGYOSH
        FROM  LIDS#LIDS154
       WHERE  JITEN=50  . . . . . . . . . 「year of 1975 」
         AND  TIIKI=27100. . . . . . . .  「code of Tokyo」
    ------------------ generated command ------------------
```

The <u>meta-scenario</u> (in Fig.7-1) is a program controlling the whole system of LIDS86. When the meta-scenario receives an input sentence like the second line of Fig.7-2, it delivers the sentence to the parser. The <u>parser</u>, a program for natural language processing, analyzes the sentence and returns a parsing tree like Fig.6-2 to the meta-scenario.

The meta-scenario judges the intention of the input sentence by the form of the parsing tree, and then sets up an appropriate scenario. A scenario is a program provided for each situation of dialogues. For this sample, the scenario for retrieving statistical data is set up. This scenario checks ambiguity in the parsing tree and, if it exists, a question is given to the user in order to resolve it. The third line in the sample is an example.

After resolving ambiguities, the scenario generates a requested data scheme like Fig.6-1 (b), and searches the conceptual file whose scheme matches the requested data scheme.

Then, it retrieves schemes of DB files underlying the conceptual file, and evaluates them according to the possibility of matching the requested data. After the evaluation, the scenario displays scheme information of the DB files in order of the evaluation scores. The output below <<data about employees>> in Fig.7-2 is an example.

Notice the message "But it is not categorized by sex" in the output. This means that the requested data "number of employees by sex in Tokyo at 1975" is not available while "number of employees in Tokyo at 1975" is available. Thus, LIDS86 can inform the user of neighboring data even when the right data fit for the request does not exist in the DB. This ability is given by the distinction of conceptual files and DB files and the evaluation of DB files as mentioned above.

If the user is satisfied with the data presented, its scheme information is sent to the command generator. The command generator generates a command for retrieving the data from the DB. An example of a generated command is shown in the bottom of Fig.7-2.

8. Conclusion

The essential points of this paper are summarized as follows:

1) A model is proposed for describing statistical data and classifications.

Problems in sharing statistical DBs mainly result from the fact that existing statistical DBs are mere collections of statistical files gathered for specific purposes.

In order to obtain data description that are neutral against specific purposes, we propose a data model describing statistical data, which has the following three features: (1) statistical data arranged by every type of objects, (2) distinctions made between a conceptual file and DB files as logical concepts for a statistical data file, and (3) relationships among classifications and categories specified independently of individual statistical data files.

(2) corresponds to the four schema approach to statistical DBs, which was proposed by [Sato 86] and [Malmborg 86]. ("Four schema" means splitting the

conceptual schema in the "three schema" proposed by [ANSI 75] into two schemata, a conceptual schema describing data conceptually obtainable and a DB schema describing data actually available in the DB.)

(3) is characterized by the distinction between a description of classification and that of categories. In our framework, classifications are sets of categories which may be chosen according to certain specific purposes, while categories and their hierarchies are independent of purpose. This distinction makes clear the concepts of the hierarchies of classifications and those of categories, which have often been confused so far.

2) Knowledge about statistical data and classifications is described on a frame system.

The complicated relationships among statistical files, classifications and categories are effectively described as links among frames representing them. This seems to indicate that an object-oriented DB planned as an enhancement to a frame system will be useful for implementing a statistical DB.

3) An experimental natural language query system is developed by using statistical data concepts.

The statistical data concepts of our data model can be used as non-terminal symbols constituting a semantic grammar for natural language query processing. In addition, the distinction of conceptual files and DB files is useful for presenting neighboring data in retrievals. These facts seem to illustrate that the concepts of our data model fits human recognition of statistical data.

Acknowledgments

This paper is based on the results of the LIDS project, in which the National Land Agency, Japan Information Service Ltd., Osaka University (the author and Professor Riichiro Mizoguchi) and Kyoto University (Dr. Toyoaki Nishida) collaborated. In addition, Professor Ryosuke Hotaka of the University of Tsukuba and Dr. Masaaki Tsubaki of Data Research Institute gave us valuable advice for our data model. We thank all those involved in the project.

References

[ANSI 75] ANSI/X3/SPARC, "Study Group on Data Base Management Systems: Interim Report," FDT (Bulletin of ACM-SIGMOD), 7(2), 1975.

[Brackman 83] R.J.Brackman, "What IS-A is and isn't: An Analysis of Taxonomic Links in Semantic Networks," IEEE Computer, Oct. 1983, pp.30-36.

[Chan 81] P.Chan and A.Shoshani, "SUBJECT: A Directory Driven System for Organizing and Accessing Large Statistical Databases," VLDB, 1981, pp.553-563.

[Cubitt 83] R.E.Cubitt, "Meta Data: An Experience of its Uses and Management," SSDBM, 1983, pp.167-169.

[Malmborg 86] E.Malmborg, "On the Semantics of Aggregated Data," SSDBM, 1986, pp.152-158.

[NLA 86] National Land Agency, Knowledge Management of Land Information, (in Japanese), Publication Bureau of the Ministry of Finance, Japan, 1986.

[Ozsoyoglu 83] Z.M.Ozsoyoglu and G.Ozsoyoglu, "An Extension of Relational Algebra for Summary Tables," SSDBM, 1983, pp.202-211.

[Reiter 78] R.Reiter, "On Closed World Data Bases," in H.Gallaire and J.Minker (eds.), Logic and Data Bases, Plenum Press, 1978, pp.55-76.

[Sato 86] H.Sato, T.Nakano, Y.Fukasawa and R.Hotaka, "Conceptual Schema for a Wide-Scope Statistical Database and Its Applications," SSDBM, 1986, pp.165-172.

[Sato 88] H.Sato, Design and Development of Statistical Databases: An Application of Data Model and Knowledge Base, (in Japanese), Ohm Co., Japan, 1988, 246 pages.

[Shoshani 82] A.Shoshani, "Statistical Databases: Characteristics, Problems and some Solutions," VLDB, 1982, pp.208-222.

[Smith 77] J.M.Smith and D.C.P.Smith, "Database Abstractions: Aggregation and Generalization," TODS, 2(2), June 1977, pp.105-133.

N.B.: SSDBM is an abbreviation of the Proceedings of the International Workshop on Statistical and Scientific Database Management.

Precision-Time Tradeoffs: A Paradigm for Processing Statistical Queries on Databases

Jaideep Srivastava †
Doron Rotem

Computer Science Research
Lawrence Berkeley Laboratory
University of California
Berkeley, CA 94720

ABSTRACT

Conventional query processing techniques are aimed at queries which access small amounts of data, and require each data item for the answer. In case the database is used for statistical analysis as well as operational purposes, for some types of queries a large part of the database may be required to compute the answer. This may lead to a data access bottleneck, caused by the excessive number of disk accesses needed to get the data into primary memory. An example is computation of statistical parameters, such as count, average, median, and standard deviation, which are useful for statistical analysis of the database. Yet another example that faces this bottleneck is the verification of the truth of a set of predicates (goals), based on the current database state, for the purposes of intelligent decision making. A solution to this problem is to maintain a set of precomputed information about the database in a view or a snapshot. Statistical queries can be processed using the view rather than the real database. A crucial issue is that the *precision* of the precomputed information in the view deteriorates with time, because of the dynamic nature of the underlying database. Thus the answer provided is approximate, which is acceptable under many circumstances, especially when the error is bounded. The tradeoff is that the processing of queries is made faster at the expense of the precision in the answer. The concept of precision in the context of database queries is formalized, and a data model to incorporate it is developed. Algorithms are designed to maintain materialized views of data to specified degrees of precision.

1. Introduction

Conventional databases have focused mainly on the efficient execution of *transaction queries*, found typically in banking and airlines applications. Such queries have following properties: *(i)* they access small amounts of data, *(ii)* their answer is sensitive to each individual data item, and most importantly, *(iii)* they can be answered from the the basic database, i.e. the stored relations. An example is,

Does there exist any account number in the bank
with a negative balance ?

† This work was done while the first author was on leave from the C.S. Division, U.C. Berkeley.

There are many applications where the overall characteristics of an entire dataset are required rather than individual data items. The required information does not exist in the database, and therefore has to be computed. An example is the need to obtain some statistical parameter, say average age of Caucasian males in California, from a demographic database. A typical query to derive this would be,

retrieve mean(*PERSON.age*)

where *PERSON.state* = *California*

and *PERSON.sex* = *male* and *PERSON.race* = *Caucasian*

Yet another example is the real-time controller of a manufacturing process, producing machine parts of specified dimensions. A typical rule for controlling the process would be,

If error(mean(*part_size*)) > mean-tolerance

or error(variance(*part_size*)) > variance-tolerance

then take some corrective action.

Applying such a rule requires computing the mean and the variance of *part_size*. Queries which carry out such computations are called *aggregate queries*. Answering an aggregate query requires extracting some *feature* from a number of data items, which in turn leads to accessing large volumes of data. Features, aggregates, etc., measure some characteristics of a set of data items, and thus are relatively insensitive to the value of individual data items. Hence, estimates of such characteristics are acceptable instead of their exact values. Conventional database systems incur prohibitive costs in providing accurate answers which are not essential. We believe that the cost of processing a query can be reduced significantly if an approximate answer is acceptable. We therefore conclude that efficient processing of aggregate queries is feasible if data management and query processing techniques are designed to accept *degree of precision* as a query parameter.

Researchers in the area of Statistical Databases [OLKE 86], [GHOS 85], [SHOS 82], [VITT 84], have faced the problem of data access bottleneck, since they had to deal with queries that calculated statistical aggregates of a set of data items. They realized the power of estimation, and introduced the statistical technique of *sampling* [COCH 53] as a database operator. This has been very successful for statistical applications, dealing with numeric data. We propose an alternative approach to this problem. As we indicate in a later section, the usefulness of our approach lies in the fact that it can also be applied to the framework of *approximate reasoning*.

We propose the idea of *materialized view maintenance* as a technique to provide efficient support for the applications mentioned above. The stored database, for example the tuples of a relational database, is the *basic* database. All knowledge that can be obtained from it is called its *derived information*.

If the entire derived information is stored at all times, processing queries that require it will be extremely fast. However, all possible derived information for any database is extremely large, making its storage infeasible. Our idea is to store the derived information that is interesting to a user, and process his/her queries on it. On top of the basic database, *views* [ULLM 82] are defined. A copy of each view, called its materialization (or *MatView*), is stored. Views are defined by the user and are specific to his application. They are defined by predicates which can be both logical and arithmetic.

Maintaining MatViews is a critical issue. Since the basic database is constantly updated the derived information changes as well, and the MatView, which is a part of the latter, has to be updated. A simple approach is to consider each individual change to the basic database and propagate its effects to all the MatViews that depend on it. However, this is prohibitively expensive due to the large number of updates to the basic database, and a correspondingly larger number of updates to the MatViews. Our approach to this problem is to maintain *approximate* MatViews. The definition of each MatView specifies not only the predicate defining it, but also a *degree of precision*, which determines its accuracy. The main idea is that updates to MatView need not be made as long as it stays within the bounds of the degree of precision attached to it. This approach reduces the total number of updates to the MatView drastically and brings it in a manageable range.

We present the tradeoff between precision and time as a paradigm for database management and query processing. MatViews are maintained at their specified degree of precision and are used to answer queries. The principal idea is that it costs less to maintain data that is less precise, and hence queries which do not require a very high degree of precision will not be forced to pay the price for it. This paradigm affects all components of a database system, namely the query language, data storage and access methods, query processing and optimization, and data distribution and replication.

There has been some past work on the idea of providing approximate answers to aggregate queries. Most notable is that of Rowe [ROWE 83] and Hebrail [HEBR 86], both of which are applicable only to static summaries (the summaries are derived once and then used for subsequent calculation). Koenig [KOEN 81] provides a framework for dynamic maintenance of summaries, but does not address the issue of how to do so efficiently. Our approach differs from all of the above in that it provides efficient ways to maintain summaries dynamically. The efficiency is gained by sacrificing some precision, thereby leading to a *precision-time tradeoff*. We also provide an analytical model of the problem.

In this paper we focus on the problem of maintaining derived information to its desired degree of precision. Specifically, we discuss the following: *(i)* a definition for precision of data and queries, *(ii)* materialized views as a mechanism for providing data with a specified degree of precision, and *(iii)* efficient automatic maintenance of views to their specified degree of precision. For clarity in exposition the examples in this paper define precision only in terms of the number of tuples of the database. However, our methods are applicable to other statistical parameters as well. Yet another application of these techniques is in the domain of approximate reasoning and deduction, which is based on *fuzzy logic* [ZADE 65]. This area has been rapidly gaining interest in the recent past.

A Real-Life Example: The idea of automatic maintenance of materialized views, defined in an application dependent manner, is especially attractive in a distributed environment. Each site stores only the views defined by the applications residing on it. The degree of precision of each view depends on the nature of the application.

We now present a real-life example *Fig. 1.1* to introduce the problem that we shall be formalizing and addressing in subsequent sections.

Consider the population data collected by the Census Bureau every time a census is carried out. Data about individuals is collected at the district level and is precise since it contains the individuals' exact age, height, weight, income, etc. Districts are grouped together to form cities, cities to form counties, counties to form states, and states to form the country. The lowest level of the hierarchy in *Fig. 1.1* may consist of a standard database, which is used for all sorts of local applications. However, the applications at the higher levels seek aggregate features of the data rather than the raw data itself. These various levels of geographic aggregation are logically equivalent to levels of information aggregation. From an information content viewpoint, it is not necessary to store any data at any level other than the lowermost. Higher levels of information can be derived from it. However, as we go upwards, the multiplicative increase in the volume of data makes it impossible to process raw data to obtain the required information on demand.

Our solution is to store, at each level of the hierarchy, the information required. Information at all levels but the lowermost, consists of *views*, which are defined in terms of the raw data, or other views, by ussing aggregation operations. Thus, we model the various levels of geographical aggregations, i.e. district, city, county, state and country, seen as a hierarchy of views defined on top of each other. The information that flows upwards is only the aggregate characteristics of the original data.

An important feature of the propagation of aggregates in such a manner is that the degree of precision required at higher levels is less than that at lower levels. The error introduced in the information in going up each level is modeled by the notion of *degree of precision* in our model. The advantage of the model presented is that it enables us to come up with efficient algorithms to maintain the various views in the hierarchy to their desired degrees of precision, at a minimum possible expense. Thus, updates need

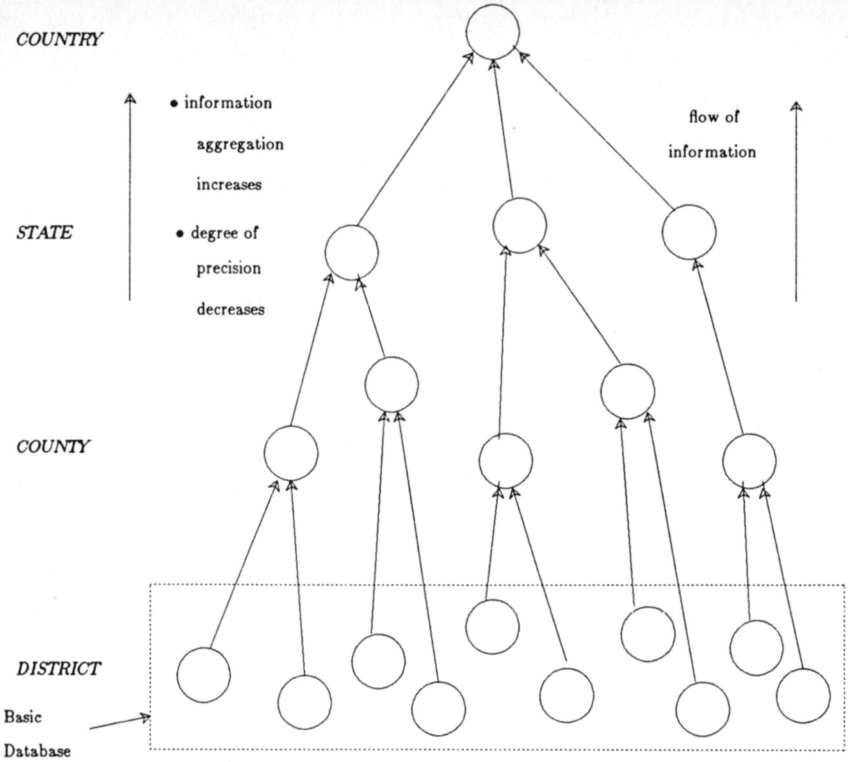

Fig. 1.1 *Hierarchy of Views in Census Data Collection.*

not propagate upwards, and flood the system, unless they are absolutely necessary.

This paper is organized as follows. Sections 3 and 3 discuss our definition of precision and the logical data model. Section 4 introduces a stochastic model for the problem of materialized view maintenance, and derives policies that maintain a MatView to its specified degree of precision. Section 5 provides a summary of the issues that future research efforts need to address, and section 6 lists our conclusions.

2. Definition of Precision

This section discusses the definition of precision which shall be used as a running example to illustrates our view maintenance algorithms. However, the approach presented here does not depend on the definition of precision chosen. To make the discussion more concrete, we illustrate our algorithms by considering the higher level abstraction to be a counter. As an example we assume the counter under consideration is the size of a view. The theory developed, however, is applicable to any counter.

We use the following notation:

N_0: Size of the view after last refresh.

N: Size of the view at some later instant.

p: Degree of precision.

q: Degree of confidence.

Degree of precision of data: N has (p,q)-degree of precision, with respect to N_0, if the probability of the difference between N and N_0, expressed as a fraction of N_0, being less than $(1-p)$ is at least q.

Formally,

$$P\{\frac{|N - N_0|}{N_0} \leq 1-p\} \geq q$$

Note: Intuitively, p is the precision of N (w.r.t. N_0)and$q$ is the *degree of confidence* in the statement,

$$\frac{|N - N_0|}{N_0} \leq 1-p.$$

Usually, q will be fixed at some high value, say 0.99, so as to have a high degree of confidence in the answer obtained. The precision requirement of a query is defined as follows.

Precision Requirement of a Query: A query with precision (p,q), requires that it be processed using copies of data having degree of precision $\geq (p,q)$. †

Given a query and the set of data items it accesses, there are many ways to process it. The decisions that have to be made are, *(i)* in what order should the subtasks of the query be performed, *(ii)* what algorithm should be used to perform each subtask, *(iii)* how should a particular data item be accessed, and *(iv)* how should the data be buffered. If data items are replicated, the additional decision of which copy to access has to be made. A set of decisions, one along each dimension, called a *query plan*, is a unique way of executing the query.

Given the above definitions, we can identify two kinds of queries.

Let,

Q: A query.

(p,q): Precision requirement of Q.

qp_i: i^{th} query plan for Q.

† $(p',q') \geq (p,q) \equiv (p' \geq p)$ and $(q' \geq q)$.

$QP=\{qp_1, qp_2, ..., qp_n\}$: Set of possible query plans to execute Q.

$cost(Q, qp_i)$: cost of executing query Q using query plan qp_i.

Precision Query: Answer query Q, in the shortest possible time, such that each copy of data used to process it has degree of precision $\geq (p,q)$,

i.e.,

$$\min_{qp_i \in QP} cost(Q, qp)$$

$$subject\ to \quad \text{for } each\ data\ copy\ D_i\ used,\ (p_i,q_i) \geq (p,q)$$

Deadline Query: Answer query Q, using maximum precision data, such that it takes no more than T time units to process it.

i.e.,

$$choose\ qp \in QP\ such\ that,\ data\ copies\ with\ maximum\ possible\ precision\ are\ chosen$$

$$subject\ to \quad 0 \leq cost(Q, qp) \leq T$$

3. Logical Data Model

The paradigm of tradeoff between precision and time is independent of the data model chosen. However, exemplifying our techniques requires us to make a specific choice. Further discussions assume the underlying data model to be the relational model with some extensions. Our choice of the model is motivated by the fact that it is both well accepted and easy to understand.

Any particular application finds only a subset of the database *interesting*. We further assume that such an interesting subset can be specified by a *view* definition.

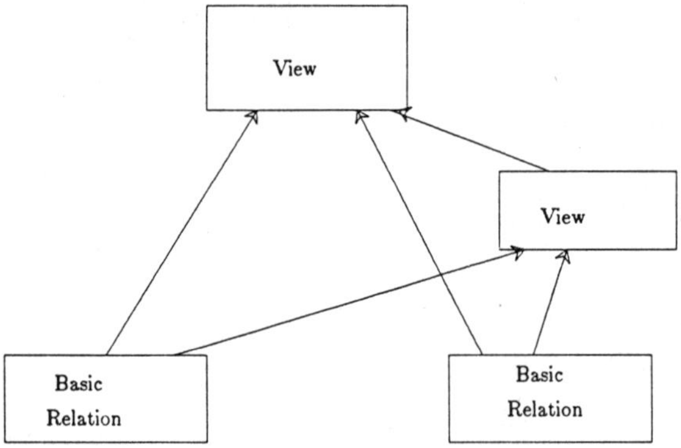

Fig. 3.1 *View Hierarchy.*

A view can be defined in terms of the base relations, in terms of other views, or both. This creates an entire hierarchy of views as shown in *Fig. 3.1.* We also assume that views are limited to those definable by the following operators.†

Relational Algebra: Relational algebra operations like *selection, projection, join, union, difference*, etc.

Example: *(A view definition using relational algebra):* Let,

$$EMP = (E.Id\#, E.name, E.age, E.sex, E.salary, E.med\_hist)$$

be a relation. A view defined using relational algebra is,

$$EMP\_HEALTH = \pi_{E.Id\#, E.age, E.med\_hist}(\sigma_{20K \leq E.salary \leq 40K}(EMP)).$$

Aggregation: Aggregation Operations like *count, sum, square-sum, data-grouping* etc.

Example: *(A view definition using aggregate functions):* Considering the same *EMP relation as above,*

$$SUM\_EMP = (sum(E.age), sum(E.salary))$$

Each view has a degree of precision attached to it, which is is one of the parameters specified during view-definition. The degree of precision depends on the requirements of the application defining the particular view. At all times, the degree of precision of the view has to be maintained within the limits specified during view-definition.

4. Creation and Maintenance of Materialized Views

This section describes the maintenance of materialized views and develops a stochastic model for it. It also describes periodic policies for the maintenance of a view to its specified degree of precision.

4.1. Maintenance of Materialized Views

A materialized view (*MatView*) is a stored copy of the view which is created at the time of view definition. Any changes made to the base relations have to be reflected in the MatView. This is done by means of a periodic maintenance process or *refresh process*. The mechanics of MatView maintenance can be described by the files that exist in the system and the processes that manipulate them.

Files: *Fig. 4.1(a)* shows the files existing in the system. The base relations from which the view is derived are stored in the file *BaseRel* and the materialized view in the file *MatView.* If MatView is refreshed as soon as a change is made to BaseRel, then these are the only files required. However,

† Our choice of the set of operators allowed for defining views is more powerful than the relational model, since it allows aggregation operations. Thus our assumption is not simplistic or unrealistic

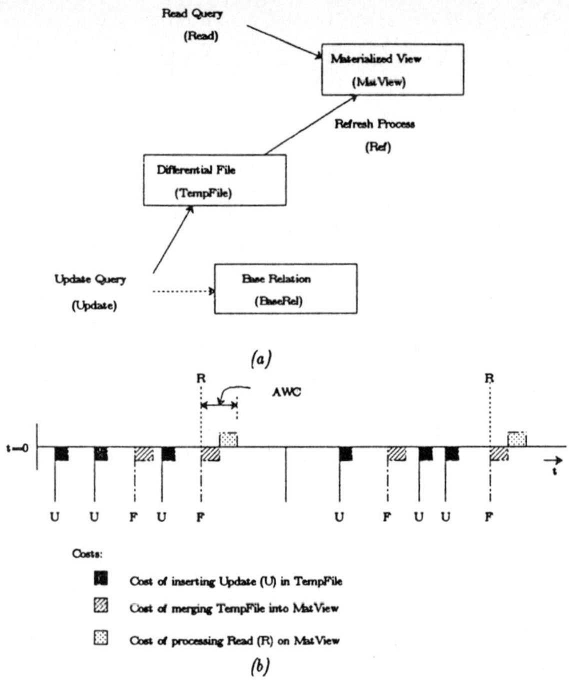

Fig. 4.1 *Mechanics of Materialized View Maintenance.*

deferring the refreshing of MatView has advantages of being more efficient under certain conditions, [ROUS 86, HANS 87]. Thus another file, the *TempFile*, is required to store the changes between successive refreshes to the MatView.

Processes: There are three kinds of processes in the system which are of interest to us, as shown in *Fig. 4.1(a).* First are Read Queries (R), that are directed to the MatView and processed using it. Second are the Update Queries (U), that can be directed either to the MatView or to the BaseRel. These are handled by making appropriate changes to the BaseRel † and also recording it in the TempFile. If the Update is an insert, delete, or change on the MatView, the appropriate action is taken on the BaseRel and the change is recorded in the TempFile. Handling Updates directed to BaseRel requires more care. Only some of these Updates affect MatView and are the only ones that have to be accounted for. This is done by *screening* each Update to the BaseRel against a *filter* (i.e. the logical predicate defining the view), to determine if it does indeed change the view too, [BLAK 86]. The ones that do so are recorded in

† Changes to BaseRel are done anyway and *not* because of maintaining materialized views.

the TempFile in addition to being changed in the BaseRel. From now on we will only consider those updates that affect the MatView. The third kind of processes are the Refreshes (F), which are executed periodically and whose function is to refresh the MatView using the contents of TempFile and bring it up-to-date with respect to BaseRel. This involves changing all tuples that appear in TempFile. No changes have to be made to BaseRel when a Refresh occurs, since the former is always up-to-date. Also, every occurrence of Refresh updates MatView with the contents of TempFile, and the latter is deleted.

4.2. A Stochastic Model of View Maintenance

Operation of the materialized view maintenance mechanism is modeled by the arrival of stochastic processes. The arrival of a Read or an Update, or the creation of a Refresh is the arrival of a job for service, and the times required for their execution are their service times, respectively. *Fig. 2.1(b)* gives a pictorial representation of the model.

Nature of the Processes: The Reads and Updates are both assumed to be stochastic processes. The Updates are assumed to have the Poisson † distribution [ROSS 85] while the Reads can have any distribution. The arrival rate of Updates is λ_U. Thus,

Read $\approx$ G (general distribution).

Update $\approx$ Poisson (λ_U).

Costs associated with processes: The cost associated with a process is the number of disk accesses made during its execution. For a fair comparison of materialized view maintenance algorithms we have developed a cost model that measures precisely the *extra overhead* that a database system incurs in supporting the materialized view, i.e. maintaining MatView and TempFile. Consider the effort required to handle insertions/ deletions to BaseRel. This would be required even if there were no materialized views, and thus we do not include its cost. There is a cost associated with *screening* each tuple to decide if it affects the view; and if it does, there is the additional cost of inserting it in the TempFile. Given below are the costs associated with the three processes in our model.

Cost of Read: The cost of accessing a single record ‡ from MatView.

Cost of Update: If the update is to MatView, the only relevant cost is of recording it in TempFile. When the update is to BaseRel, it also has to be screened to see if it affects MatView. From our point of view only the *relevant updates* [BLAK 86], i.e. the ones that affect the view, are of interest.

---

† The underlying assumption here is that the updates are independent. For large multiuser databases this is a good assumption and is widely used by researchers in the analysis of stochastic models.

‡ For simplicity we assume exactly one record is accessed by a Read.

Since screening is done in main memory, its cost is negligible and is henceforth ignored. Thus, the cost associated with an Update is that of recording it in TempFile regardless of whether it is directed to MatView or BaseRel.

Cost of Refresh: The cost of updating MatView with the contents of TempFile to bring the former up-to-date.

The work done in maintaining the files MatView and TempFile is exclusively for the purpose of view materialization. Thus the extra cost incurred by the database system should be borne by the queries made to the view. Hence, the cost of a Read is augmented by the expected cost of handling all the Updates since the previous Read. Table 4.1 lists the various parameters of the model.

Symbol	Meaning
$Read(R)$	Read Process (Query)
$Update(U)$	Update Process (Query)
$Refresh(F)$	Refresh Process
$\lambda_R, \lambda_U, \lambda_F,$	arrival rates of processes R, U, F respectively
p_U, p_F	probabilities of arrival of U, F respectively
N_0	value of counter (MatView size) at last refresh
N	value of counter (MatView size) at a later instant
p	degree of precision
q	degree of confidence

Table 4.1 *Summary of Notation.*

4.3. Materialized View Creation

When a user defines a view, its materialized copy (MatView), is created using the basic database. It consists of records (if relational algebra operators are used in its definition), or aggregate information (if aggregation operators are used). It is then stored at the site of the user creating it. As shown in *Fig. 4.2*, updates made to the main database have to be propagated to the view. If the view was required to be completely precise, i.e. reflect all changes in the main database, there would be no choice but to propagate each update on the main database immediately to the view. This is called the *immediate refresh policy*. However, for the applications we are looking at, some imprecision in data is tolerated. This enables us to design periodic refresh policies which propagate updates from the main database to the view periodically. *Table 4.2* compares periodic refresh policies with immediate ones.

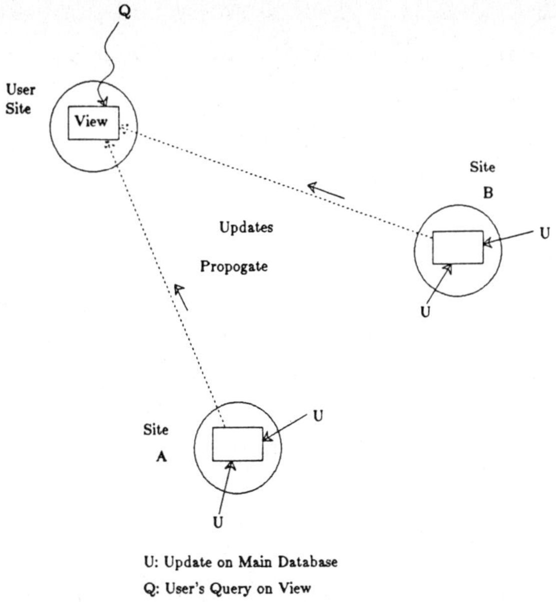

Fig. 4.2 *Propagation of Updates Between Copies.*

Cost Factor	Immediate Policies	Periodic Policies
Transmission Cost of Data	High	Low
Contention on Data	High	Low

Table 4.2 *Comparison of Refresh Policies.*

4.4. Nature of View Updates

In this paper we have assumed that all the updates to the MatView are insertions. Since the precision of a materialized view has been defined as the fraction of tuples that it is out-of-sync with the main database, it suffices to consider each update as an arrival of a random variable, U_i, with a constant value 1. Formally the Update process can be specified as,

$$P(U_i = 1) = 1; \;\; i = 1, 2, 3, \cdots$$

If T_i is the time of arrival of U_i, then

$$(T_{i+1} - T_i) \approx Exponential(\lambda_U), \;\; i = 1, 2, 3, \cdots$$

Thus, U_i's are *Poisson* arrivals.

One way to model a database with both additions and deletions is to consider them separately, and then combine the effect by adding up the errors. However, addition of errors leads to bounds that are weak and do not adequately capture the quality of the data. A better model for such a database is to consider the updates U_i as both increasing or decreasing the counter (database size). We are currently working with the following model for U_i's.

$$P(U_i = 1) = r, \; P(U_i = -1) = 1-r; \quad i = 1, 2, 3, \cdots$$

$$(T_{i+1} - T_i) \approx Exponential(\lambda_U), \quad i = 1, 2, 3, \cdots$$

4.5. Periodic Time Refresh Policy

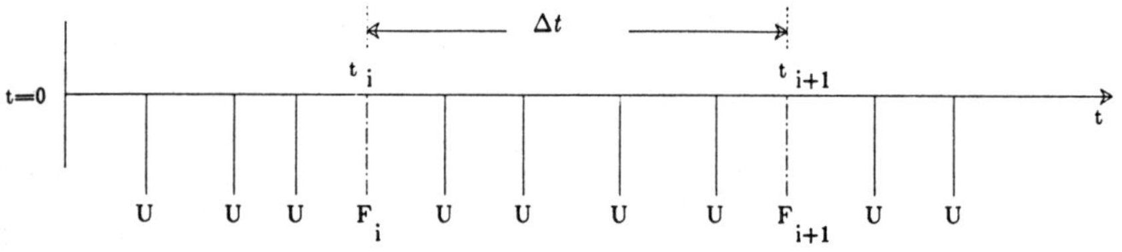

Fig. 4.3 *Updates Collecting Over Time.*

In this section we consider the design of a periodic time refresh policy which maintains the view MatView to a *desired degree of precision*. *Fig. 4.3* illustrates the arrival of various processes. From the definition of *degree of precision* we have,

$$P\left(\frac{N - N_0}{N_0} \leq (1-p) \right) \geq q$$

$$<=> P(N - N_0 \leq (1-p)N_0) \geq q$$

$$<=> \sum_{i=0}^{(1-p)N_0-1} \frac{e^{-\lambda_U \Delta t} (\lambda_U \Delta t)^i}{i!} \geq q$$

The expression on the left hand side can be approximated by the *Normal distribution* for large values of $\lambda_U \Delta t$ [FELL 68]. Formally,

$$\frac{(N - N_0) - \lambda_U \Delta t}{(\lambda_U \Delta t)^{1/2}} \approx Normal(0,1)$$

$$<=> P\left\{ \frac{(N - N_0) - \lambda_U \Delta t}{(\lambda_U \Delta t)^{1/2}} \leq \frac{(1-p)N_0 - \lambda_U \Delta t}{(\lambda_U \Delta t)^{1/2}} \right\} \geq q$$

$$<=> \frac{(1-p)N_0 - \lambda_U \Delta t}{(\lambda_U \Delta t)^{1/2}} \geq F^{-1}(q)^{\dagger}$$

$$<=> (\Delta t)^2 - \frac{(F^{-1}(q))^2 + 2(1-p)N_0}{\lambda_U}(\Delta t) + \left\{ \frac{(1-p)N_0}{\lambda_U} \right\}^2 \geq 0 \quad (4.1)$$

† $F^{-1}(q)$ is the inverse function of the Normal distribution that gives the value of the *standard normal random variable*, to the left of which the probability is q.

$$<=> \Delta t = \frac{1}{2\lambda_U}[\ 2(1-p)N_0 + (F^{-1}(q))^2 \pm [(F^{-1}(q))^4 + 4(1-p)N_0(F^{-1}(q))^2\ ]^{1/2}\ ]$$

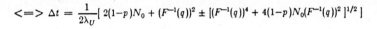

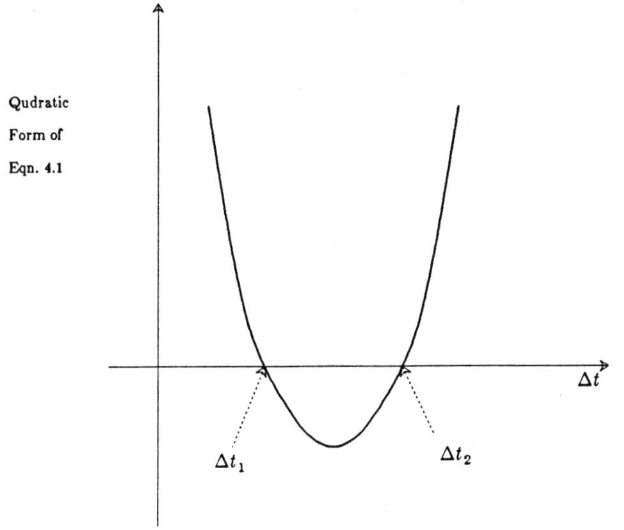

Fig. 4.4 *Roots of Eqn. 4.1*

As shown in *Fig. 4.4*, the quadratic form of *Eqn. 4.1* is non-negative in the regions $(-\infty, \Delta T_1]$ and $[\Delta t_2, \infty)$. The value of Δt (call it T) required is such that the quadratic form of *Eqn. 4.1* is non-negative for all $\Delta t \leq T$. This limits the choice to be,

$$\Delta t \in (-\infty, \Delta t_1]$$

The minimum cost refresh policy is to choose the largest possible value for Δt, i.e.

$$\Delta t = \Delta t_1$$

$$<=> \Delta t = \frac{1}{2\lambda_U}[\ 2(1-p)N_0 + (F^{-1}(q))^2 - [(F^{-1}(q))^4 + 4(1-p)N_0(F^{-1}(q))^2\ ]^{1/2}\ ]$$

Therefore, the optimal refresh algorithm is a periodic time algorithm. Every Δt_1 time unit it is executed to refresh MatView by installing all the Updates that have arrived since the last refresh.

Example: Time periods of the minimal policies were calculated for some values of N_0, p, and q. The results are presented in *Table 4.3(a)* and *Table 4.3(b)*.

As seen in *Table 4.3(a)*, Δt, i.e. the frequency of refreshes, decreases with increase in p, and also with increase in N_0. As observed in *Table 4.3(b)*, Δt decreases with increase in N_0, and with increase in q. For

p / N_0	0.90	0.95
100	0.5279	0.2054
1000	8.1453	3.7287
10000	93.7093	45.6112

$q = 0.98$, $\lambda_U = 10/s$

Table 4.3(a) *Change in Δt with N_0 and p, fixed q.*

q / N_0	0.95	0.98
100	0.2434	0.2054
1000	3.9643	3.7287
10000	46.4545	45.6112

$p = 0.95$, $\lambda_U = 10/s$

Table 4.3(b) *Change in Δt with N_0 and q, fixed p.*

$N_0 = 1000$; $p = 0.90$ and $q = 0.98$,

$\Delta t = 8.1453s$

$\lambda_U = 10/s$ means that on the average, 81.453 updates are allowed between successive refreshes. Since the precision required is only 0.90 it would seem that $(1-p)N_0 = 100$ updates should be allowed between refreshes. However, the refresh rate is higher (Δt is smaller) because $\lambda_U \Delta t$ represents only the *average* number of updates in the interval Δt. In fact there is a non-negligible probability of having more arrivals. Since our aim is to provide 0.90 precision in the worst case (actually not quite since it is being done only with 0.98 confidence), the refresh rate is slightly higher.

4.6. A Stochastic Count Refresh Policy

The idea behind a stochastic count policy is to design a refresh process, *Refresh,* based on the number of the updates (count) between successive refreshes. The refresh process will be designed as a Poisson process. The important criterion here is that the probability that the count of updates between successive refreshes exceeds a certain threshold is extremely small. From the definition of *degree of precision* we have,

$$P\left(\frac{N - N_0}{N_0} \leq (1-p)\right) \geq q$$

Considering the mixture of Update and Refresh processes we have a process in which the probability of any arrival being Update or Refresh is fixed, and is given by

$$p_U = \frac{\lambda_U}{\lambda_U + \lambda_F}; \; p_F = \frac{\lambda_F}{\lambda_U + \lambda_F}$$

The criterion mentioned above is equivalent to saying that the probability of the number of Update arrivals between successive Refreshes being more than $(1-p)N_0$ is smaller than $1-q$. Thus,

$$P(\#\ \textit{Updates between successive Refreshes} > (1-p)N_0) \leq 1-q$$

$$<=> P(\#\ \textit{Updates between successive Refreshes} \leq (1-p)N_0) \geq q$$

$$<=> p_F + p_F p_U + p_F p_U^2 + \cdots + p_F p_U^{(1-p)N_0} \geq q$$

$$<=> 1 - p_U^{(1-p)N_0+1} \geq q$$

Now, $p_U = \dfrac{\lambda_U}{\lambda_U + \lambda_F}$

Thus,

$$\lambda_F \geq \lambda_U((1-q)^{\frac{-1}{(1-p)N_0+1}} - 1) \quad (4.2)$$

Thus, the optimal refresh process is a Poisson process that has the rate λ_F given in *Eqn. 4.2* above.

Example: Update rates for the minimal refresh policies were calculated for some values of N_0, p, and q. The results are presented in *Table 4.4(a)* and *Table 4.4(b)*.

p / N_0	0.90	0.95
100	4.271	9.194
1000	0.395	0.797
10000	0.0392	0.0784

$q = 0.98$, $\lambda_U = 10/s$

Table 4.4(a) *Change in λ_F with N_0 and p, fixed q.*

q / N_0	0.95	0.98
100	6.475	9.194
1000	0.605	0.797
10000	0.060	0.0784

$p = 0.95$, $\lambda_U = 10/s$

Table 4.4(b) *Change in λ_F with N_0 and q, fixed p.*

As observed in *Table 4.4(a)*, the rate of the refresh process, λ_F, decreases on lowering the precision requirement, p, while moving along a row. Also, the rate decreases along a column as N_0 increases. As observed in *Table 4.4(b)*, λ_F *increases with* q. *For*

$N_0 = 1000$; $p = 0.90$ *and* $q = 0.98$,

$$\frac{\lambda_U}{\lambda_F} = \frac{10}{0.395} = 25.317$$

This rate is higher than one would expect, reasons for which being similar to that for the Periodic Time Refresh Policy. With the solution derived above there is a non-zero probability of successive refreshes with no updates in between. A design which makes it mandatory for some updates to occur between successive refreshes decreases λ_F, thus making the refresh policy cheaper. However, the analysis of such a policy is quite involved, and is not presented here.

5. Other Considerations in Precision-Time Tradeoff

Many challenging problems remain, and our future efforts will be directed towards them. Specifically the following problems are important:

Logical Model: Extensions to the query language have to be provided to *(i)* enable the user to define views, i.e. data copies, with desired degrees of precision, and *(ii)* express the precision requirement of a query.

Query Processing & Optimization: Query processing and optimization strategies have to take into account the existence of data copies with differing degrees of precision. Processing queries from a copy with less precision will be cheaper due to lesser amount of data contention on it. The query optimizer should generate an access path such that the data accessed has no more degree of precision than is necessary.

Physical Model: The techniques, i.e. data structures and algorithms, developed have to be implemented to see how well they perform on real data. The queueing model we have developed for view maintenance has to be extended to include replicated copies of views. New data storage techniques have to be devised, since conventional ones are oriented towards small and simple queries.

Data Distribution: The problems of data fragmentation, i.e. does a database need to be broken up into parts, and data placement, i.e. where should the units be placed, have to be revisited. Data of different degree of precision may be located at different nodes. A query arriving at a node is checked to determine its precision requirements, and sent to the appropriate node to be processed.

Data Replication: How many data copies of each degree of precision are required? It will depend on the requirements of the environment.

Fault Tolerance: If multiple copies of data exist in a distributed conventional database, the failure of sites holding more than half the copies can cause the whole system to stop. The reason for this is the need for complete precision of all copies. This approach is called the pessimistic or the careful approach. Since we allow the precision of some of the data copies to be less than complete, such a system can still continue operation. A challenging problem is to model the tradeoff between the availability of the system (i.e. how long after the failures the system is still operational), to the degree of imprecision accumulated in the inactive data copies.

Extension to Knowledgebases: Our emphasis so far has been on views defined in terms of relational and aggregate operations. Recent years have seen the rapid development of *fuzzy logic* [ZADE 65] and *possibility theory* [ZADE 78], and their applications to *approximate reasoning*. In this logical framework the *truth* or *falsehood* of a proposition is not a binary value. Instead it is a real number, α, between 0 and 1, where α and $1-\alpha$ are the *degree of truth* and *degree of falsehood* of the proposition, respectively. The value α *depends on the evidence collected so far, both for and against the proposition. Each new piece of information increases or decreases* α *depending on whether it provides evidence for or against the proposition.*

Our model of abstraction maintenance is very well suited to support such a framework of logic. Consider a set of propositions, representing fuzzy facts about some domain, which are defined in terms of some premises, i.e. existing pieces of evidence. The former correspond to the abstractions in our model while the latter correspond to the base data. As more knowledge about the domain is gathered, additional evidence about the facts is available and the corresponding α values have to be revised. For a specified degree of precision in the α's what is the most efficient way of maintaining them, is precisely the problem that our model provides an answer to.

6. Conclusions

Conventional databases are often not suitable for statistical applications, since they do not fulfill many of the important needs of the latter. Specifically, during calculating statistical parameters, they provide results that are completely precise, and spend prohibitive amount of resources in doing do. Almost all analysis in statistics can be done in the presence of imprecise data, as long as the error is bounded. Bearing this in mind we present the idea of *precision-time tradeoff* as a new paradigm for processing queries that evaluate statistical parameters from a database. The notions of *precision of data* and *precision-requirement of query* were defined. The main idea is that the cost of processing a query should be proportional to its precision-requirement, i.e. highest if full precision is required. We described a data model which allows data with varying degrees of precision. We next described *materialized views* as a mechanism (i.e. the physical model) to support data with a specified degree of precision. We used a stochastic model to analyze the problem of view maintenance, and derived two efficient maintenance policies. Finally, we outlined the various issues that the new paradigm raises, and that need to be addressed.

7. References

[ASTR 77] Astrahan, M.M., "System R: A Relational Database Management System", IBM Research Report.

[BLAK 86] Blakeley, J.A., P.Larson and F.W.Tompa, "Efficiently Updating Materialized Views", Proc. of the 1986 ACM-SIGMOD Conf. on Management of Data, Washington DC, May 1986, 61-71.

[COCH 53] Cochran, W.G., "Sampling Techniques", John Wiley Sons, New York, USA, 1953.

[FELL 68] Feller, William, "An Introduction to Probability Theory and Its Applications", John Wiley & Sons, Inc., New York 1968.

[GHOS 85] Ghosh, S.P., "SIAM: Statistics Information Access Method," IBM RJ 4865 (51295).

[HANS 87] Hanson, Eric N. "A Performance Analysis of View Materialization Strategies," Proc. of the 1987 ACM-SIGMOD Intl. Conf. on the Management of Data, San Francisco, CA, May 1987.

[HEBR 86] Hebrail, G., "A model for summaries for very large databases," 3rd Workshop on Statistical & Scientific Databases, 1986.

[HOEL 71] Hoel, P.G., S.C.Port and C.J.Stone, "Introduction to Probability Theory," Houghton Mifflin Company, Boston, 1971.

[HOU 87] Hou, Wen-Chi, G. Ozsoyoglu, B.K. Taneja, "Statistical Estimators for Relational Algebra Expressions," Deptt. of Comp. Sc., Case Western Reserve University, 1987.

[KOEN 81] Koenig, S. and R. Paige, "A Transformational Framework for the Automatic Control of Derived Data," Proc. of the VLDB Conference, 1981.

[OLKE 86] Olken, F. and D.Rotem, "Simple Random Sampling from Relational Databases," Proc. of the Conf. on VLDB, Kyoto, Japan, August, 1986.

[ROSS 85] Ross, Sheldon M., "Introduction to Probability Models", Academic Press, Inc., Orlando, Florida, 1985.

[ROUS 86] Roussopoulos, N. and H.Kang, "Principles and Techniques in the Design of ADMS+/-", Computer, December 1986.

[ROWE 83] Rowe, N.C., "Rule-Based Statistical Calculation on a Database Abstract," Rep. STAN-CS-83-975.

[SHOS 82] Shoshani, A., "Statistical Databases: Characteristics, Problems, and Some Solutions." Proc. 8th Intl. Conf. on VLDB, 1982, pp 208-222.

[SRIV 87] Srivastava, J. and Doron Rotem, "Analytical Modeling of Materialized View Maintenance," Lawrence Berkeley Laboratories Tech. Rep., 1987.

[ULLM 82] Ullman, J.D., "Principles of Database Systems," Computer Science Press, 1982.

[VITT 84] Vitter, Jefferey S., "Faster methods of Random Sampling," CACM 27(7):703-718, July 1984.

[ZADE 65] Zadeh, L.A., "Fuzzy Sets", Information and Control 8, 1965, pp. 338-353.

[ZADE 78] Zadeh, L.A., "Fuzzy Sets as a basis for a theory of possibility." Fuzzy Sets and Systems, 1, pp. 3-28, 1978.

INTERPRETATION OF STATISTICAL QUERIES TO RELATIONAL DATABASES

Alessandro D'Atri, Fabrizio L. Ricci

Alessandro D'Atri: Dipartimento di Ingegneria Elettrica, Università dell'Aquila, Poggio di Roio, 67040 L'AQUILA (ITALY). datri@irmiasi.bitnet

Fabrizio L. Ricci: Istituto di Studi sulla Ricerca e la Documentazione Scientifica del CNR, via C. De Lollis 12, 00185 ROMA, (ITALY).

ABSTRACT

A research in progress concerning the problem of extending the logical independence approach to statistical query answering is described. The main ideas and limitations of a system which assists a user in obtaining a statistical table from a relational database by describing only elements that compose the table are discussed. The proposed system assists the user in formulating statistical queries by means of a universal relation interface, i.e., independently from the underlying database structure. To generate the result, an extended relational algebra is needed to process aggregations in queries function; heuristics for the interpretation of queries and their translation in terms of database primitives are the basis of the proposed approach.

1. INTRODUCTION

Statistical/scientific databases (SSDBs) are different from traditional databases in several aspects.Usually, they are more static (they represent consolidated events), i.e., update operations are infrequent and are applied to a small percentage of the data; on the other hand, a statistical query is, in general, more complex and involves a large amount of data [Shoshani 85].Tipically, it is harder for the user to be formulate and more expensive for the system to evaluate.

There are two kinds of SSDB: *micro* and *macro* SSDBs [Wong 84]. Micro-SSDBs handle information about single events or individual entities and their statistical features are embedded in the query system; on the other hand, macro-SSDBs contain mainly summary data of micro-SSDBs, i.e., the results of statistical operations such as cross tabulation or regression.

In this paper micro-SSDBs and queries to obtain statistical tables from a relational database are considered. Even if not all statistical queries are categorical in nature and not all statistics has to do with tabular output, we have restricted our attention to such cases of statistical analysis, since we consider this as a first step towards a more general approach to the problem.

The extensions of relational query languages proposed in the literature to handle statistical applications provide a formal environment for users that are sophisticated in terms of knowledge about the application domain and the task to be performed and expert in terms of the way in which the system works. In particular, it is required that users know: the data model, that is used to formalize the realm of interest, the syntax and semantics of the query language, and the way in which the application is represented in the system (the database scheme). In absence of one or more of the above requirements, query formulation becomes very hard.

For commercial database systems various interaction techniques have been proposed to take care of the way in which naive users work. One technique is based on the construction of database *views*. Views are designed and implemented by intermediary expert persons (application programmers), hence, they have a "static" nature, and are intended for regular users, whose interactions are predefined and frequently activated. In this way naive users can obtain personalized views on the database that are easier to understand and to use, but cannot modify easily their interaction goals or style. This does not seem to be suitable for statistical users of micro-databases that do not have, in general, a predefined task to perform and cannot interact with the model of the realm through an intermediary.

Another kind of interaction technique intended for naive users is the *navigation* technique implemented by browsers (see, for example, [Motro 86] and [Motro 88]) through the database. These tools are very simple to use, since they are based on a simplified data model and a set of elementary operations that move step-by-step the observation point in the database. Unfortunately, browsing is limited in its capability of performing complex tasks, and it is not clear how it can be used to specify a statistical query.

An intermediate interaction technique is query answering by finding "*conceptual connections*" between a set of user defined "goals" under the hypothesis that the database scheme is not shown to the user [Ausiello 86]. This paper concerns the problem of extending to the statistical environment a special case of this approach: the *logical independence* approach to query answering in relational databases, and in particular the possibility of using universal relation interfaces for statistical queries.

A universal relation interface [Ullman 83] allows the user to see a database as a single (universal) relation, by assuming that attributes have a global meaning, not limited to the relation they belong. Therefore, the real world is seen by the user as entities and associations which can be obtained by composition of attribute names. Such interfaces offer the advantage of a simpler formulation of a query (it is not necessary to specify the relation to which the attributes belong: "logical independence"). A correspondence that allows one to obtain from a relational database and a given set of attributes (query scheme) a relation instance on these attributes, used to compute the answer to the query, is called a *window function* [Maier 83]. There are various kinds of window functions that differ both in their assumptions and in the algorithms that are proposed to calculate the resulting relation.

To generate a statistical table from microdata stored in a relational DBMS, a relation has to be obtained on which an "aggregation operator" can be applied (construction phase of the query process), then the operator is computed and the result is presented in tabular form (presentation phase). The construction activity can be done by an "extended" relational algebra (e.g., with set-valued attributes) with emphasis on management of data. On the other hand, in the presentation phase the emphasis is on computing and on displaying results. Therefore, to formulate a statistical query, two expertises are needed: on data management and on statistical analysis. To handle this problem, commercial systems are based on either of three different approaches:

1) systems consisting of a statistical package with some facilities of database management systems (i.e., [Buhler 81]);
2) database management systems specialized to find solutions to particular statistical problems (i.e., [McCarthy 82]);
3) interfaces between statistical computing systems and database management systems (i.e., [Hilhorst 87]).

Since several problems arise when the facilities of a system are extended, we propose to make the data management activity transparent to the user, i.e., to have it handled by the system by extending its universal relation interface. Hence, the system we propose (the July system) is based on the assumption that the underlying database management system contains a universal relation interface; using this interface the user query can be simplified. In fact, it only handles the statistical part of the query in terms of attributes. Therefore the extension of the logical independence approach to SSDBs allows one to perform a statistical query without any knowledge about the database structure. This approach is based on a proposal in [D'Atri 87], where in an ambiguous situation the system shows only that portion of the database structure which is needed to resolve the ambiguity.

In the following section we intrduce the hypotheses, definitions, and heuristics on which our proposed system is based. The system is capable to obtain a statistical table by describing only those elements that constitute the table. This approach is compared with similar proposals in the literature, and limitations and possible extensions of our proposal are discussed.

2. MAIN CHARACTERISTICS OF THE PROPOSAL

The main components of the data model that has to considered in order to process a statistical query are now introduced.

First, the attributes of a relational database can be classified according to the following three types (that are, in general, trasparent to the user):

- primitive, if they belong to the underlying database structure;
- derived, if they are associated with derivation rules applied to single tuples (the value of the derived attribute is obtained from the values in the same tuple according to the rule).
- summarized, if they are obtained by means of an aggregation operator and the evaluation of an aggregation function (applied to tuples grouped according to the values that appear in a given set of attributes).

Since the relation schemes considered in universal relation interfaces are only composed by primitive attributes, it follows that a mapping (and the corresponding sequence of derivation rules and/or aggregation operators) to an attribute used by the aggregation function from its underlying primitive attributes have to be determined by the system.

Statistical queries are more complex than normal database queries, in fact the relational model has to be extended by introducing unnormalized relations, aggregation functions, and operators that perform elementary aggregation functions for the generation of statistical tables [Ozsoyoglu 87]. In particular, two aggregation operators have been proposed in the literature to handle the generation of statistical tables: the *aggregate formation* [Klug 82], that in a first step partitions the set of tuples of a relation (according to the values they assume on a subset of their attributes) and, in a second step, it applies an aggregation function to each partition (e.g., count, sum, average) that takes as input a set of relation tuples and computes a numeric value; and the *aggregation by template* [Ozsoyoglu 87], that is based on the concept of grouping according to an appropriate criterion, contained in a template relation, and applying to each group an aggregation function. It has to be noted that the grouping concept produces unnormalized relations [Schek 86]. Since we restrict our proposal to descriptive statistics, these two operators are sufficient to produce the information for the statistical tables.

The second aspect we considered is the logical organization of statistical tables, where the basic structures to describe tables (e.g., GRASS [Rafanelli 83], Mefisto algebra [Fortunato 86], SAM* [Su 83], Subject [Chan 81]) are:

- the summary attribute, that characterizes the final result of a statistical analysis and summarization;
- the summary type, i.e., the aggregation function used for computing the summary attribute;

- a set of category attributes, which provide an identification of the summary attribute, i.e., a combination of values of these attributes identify each value of a summary attribute (this set of attributes can be considered as a composite key for the summary attribute [Shoshani 82]);
- the specific domain of each category attribute, i.e., its active set of values (and not the admissible set of values, as it is usually done in the relational model). Notice that such values can also be sets of values in the domain of the correspondent database attribute;
- the selection conditions, that are used to restrict the analysis only to certain events of a phenomenon and that can be also applied to summarized (and derived) attributes after a summarization (and/or derivation) step [Johnson 81].

The differences between relations that can be obtained from a traditional universal relation interface and relations needed to compute an aggregation function are the following.

Let a minimal relation with respect to a statistical query be a strictly necessary relation for the application of an aggregate formation or an aggregation by template. A minimal relation is composed by a first set of attributes, needed for obtaining the values of category and summary attributes, and a second set of attributes, called key attributes, that allow the identification of distinct instances of the phenomenon considered in the statistical query (note that the first set of attributes is often functionally dependent on the second set of attributes). Since the summarized attributes are obtained by aggregate formation, we find that a minimal relation has to be associated with the summary attribute in the output table and with each summarized attribute. Let the window relation be the relation obtained from the micro-database management system by means of the universal relation interface (this relation is composed of primitive attributes only).

Similarly to minimal relations, in the window relation three types of attributes can be identified: those from which the values of category and summary attributes are computed, the key attributes, and the selection attributes.

A special kind of selection condition in a query is that which is applied to a category attribute of the output table (this selection is used with the operators of aggregate formation and aggregation by template). If the selection is applied to primitive attributes then it is performed by the universal relation interface, else it has to be performed by a selection operator. This selection is applied to summarized and/or derived attributes; hence, in this case, the selection attributes of the window relation are the primitive attributes needed.

Hence, a minimal relation can be different from its corresponding window relation, in fact:

1) a minimal relation may also contain non-primitive attributes (corresponding to derivation rules or aggregation functions);
2) a window relation may contain selection attributes.

Finally, if in a minimal relation there are summarized attributes then more than one window relation are needed.

3. THE QUERY INTERPRETATION PROCESS

Starting from a simple and possibly incomplete description of the structure of a statistical table that the user wishes to obtain by his statistical query, the July system generates this table by giving a complete specification of it and by determining the evaluation procedure. In order to obtain this result, it analyzes the description of the elements that compose the structure, and starts a dialogue with the user, under the assumption of logical independence, to resolve the interpretation ambiguities. The structure given by the user includes not only a "logical" description of the table, but also its "layout" description, i.e., its display parameters on the output devices as the position of category attributes in rows and columns or the ordering of their values. Such information can be exchanged by using several distinct kinds of interaction techniques [D'Atri 85]. Among these techniques we have selected a suitable extension of the dynamic query interpretation approach (proposed in [D'Atri 87]).

The query interpretation problem for universal relation interfaces arises from the consideration that several distinct answers (due to distinct window functions) may be associated with the same query, because the user ignores the database scheme (logical structure). There are two possible ways to solve this problem: the *static* approach, that assumes that a unique "basic" interpretation exists and can be computed following a suitable procedure that does not involve the user, and the *dynamic* approach, that bases the interpretation process on a user-system dialogue that does not show the whole logical structure, but consists in a sequence of user decisions about the relevance of an attribute with respect to his/her desiderata. The July system extends the dynamic approach to the statistical environment by considering the concepts given in the previous section.

Besides displaying the table, the July system allows also to store it in a (macro) SSDB, or to store the formal description of the table in its data dictionary. In this way the system can also be used to recognize and to handle the relationships between its micro and macro data.

The July system (see Figure 1) manages a predefined set of aggregation functions and a set of interpretation techniques (which will be briefly described later). A first interpretation step is devoted to the decomposition of the statistical query in its aggregation operators (aggregate formation and aggregation by template); when such operators are identified, for each of them an appropriate minimal relation is obtained. Hence, there are two main problems to solve in this phase:

1) the query interpretation from the statistical-mathematical point of view;

2) the construction of minimal relations according to the described procedures.

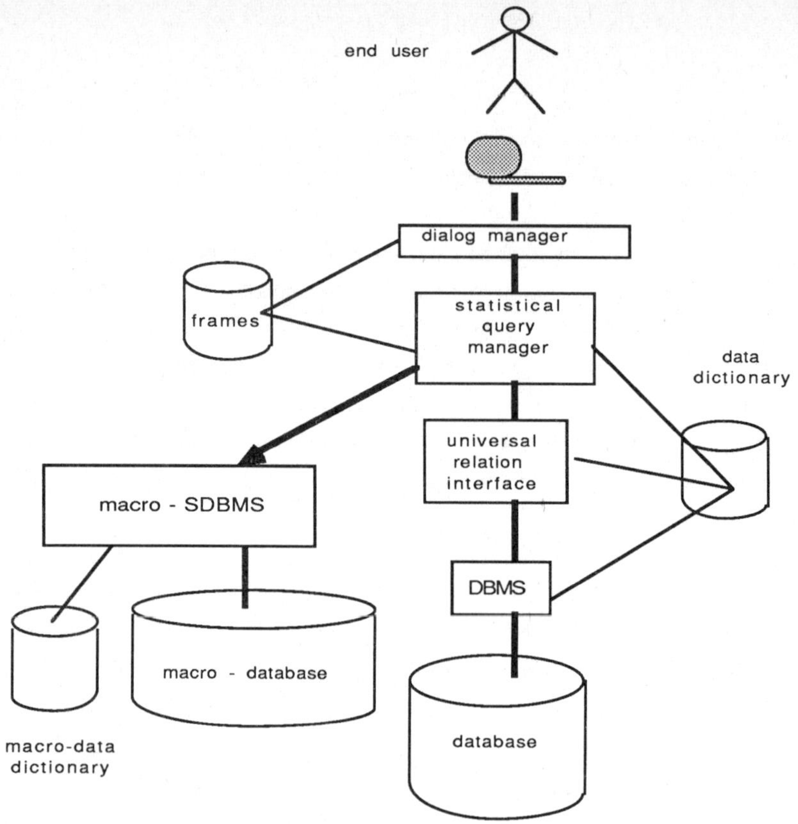

Fig. 1. The architecture of the July system.

The second phase concerns the mapping between minimal relations and universal relations, by selecting for each minimal relation a corresponding window relation (or relations). In this phase two interpretation problems arise:

1) the identification of the window relation to submit to the universal relation interface;
2) the synthesis of the minimal relations, associated to each statistical function, starting from the window relations.

The third phase is devoted to obtain the window relation from a set of primitive attributes and a selection on them. This phase handles the final logical independence level between the underlying database scheme and the universal relation interface.

A friendly, logically independent and dialogue-based interpretation of a statistical query is a very complex process, since it requires that the system cooperates with the user in a suitable environment to assist him both in formulation and in disambiguation phases.

The July system performs such a process by a module that works as an expert assistant in statistical querying, i.e., an expert system that contains the expertise of a sophisticated user of SSDBs. A similar approach works in the KIWI system [Saccà 86] for (non-statistical) database querying. In the KIWI system an Advanced Database Environment (based on relational databases and logic programming facilities) interacts with a Knowledge Handler; this Knowledge Handler manages the user profile and the expertise in database querying which are stored in an (object oriented) knowledge base.

The knowledge base of the July system is described by frames [Minsky 75], that are related to expectations, generated by experts, and that work according to the experience needed for statistical query interpretations. In order to answer a query, the system begins by instantiating a frame which analyses the output table and then activates other frames (that are specified in the slots of the instantiated frame), in which the analysis technique is coded by heuristic procedures. At the same time a corresponding sequence of operators is determined and applied to minimal relations to compute the resulting table.

4. DISCUSSIONS

The main criticism of the universal relation approach (see [Atzeni 82], [Kent 81]) is that the data model shown to the user is too poor. In fact, it is a simplification of the relational model, which is too flat in itself (it does not contain aggregation and generalization primitives [Smith 77] which play a fundamental role for representing and understanding the database content, and do not provide the user the "complete" power of a relational query language to retrieve information).

This approach is tailored to naive users and can be useful to implement natural language interfaces to relational databases (as proposed in [Ullman 88]); the universal relation can also be useful for implementing graphical interfaces, where the attributes (or perhaps relations) are displayed and the user indicates those attributes among which he wishes to see the logical connections.

We believe that logical independence benefits also the scientific/statistical user. The arguments for using the logical independence approach in SSDBs are the following. Firstly, it is unreasonable to assume that the user knows the detailed database structure; in fact, normalization and performance criteria (and not only semantic ones) drive the database design. Secondly, in query formulation, the user has to specify "connections" when jumping from relation to relation (logical navigation by join), but this join operator is frequently difficult to use and it does not help the user to better understand the query semantics. From a statistical point of view, e.g., in questionnaire analysis, the user wishes to see the database structure as a set composed by few and very abstract concepts and a list of variables [Falcitelli 87]. Finally, we note (also according to our experience in using SSDBs for epidemological analysis [De Rosis 84]) that queries to obtain minimal relations are often simple PSJ (projection-selection-join) queries, if there are no summarized and derived attributes (in this case aggregate formation and derivation rules are needed); hence, the query language does not need to be complete and the strategies to obtain minimal relations can be forecast.

The reason for using a special approach for statistical queries is that, even if the commercial relational DBMS use aggregation primitives (e.g., grouping, partition, partition selector, and duplicate), a relational query language based on the extended relational algebra [Ozsoyoglu 87] is more powerful, because it achieves this power through the simple idea of subquery with parameters. Several statistical query languages have been proposed in the scientific literature and there are various criteria for their evaluation [Ozsoyoglu 85]. In order to compare them with our proposal, it is necessary to restrict the analysis to languages based on the relational model and those that pay particular attention to the man-SSDB interface.

Relational SSDB systems can be classified in the following way: systems that are based on relational algebra (e.g., RAPID [Turner 79] and CAS SDB [Kohji 83]), systems that are based on relational calculus (e.g., Abe [Klug 81] and STBE [Ozsoyoglu 82] that use languages similar to QBE [Zloof 77]), and systems that are extensions of SQL (e.g., SIR [Anderson 83], GENISYS [Maness 81], CANTOR [Karasalo 83]). JANUS [Klensin 83] can also be considered a relational SSDB systems since it uses objects similar to relations. A compact method for defining tables is proposed by Ikeda and Kobayashi [Ikeda 81] that expands the facilities of a relational DBMS to allow the creation and management of tables. These systems do not consider simple interaction techniques, such as logical independence from the scheme, hence, they may become "unfriendly" for a naive user when the statistical query is complex.

GUIDE [Wong 82] (a system based on the Entity-Relationship model) assigns to each entity and to each attribute in the scheme a value that gives the "relevance" of this object. It frees the user from havig to remember too many database scheme details and it allows him to focus his attention to other aspects. Although the authors analyze several interesting problems concerning man-machine interaction, they do not propose a powerful solution, since the user can find himself navigating "blindly" through metadata without any system aid.

GENISYS [Maness 81] is a system that has been proposed in order to simplify the interaction with the underlying relational database, by using an approach similar to universal relation interfaces, but this approach works only with predefined joins. Still closer to a universal relation interface is TPL [Weiss 81] that has a nonprocedural language which generates tables starting from their description by the user. The system stores a set of predefined views and the user, in defining the table, specifies the name of a view on which he intends to work. The difference between the universal relation approach and these proposals is the static nature of predefined views. Finally, a system based on a universal relational interface that handles summarized attributes in the output relation is SMARTY [Stecher 86]; this system does not consider statistical tables (it works within a traditional relational algebra).

5. CONCLUSIONS AND FURTHER RESEARCH

A universal relation interface to a traditional database system simplifies query formulation, since the query does not refer to the underlying database structure, and the logical organization of the scheme is not considered by the user. Within the realm of statistical querying, the approach of logical independence implies also the independence of the user interaction from the underlying possible aggregations of data, which are related to the studied phenomenon. Hence, derived and summarized data are indistinguishable from primitive data, with respect to the user's point of view. The query language is statistically oriented; thus, the user formulates a query by giving only information needed to describe output tables.

The July system runs in a preliminary version on MS-DOS environments. It mediates a weak interaction between a traditional universal relation interface and a separate knowledge-based module that handles statistical queries according to a logical independence approach. In order to obtain a stronger integration or to have a unique universal relation interface suitable for statistical queries, several problems have to be studied. In particular, a first methodological aspect to be considered is the extension of the logical independence theory to a richer relational data model (for instance with the capability of dealing with unnormalized relations, and containing primitives for structuring metadata); furthermore, it can be useful to include in the system (e.g., by frames) more statistical concepts and information about users. These two aspects are both important in order to allow sophisticated statistical analysis. The universal relation interface has to be able to learn about the user with whom it interacts, as well as to receive (or to deduce) definitions of new macro structures such as other types of "macrodata", and new aggregation functions.

Acknowledgement. The authors would like to thank P. Di Felice, P. Svensson and the unknown referees for their comments to the paper and their useful suggestions.

REFERENCES

[Anderson 83] G.A. Anderson, T. Snider, B. Robinson, J. Toporek, "An integrated research support system for inter-package communication and handling large volume output from statistical database analysis operations", Second International Workshop on Statistical Database Management, 1983.

[Atzeni 82] P. Atzeni, D.S. Parker, "Assumptions in relational database theory", ACM SIGACT-SIGMOD Symposium on Principles of Database Systems, 1982.

[Ausiello 86] G. Ausiello, A. D'Atri, M. Moscarini, "Chordality properties on graphs and minimal logical connections in semantic data model", Journal of Computer and System Science, 33, 2, 1986.

[Buhler 81] R. Buhler, "Data manipulation in P-STAT", First LBL Workshop on Statistical Database Management, 1981.

[Chan 81] P. Chan, A. Shoshani, "Subject: a directory driven system for organizing and accessing large statistical databases", Very Large Data Bases Conference, 1981.

[D'Atri 85] A. D'Atri (ed.), "Comparative analysis of presently available languages and methods for supporting user interaction with data and knowledge base systems", Report on Task A2, Esprit Project 641, 1985.

[D'Atri 87] A. D'Atri, P. Di Felice, M. Moscarini, "Dynamic query interpretation in relational databases", ACM SIGACT-SIGMOD Symposium on Principles of Database Systems, 1987.

[De Rosis 84] F. De Rosis, A. Franich, S. Pizzutilo, F.L. Ricci, "Handling and analyzing epidemiologial data by microcomputer", Lecture Notes in Medical Informatics, 24, Springer-Verlag, 1984.

[Falcitelli 87] G. Falcitelli, "Stipam: a system for preliminary management of statistical information oriented towards multivariate analysis", Cinquièmes Journées Internationales Analyse des Données et Informatique, 1987.

[Fortunato 86] E. Fortunato, M. Rafanelli, F.L. Ricci, A. Sebastio, "An algebra for statistical data", Third International Workshop on Statistical and Scientific Database Management, 1986.

[Hilhorst 87] R.A. Hilhorst and others, "ROCHEFORT: research on creating a human environment for on-line research tools", Statistical Software Newletter, 13, 2, 1987.

[Ikeda 81] H. Ikeda, Y. Kobayashi, "Additional facilities of a conventional DBMS to support interactive statistical analysis", First LBL Workshop on Statistical Database Management, 1981.

[Johnson 81] R.R. Johnson, "Modelling summary data", ACM-SIGMOD International Conference on Management of Data, 1981.

[Karasalo 83] I. Karasalo, P. Svensson, "An overview of CANTOR - A new system for data analysis", Second International Workshop on Statistical Database Management, 1983.

[Kent 81] W. Kent, "Consequences of assuming a universal relation", ACM Trans. on Database Systems, 6, 4, 1981.

[Klensin 83] J.G. Klensin, "A statistical database component of a data analysis and modelling system: lessons from eight years of user experience", Second International Workshop on Statistical Database Management, 1983.

[Klug 81] A. Klug, "ABE - A query language for constructing aggregate-by-example", First LBL Workshop on Statistical Database Management, 1981.

[Klug 82] A. Klug, "Equivalence of relational algebra and relational calculus query languages having aggregate functions", J. of ACM, 29, 3, 1982.

[Kohji 83] H.Kohji, H. Sato, "Statistical database research project in Japan and the CAS SDB project", Second International Workshop on Statistical Database Management, 1983.

[Maier 83] D. Maier, "Windows on the world", ACM-SIGMOD International Conference on Management of Data, 1983.

[Maness 81] A.T. Maness, S.A. Dintelman, "Design of the genealogical information system", First LBL Workshop on Statistical Database Management, 1981.

[McCarthy 82] J. McCarthy and others, "The SEDIS project; a summary overview of social, economic, environmental, demografic information system", L.B.L. report, Un. of California, Berkeley, 1982.

[Minsky 75] M. Minsky, "A framework for representing knowledge", in "The psychology of computer vision", McGraw-Hill, 1975.

[Motro 86] A. Motro, "Baroque: a browser for relational databasee", ACM-SIGMOD International Conference on Management of Data, 1986.

[Motro 88] A. Motro, A. D'Atri, L. Tarantino, "The design of KIWI: an object oriented browser", Expert Database Systems Conference, 1988.

[Ozsoyoglu 82] Z.M. Ozsoyoglu, G. Ozsoyoglu, " STBE - Database query language for manipulating summary data", Report CES-82-2, Case Western Reserve University, Cleveland, 1982.

[Ozsoyoglu 85] G. Ozsoyoglu, Z.M. Ozsoyoglu, "Statistical database query languages", IEEE Trans. on Software Engineering, SE-11, 10, 1985.

[Ozsoyoglu 87] Z.M. Ozsoyoglu, G. Ozsoyoglu, V. Matos, "Extending relational algebra and relational calculus with set-valued attributes and aggregate functions", ACM Trans. on Database Systems, 12, 4, 1987.

[Rafanelli 83] M. Rafanelli, F.L. Ricci, "Proposal of a logical model for a statistical database", Second International Workshop on Statistical Database Management, 1983.

[Saccà 86] D. Saccà, et al., "Description of the overall architecture of the KIWI system", in "ESPRIT 85: Status Report on Continuing Work", Elsevier Science Publishers B.V. (North-Holland), 1986.

[Schek 86] H.J. Schek, M.H. Scholl, "The relational model with relational-valued attributes", Information Systems, 11, 2, 1986.

[Shoshani 82] A. Shoshani, "Statistical databases: characteristics, problems and some solutions", Very Large Data Bases Conference, 1982.

[Shosani 85] A. Shoshani, H.K.T. Wong, "Statistical and scientific database issues", IEEE transaction on Software Engineering, SE-11, 10, 1985.

[Smith 77] J.M. Smith, D.C.P. Smith, "Database abstractions: aggregation and generalization", ACM Trans. on Database Systems, 2, 2, 1977.

[Stecher 86] P. Stecher, P. Hellemaa, "An intelligent extraction and aggregation tool for company data bases", Decision Support Systems, 2, 1986.

[Su 83] S.Y.W. Su, "SAM*: a semantic association model for corporate and scientific-statistical databases", Information Science, 29, 2 and 3, 1983.

[Turner 79] M. Turner, R. Hammond, P. Cotten, "A DBMS for large statistical databases", Very Large Data Bases Conference, 1979.

[Ullman 83] J.D. Ullman, "Universal relation interfaces for database systems", IFIP Conference, 1983.

[Ullman 88] J.D. Ullman, Principles of Database and Knowledge Base Systems, Computer Science Press, 1988.

[Weiss 81] S.E. Weiss, P.L. Weeks, N.J. Byrd, " Must we navigate through databases?", First LBL Workshop on Statistical Database Management, 1981.

[Wong 82] H.K.T. Wong, I. Kuo, "GUIDE: graphical user interface for database exploration", Very Large Data Bases Conference, 1982.

[Wong 84] H.K.T. Wong, "Micro and macro statistical/scientific database management", IEEE Conference on Data Engineeering, 1984.

[Zloof 77] M.M. Zloof, "Query-by-Example: a data base language", IBM Systems Journal, 16, 4, 1977.

GQL, A Graphical Query Language for Semantic Databases ψ

Manoochehr Azmoodeh
Hongbo Du

Department of Computer Science
University of Essex
Wivenhoe Park
Colchester CO4 3SQ.
U.K.

Abstract

GQL is a formal query language for graphically manipulating a semantic database. The language provides a set of pattern images and a set of simple rules for constructing pattern graphs. Users describe their queries by drawing pattern graphs against the graphical schema of the database. This paper describes various aspects of GQL: the underlying model, the pattern images and graphs, and the representation of operations. Implementational issues of a GQL interpreter and its relation to other query languages are also discussed.

Key words: Database Query Language, Data Model, Database Schema, Graphics, Pattern Image.

1. Introduction

Database query facilities provide users with a formalism for manipulating a database (throughout this paper, "query" is a general term referring to data retrieval, data manipulation or data definition). Different techniques have been used to design a query facility which should be easy to use for database end users. A query facility using graphics supports a two dimensional syntax. It allows users to directly specify queries using the graphical representation of the database.

A number of graphical query languages and graphical database user interfaces have been described in the literature (QBE, ISIS [Kenn 85], GUIDE [Wong 82], SNAP [Bryc 85], BRMQ [Azmo 85], etc). Like textual query language, graphical query facilities are defined on the basis of an underlying data model (e.g. QBE on a relational data model and ISIS, SNAP, BRMQ on a semantic data model). The model determines the expressive power of the query facilities. The concepts of existing languages and systems mainly follow either of two strategies. One is called the "live-inside-database" strategy.

ψ A shorter version of this paper was presented at the Computer Graphics International Conference in Geneva, May 1988.

Users build queries as they explore the database values and schema using graphical tools. Queries are composed by a sequence of "up-to-now" sub-queries. A typical example of this approach is ISIS. The other strategy is that users describe entire queries at the database schema level by using the graphical specification method of the language. GUIDE and SNAP are examples of such systems.

The "query as explore" idea of the first strategy defines an informal and procedural way of making a query. The query systems using this strategy are designed for users who wish to specify their queries while exploring the database. However, in order to keep track on "up-to-now" sub-queries, systems such as ISIS have to use derived data definitions for them. Another problem with this strategy is that the idea of "query as explore" may help users to understand the database, but at the same time may mislead users from their original query goals. In contrast, the second strategy allows users to specify their entire queries at the conceptual schema level. Therefore, this strategy defines a general and descriptive way of specifying a query which is believed to be natural for users in general. This latter strategy is the approach taken in our query language.

The definition of a query language determines its expressiveness in a number of ways, e.g., the qualification and quantification of data, the specification of query operators, and the representation of operations for query construction. Many existing graphical query facilities are prototype systems, such as GUIDE and SNAP. They lack full consideration of these issues, and their expressiveness is therefore limited. Some languages, like QBE and BRMQ, still have a text-oriented syntax although they are called graphical languages.

In SNAP and GUIDE, queries are represented as a highlighted segment of the database schema. Because the schema itself only shows the conceptual structure of the database, it is not sufficient as a general tool to represent manipulations of the database. SNAP and GUIDE tried to attach some qualification conditions on the segment graph, but both of them run into difficulty when representing quantifiers, like "for all", "exist some", and general operations, such as arithmetic operations and aggregate functions which usually appear in a user's query.

In this paper, we present a Graphical Query Language (GQL) which is defined on a subset of the first order predicate calculus. The language provides a graphical method for specifying queries. It defines a set of pattern images and a set of rules for constructing pattern graphs out of the images. To make a query in GQL, users just use the GQL pointer device (mouse, light pen, etc) to draw pattern graphs and to call necessary operations against the schema network of the database displayed on the screen.

The structure of the paper is as follows. In section 2, we introduce the underlying data model of GQL and the graphical representation of the model. Section 3 discusses the basic set of pattern images, pattern graphs, and their underlying meanings. Section 4 shows the methods for representing and calling operations. Section 5 is an example session. Finally, we briefly discuss the implementational issues of GQL and its relations with other query languages.

2. The Underlying Model and Its Graphical Representation

GQL underlying data model is a general semantic data model developed from existing models E-R [Chen 76], SBRM [Azmo 84], QBRM [Jian 85] and IFO [Abit 84]. The model describes the real world as "objects" and various "relationships" between objects. An object represents an element in the world and a relationship represents a property of an object. We first give the informal description of the model and then the graphical representations of the model will follow.

2.1. Objects and Object Types

Every object has a value and belongs to a particular object type. In this model, an object is either an "atom" which is a string, an integer, a floating number or a boolean, or a "structure" which is composed of some atomic objects and/or other structures. A structured object represents an abstract concept. All the object values of the same type form the "domain" of the type that can be considered as the set of values of that type.

$$D \equiv \{ v_1, v_2, \ldots , v_N \}$$

If domain D is of an atomic type, the values are either strings, or integers, etc; otherwise, the values are abstract identifiers (they are unprintable) which define the references to the structures.

Every object type has a name given by the user. In logic terminology, object type P defines a "member" predicate P(x). P(x) is true iff term x refers to an object in the domain of P; otherwise P(x) is false.

2.2. Relationships and Relationships Types

In this model, all relationships are binary , describing associations between two objects. Relationships with the same conceptual meanings are grouped into a relationship type which represents a type of associations between two types of objects in abstract form. Relationship types can denote one-one, one-many, many-one or many-many mappings. Some relationship types can be conceptually categorised according to their functionalities :-

i) Relationships for composing complex structures. Such relationships construct a structured object and define the "aggregation_of" hierarchy between the structured object and its component objects. These component objects are known as the attributes of the structured object. Such relationship types only represent many-one mappings.

ii) Binary relationships can specify associations not only between two objects but also among several objects. The model adopts QBRM's idea of qualified binary relationships to define the latter. Namely binary relationships can be nested in such a way that an argument of a relationship can be another relationship.

iii) Relationships for representing a hierarchy of object types. This type of relationships are defined between object types. It shows an object type "is_a_subtype_of" another. Along the IS_A hierarchy, objects have property inheritance, i.e. the properties that a supertype has, are also properties of the subtype. An example of property inheritance is "operation". Operations for object types are classified along IS_A hierarchy so that operations for a super type are also available for sub-types.

iv) Relationships for representing functions. Generally, relationships denote the associations of existing objects. In order to represent functions, we need to define the results of functions as an argument of the relationships. Therefore the relationships represent the mappings from function arguments to function results.

From the point of view of logic, a relationship corresponds to a binary predicate. The predicate R(a,b) is true iff the relationship R exists between objects a and b; otherwise false. The logical meaning of nested binary relationships R1(R2(x,y),z) is that predicate R1(R2(x,y),z) is true iff R2(x,y) as a predicate is true and R2(x,y) as an object makes R1(R2(x,y),z) true.

2.3. Graphical Representation of the Model

In our graph representation of a database model, nodes stand for objects and arcs for relationships. We use several node and arc images to represent the conceptual items of the model graphically. Figure 2-1 shows these images.

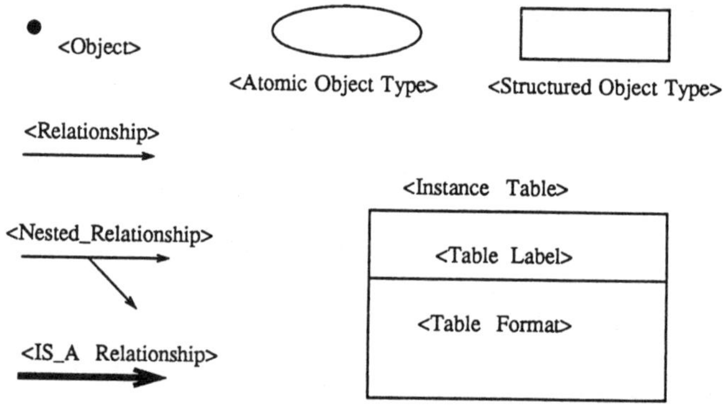

Figure 2-1. Graphical Images for the Underlying Data Model

We use a dot to represent an object or a variable denoting an object. We define a table form to show the domain of an object type and the instances of a relationship type. The table is single-columned for an atomic object type, multiple-columned for a structured type and two-columned for relationships. Table forms for qualified relationships are nested.

We define oval and rectangle images for atomic and structured object types respectively. The image for "IS_A" relationship is a heavy-line directed arc whose arrow indicates the "is_subtype_of" mapping from a subtype to a supertype. The image of other relationship types is a single-line directed arc. The arrow of the arc represents the direction of the mapping. We shall introduce graphical images for operations in a later section. All images for object types and relationships have type names as labels.

We do not define different images for a relationship and a relationship type because users can directly tell if an arc represents a particular relationship or a general description of a relationship type from the context of a graph.

The database schema is designed in terms of this graphical representation of the underlying data model. Figure 2-2 gives a sample database schema which describes the supplies and the employment in a company.

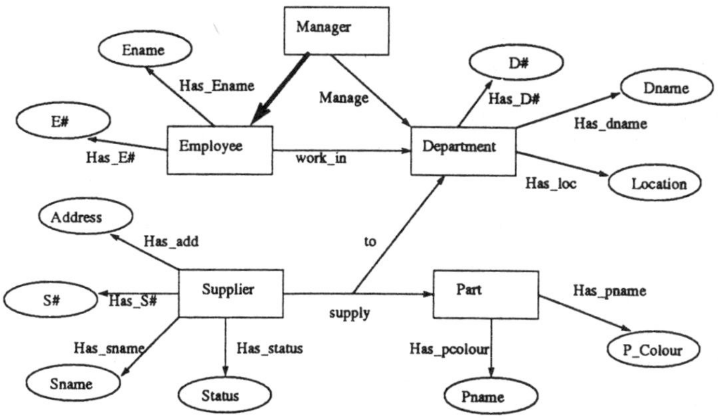

Figure 2-2. A Sample Database Schema Network

3. Overview of GQL

With the graphical images of the model, we are now able to display the structure of a database (schema). From now on, we start to define a data manipulation language operating on the schema. In this section, we present the basic pattern images, pattern graphs and the construction and interpretation of query graphs. However, we first introduce the concept of a GQL pointer.

In a textual language, users describe queries as a sequence of textual terms according to its "linear" grammar. The definition of such a language does not need to know what device is to be used for specifying the terms. However, in the definition of a graphical query language with two dimensional syntax, a two dimensional "locator" is desirable. A GQL pointer is such a locator. It is an abstraction of devices such as light pen or mouse. Associated with a set of access keys, the pointer is used for specifying data types, drawing pattern images and calling operations.

3.1. Basic Pattern Images

The GQL basic pattern images include those for representing an object, a relationship, a set of objects, a set of relationships, a "member_of" relationship, and those for query qualification and operations. Figure 3-1 lists some of them, where P and R respectively stand for an object type and a relationship type. The rest will be introduced in later sections.

The dot image denotes an object x such that x ∈ P. The arc image indicates a particular relationship R(x,y) if the arc links two dots, or a set of relationships R(x,y)s if the arc links to a subset (see section 3.2). The subset image indicates subset P' ⊆ P. The membership image shows that subset P' is composed of objects x (a dot), and object x is a typical member of subset P'. Window images are used when specifying operations over objects and relationships, and therefore associated with images for objects and relationships. Windows can be used for query qualification/quantification, and for general operations.

A pattern image can be created and erased by using the GQL pointer with "specify" and "abort" keys respectively. In addition, a window can be opened and closed by the pointer with keys "open" and "close".

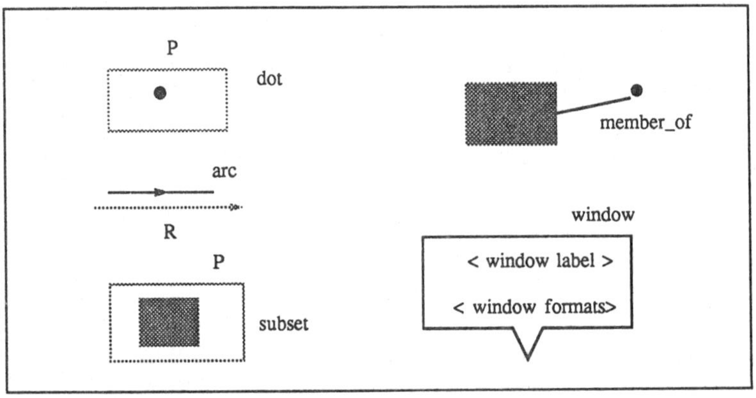

Figure 3-1. GQL Basic Pattern Images

3.2. Q/Q Windows and Pattern Graphs

3.2.1 Windows for Qualification and Quantification

The most frequently used window images are those for qualification/ quantification (Q/Q windows). Q/Q windows have different formats according to what images the windows are associated with. The window used for objects ("o") has a menu format. It takes the name of the object type as the window label and the attribute object type names as the entries of the menu. Because the attribute of an atomic object is the object itself, the window for it takes the type name both as the label and the only menu entry. Users textually input noun-phrases like values, the results of functions, etc. against menu entries for qualification/quantification purposes.

The window for relationships "-->--" has format "___rel_name___". A qualification operator over the relationship is applied to the left underlying position while a quantifier on the second argument of the relationship is applied to the right. The formats of Q/Q windows are illustrated in Figure 3-2 where "QL" and "QL_i" stand for qualification operators and "QN" for quantifiers.

The window format for a structured object makes it possible to specify a query at the structure level without going further down to its attributes. Through the window, the qualification on the structured object QL is distributed into the qualifications, $QL_1, \ldots, QL_n$, to its attribute objects and $QL \equiv QL_1 \wedge \ldots \wedge QL_n$.

Creating a Q/Q window followed by closing that window defines a new qualification/quantification. Opening a Q/Q window followed by closing it provides a way of viewing or modifying the current qualification/quantification. Opening or creating a Q/Q window followed by aborting that window discards the current qualification/quantification.

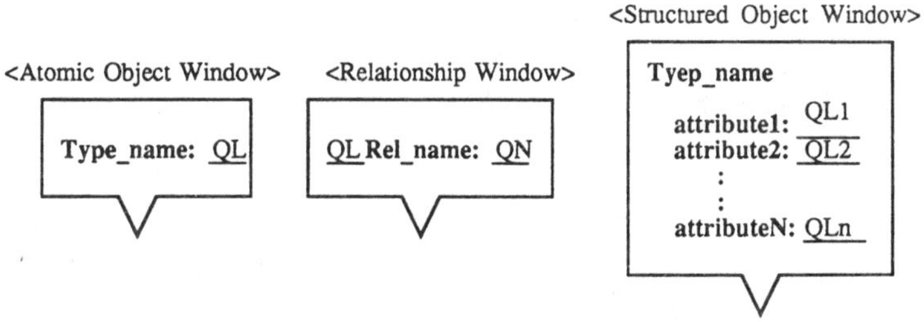

Figure 3-2. Formats of Q/Q Windows

3.2.2. Pattern Graphs

In this section, let us assume that we have :-
1) objects x and y, and object types X and Y where $x \in X$ and $y \in Y$;
2) relationship type R defined from X to Y;
3) qualification function Qual() applied on y through a window, specifying a subset $Y' \subseteq Y$;
4) the required subset X' of X.

Basic Pattern Graphs

Pattern 0 : The specification of objects.
Graph:

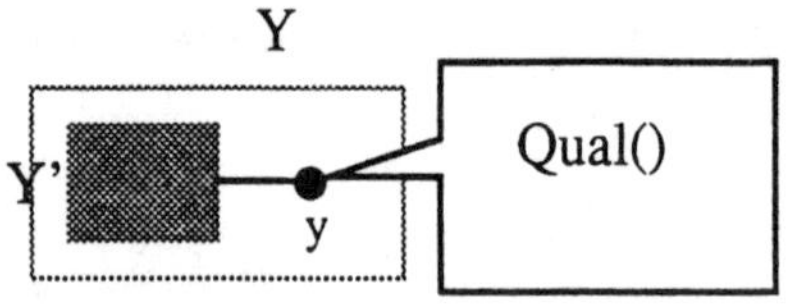

Semantics: $Y' \equiv \{y \mid (y \in Y) \wedge Qual(y)\}$

Pattern 1 : The existential reference to objects (∃)
Graph:

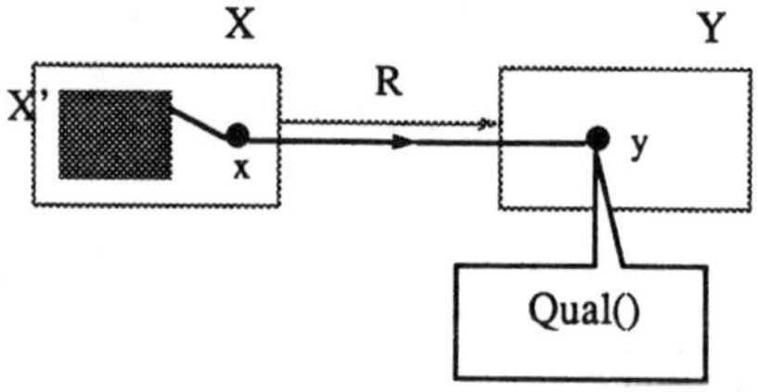

Semantics: $X' \equiv \{x \mid (x \in X) \wedge (\exists y)\ ((y \in Y) \wedge Qual(y) \wedge R(x,y))\}$

Pattern 2 : The universal reference to objects (∀)
Graph:

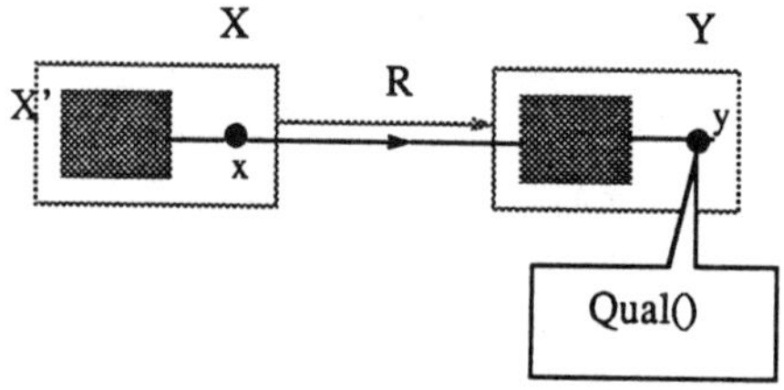

Semantics: $X' \equiv \{x \mid (x \in X) \wedge (\exists y)\ ((y \in Y) \wedge Qual(y)) \wedge (\forall y)\ ((y \in Y) \wedge Qual(y) \rightarrow R(x, y))\}$

Pattern 0 indicates that Y' consists of y's satisfying Qual(). Pattern 1 states that for every x required there exists a y in Y which satisfies Qual() and R(x,y) holds. Pattern 2 means that every x required links through R to all y's which satisfy Qual(). Note that pattern 2 takes the natural semantics from a natural language, assuming Y' is not an empty set. This assumption prevents x's from being included in the required subset X', if they do not link to any y's not satisfying Qual(), For example a query like "find suppliers who supply all the red parts to a department" implies that there must be at least one red part. If there is not any red part, we do not get any suppliers as answers.

Generalised Pattern Graphs

In order to generalise the pattern graphs and extend their expressive powers, two alternative generalised pattern graphs, patterns 3a and 3b, are defined.

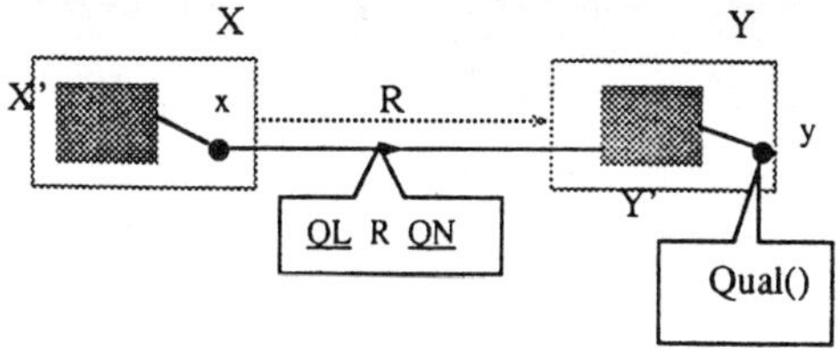

3 (a)

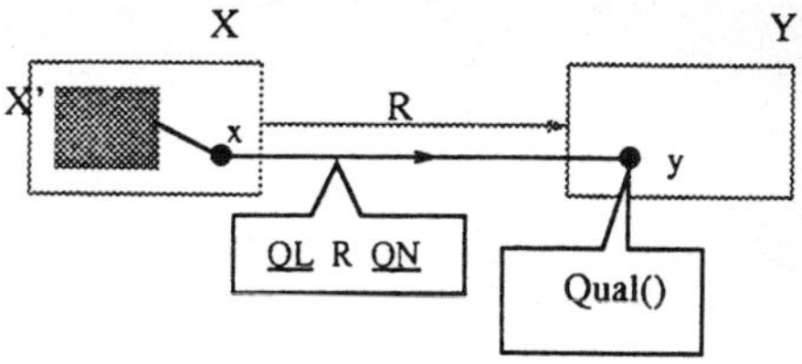

3 (b)

Pattern 3a states that object x refers, through R, to one or a collection of the objects in subset Y' which satisfy Qual(). How many such objects to be referred and how x refers to them are determined by the operators QL and QN in the window associated with the arc image for R. Pattern 3b only differs from 3a by extending the meaning of dot image y. y now represents not only a particular object, but also a collection of objects. To keep the consistency with the previous patterns, patterns 3a and 3b treat "for_all" and "exist some" as the default cases respectively.

Based on pattern 3, extended quantifiers "at_least(N)", "at_most(N)" and "exact(N)", and qualification operators "Not" and "Only" can be applied through the window on the arc R. The extended quantifiers have the following meanings.

at_least(N):

$$X' = \{x \mid (x \in X) \wedge Count(\{y \mid (y \in Y) \wedge Qual(y) \wedge R(x,y)\}) \geq N\}$$

at_most(N):

$$X' = \{x \mid (x \in X) \wedge Count(\{y \mid (y \in Y) \wedge Qual(y) \wedge R(x,y)\}) \leq N\}$$

exact(N):

$$X' = \{x \mid (x \in X) \wedge Count(\{y \mid (y \in Y) \wedge Qual(y) \wedge R(x,y)\}) = N\}$$

where function Count(S) returns N, the number of elements in set S. Operators "Not" and "Only" associates different quantifiers with different meanings.

"Not" with "for all":

$$X' = \{x \mid (x \in X) \wedge \neg(Y' \neq \{\} \wedge (\forall y)\ (((y \in Y) \wedge Qual(y)) \rightarrow R(x,y)))\}$$
$$= \{x \mid (x \in X) \wedge (Y' = \{\} \vee (\exists y)\ ((y \in Y) \wedge Qual(y) \wedge \neg R(x,y)))\}$$

"Not" with "exist some":

$$X' = \{x \mid (x \in X) \wedge \neg((\exists y)\ ((y \in Y) \wedge Qual(y) \wedge R(x,y)))\}$$
$$= \{x \mid (x \in X) \wedge (\forall y)\ (((y \in Y) \wedge Qual(y)) \rightarrow \neg R(x,y))\}$$
$$= \{x \mid (x \in X) \wedge ((Y' \neq \{\} \wedge (\forall y)\ (((y \in Y) \wedge Qual(y)) \rightarrow \neg R(x,y))) \vee Y' = \{\})\}$$

where $Y' = \{\} \equiv \neg(\exists y)\ ((y \in Y) \wedge Qual(y))$
and $Y' \neq \{\} \equiv (\exists y)\ ((y \in Y) \wedge Qual(y))$

"Only":

$$X' = \{x \mid (x \in X) \wedge (Qy)\ ((y \in Y) \wedge (Qual(y) \leftrightarrow R(x,y))\}$$

where Q stands for either "$\forall$" or "$\exists$".

The associations of "Not" and the extended quantifiers have similar meaning to "Not with exist-some", but with the negations to predicates P(Count({}), N) (P stands for a comparison operator) for "at_least", "at_most" and "exact". The associations of "Only" with the last three are not meaningful and are not allowed in GQL.

Pattern Graphs with Nested Arcs

Pattern graphs for nested arcs can be constructed by nesting pattern graphs for the component binary arcs, and have corresponding semantics. This can be proved by using pattern 3 and the semantics of nested binary relationships, and considering the applications of different qualification operators and quantifiers.

3.2.3. Pattern Graph Construction and Interpretation

The pattern graphs that we have discussed can be directly or recursively applied to construct a complex graph for the specification of a query. Using the GQL pointer and associated access keys, a user can either draw the whole pattern graph "part by part" in a sequence, or specify different parts separately and then link them together.

Find students who have name 'J.Smith' and take a course with grade 'A'.

sequential construction: **1, 2, 3, ...**

piecemeal construction: **I, II, III, ...**

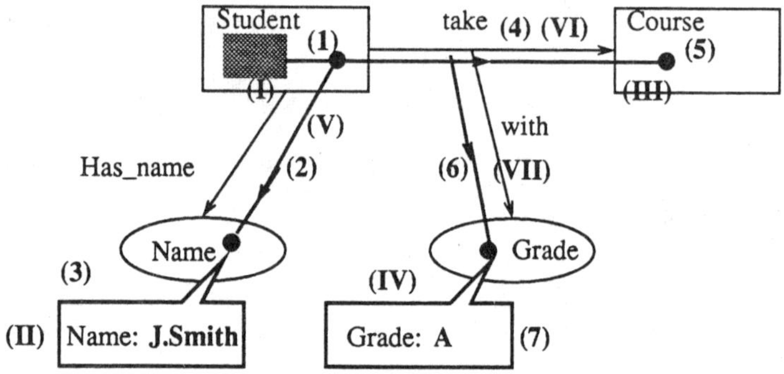

where numbers in () indicate the sequence in which the query is made.

These two specification criteria are respectively called "sequential construction" of the graph and "piecemeal construction" out of sub-graphs. The above diagram shows an example of a query and the two ways of describing it.

A pattern graph corresponds to an underlying FOPC expression. An object ("o") with specifications and without references from other objects refers to a free variable. A subset which take such objects as its members is called "required subset" that contains the answers of a query. An object with references corresponds to a bound variable. Relationships (specified by arc images) and qualification predicates (defined through windows) refer to predicates.

4. GQL Representations of Operations

Two types of operations are considered, query operations such as "retrieve" "insert", etc and aggregate functions such as "maximum", "sum", etc. As we have mentioned in section 2, operations are associated with particular types of objects and relationships. For instance, a string type object can not take part in arithmetic operations with other objects. Query operators like "retrieve" can be considered as operations associated with object type "model". GQL defines the operations as its built-in operators and functions. The operators and functions can be called by their textual names when users describe queries through windows. An operator "name_var(x)" which names a variable x for an object is provided to meet the requirement for the call of a function with parameters. GQL also provides graphical representations, operational arcs and operational windows, for the operators and functions. Users can call these by the GQL pointer device and access keys.

4.1. Operational Windows for Query Operations

Query operators include "retrieve" for selecting objects and relationships from the database, "insert", "delete" and "update" for modifying the database, and "define_new" and "define_derived" for defining object and relationship types. GQL defines operational window images with different formats to represent the query operators. The windows are then applied to images for objects and relationships. The window formats are illustrated in figure 4-1.

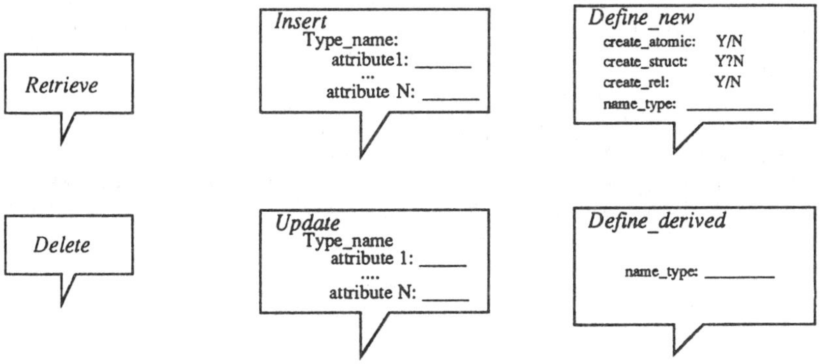

Figure 4-1. Windows for Query Operators

The windows for retrieval and deletion are the ones with the operators inside. The window for "retrieve" can be applied to a dot image and a subset image, respectively representing "retrieve a" and "retrieve all". If "retrieve" is applied to a structured object image, the answer will be complete structures which the required identifiers refer to. The windows for the insertion and the update have the same format as a Q/Q window except that they take the operators as the window labels. The new data values are the phrases that a user inputs against a window menu entry.

The window for "define_new" takes the operator as the window label. The window menu entries are three alternative commands for creating an image of a data type and a command for naming an image.

The window for "define_derived" only takes the naming command as the menu entry because the type image is determined automatically. "Define_derived" results in not only the generation of a data type, but also the generation of an IS_A link from the derived type to the original type.

To call query operators, users use the GQL pointer and the "create" key to create the windows in proper places (a place in the schema network space for "define_new" and object or relationship images for the rest), and then input textually the necessary phrases against the window entries (See the examples in section 5).

4.2. Operational Arcs

Figure 4-2 lists operational arcs for logical operators "∧" and "∨", comparison operators ">", "<", "≥", "≤", "=" and "≠", set operators "∪" (union), "∩" (intersection) and "-" (set difference), and arithmetic operators "+", "-", "x", and "÷". The arcs for set and arithmetic operators represent functions with result values, so they take corresponding images to represent the results.

Arcs for different types of operators are applied to different types of images: those for logical operators apply to windows or relationships which correspond to predicates; those for set operators apply to subset images, and those for arithmetic and comparison operators apply to objects of numerical (string) type.

Some operational arcs can be nested. Logical, set and arithmetic operations can be nested for representing complex logical, set and arithmetic connections. Figure 4-3 shows a few examples. GQL also allows a comparison arc to be added to an arithmetic arc specifying that the result of the arithmetic operation satisfies the comparison relation.

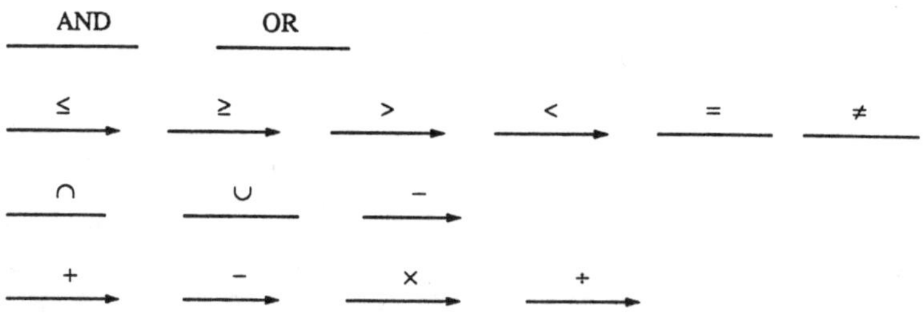

Figure 4-2. GQL Operational Arcs

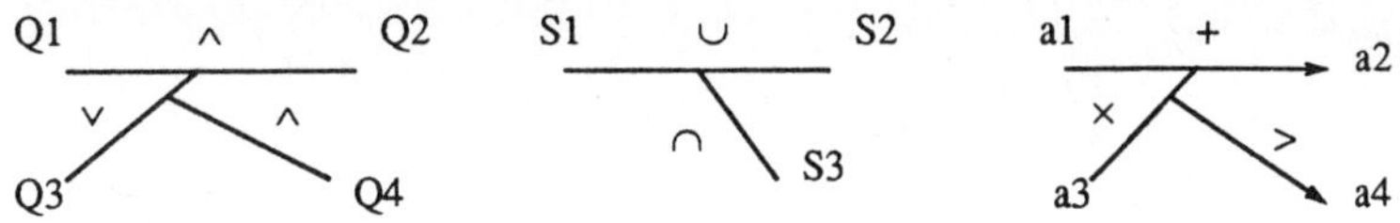

$((Q1 \wedge Q2) \vee Q3) \wedge Q4$ $(S1 \cup S2) \cap S3$ $(a1 + a2) \times a3 > a4$

Figure 4-3. Some Examples of Complex Calculations

For convenience, GQL also provides a default way of representing logical "and" and "or" in pattern graphs :-

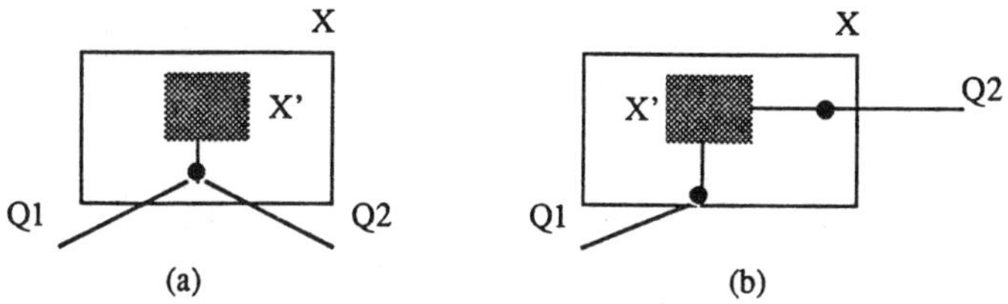

$X' = \{ x \mid (x \in X) \wedge Q1(x) \wedge Q2(x) \}$ $X' = \{ x \mid (x \in X) \wedge (Q1(x) \vee Q2(x)) \}$

In (a), the subset contains the objects each of which satisfies both Qual1 and Qual2 ("and"), while in (b) the subset contains the objects which satisfy Qual1, and the objects which satisfy Qual2. In other words, an object in the subset satisfies either Qual1 or Qual2 or both ("or").

4.3. Aggregate Functions and Grouped_by Operation

Aggregate functions "Count", "Average", "Total", "Minimum" and "Maximum" are applied to a collection of objects (mostly numerical-type), and return a single numerical value. In GQL, the aggregate functions are represented by a particular kind of operational arcs which link an image for a collection of objects to a dot image. The aggregate arcs can be applied directly to object type images because the object type can be considered as a set of objects. So are the operational arcs for set operations. Here is an example of using the aggregate function "Maximum".

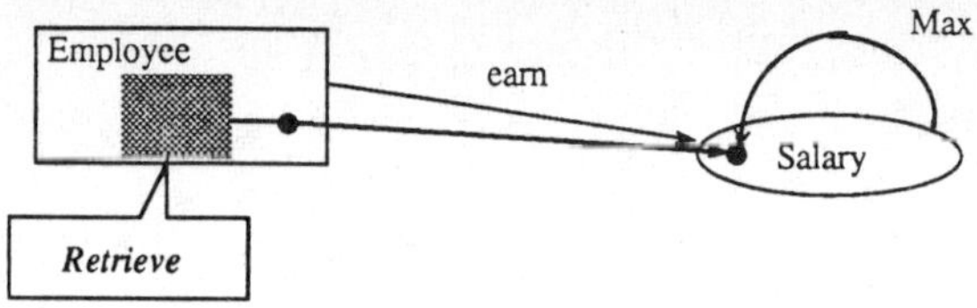

"Find employees who earn the highest salary".

In practice, users sometimes need to gather objects into different groups according to another type of objects. The aggregate functions are applied to each group. A typical example is "find employees who earn the highest salary in a department". The "Group_by" operation is available in SQL and most other query languages.

Find employees who earn the highest salary in each department.

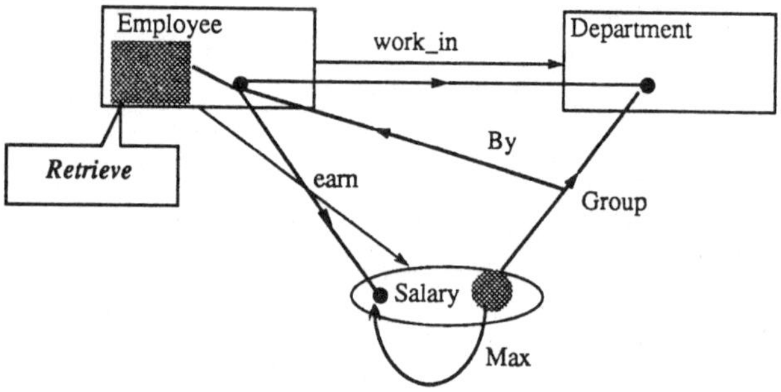

Figure 4-4. An Example of Grouping

GQL defines an operational arc for handling a fairly simple case of "grouped_by" operation. This arc can be directly applied from a new image for a collection of objects (group●) in an object type image to a dot in another object type image if there is a relationship type between the two object types. There is another operational arc called "via" available for being nested with a "grouped_by" arc. It represents the grouping via another type of objects. The meaning of "(A)grouped_by(B)_via(C)" is that objects in C are grouped according to objects in B and objects in A are grouped correspondingly according to the groups of objects in C. This nested arc is only applied in the case that grouping objects have no direct links to objects to be grouped by, but via another type of objects. The example of figure 4-4 illustrates this operation.

5. An Example Session

In this section, we present a couple of query examples described on the database schema network of Figure 2-2. We shall show the pattern graphs for the queries and how a user uses the GQL facilities to specify them. The first query shows how a retrieval operation is carried out.

Query1: "Find a supplier which supplies at least one red pen to all departments on the second floor".

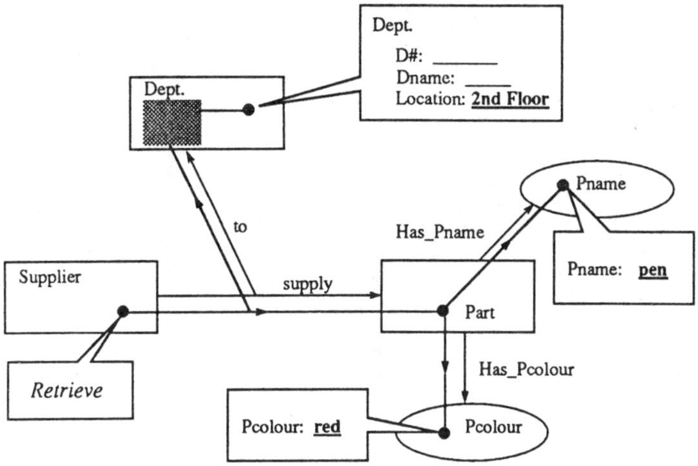

First, the user draws a dot against box "supplier", and from that dot, draws an arc along relationship type "supply" to box "part". The user then draws a dot connected to the arc, opens a window against the dot, and specifies the part to be red and have name "pen". Then he/she draws a nested arc along relationship type "to" to box "department", and creates a subset image with a "member_of" link to a dot. The subset image is connected to the nested arc. Then he/she opens a window against the dot on "department" and specifies the dot to be on 2nd floor against entry "location" of the window menu.

Now the user comes back to the original dot image on "supplier", opens an operational window with "retrieve" operator. This completes the whole query. Users may be required to close a window before moving to the next step in a particular implementation of the language.

The next example shows how the entity type "Specialised-London-Supplier" is defined as a subset of the "Supplier" entity type. This is carried out by using the operation "Define-derived".

Query2: "Define a new object type of specialised London suppliers which are either a 'limited company' or a 'Chain-shop'."

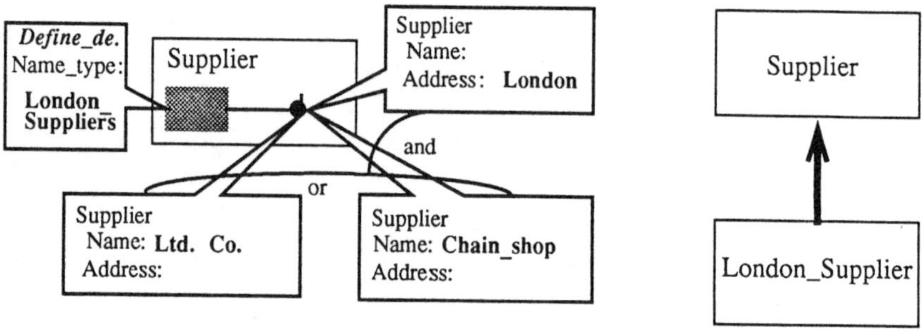

The user first creates a subset-with-a-dot image at "supplier", and then opens three windows against the dot. The windows specify "London" on entry "address", "limited company" and "Chain-shops" on "status" respectively. Then the user applies the nested arc "and" and "or" among the three windows. The user now open a definition window for derived object type against the subset image. He/She then names the derived type as "Specialised-London-supplier". Once the query is evaluated, the derived object is defined, and so is an IS_A link from the type to its supertype "supplier". The corresponding graphical images of the new type and the IS_A link will then appear in the schema network.

6. Discussions on GQL

6.1. GQL Implementation

GQL is supported by an interpreter. The interpreter checks the syntax grammar of GQL when users draw pattern graphs. It then translates a legal pattern graph into an underlying logical expression for a query. The interpreter starts to evaluate a query only when users call a built-in command "run_query". The answers of a query returned from the interpreter are either organized in a table form or are just a simple "yes/no" depending on the type of the query. If we require all the answers for a retrieval query by specifying the required subset image, the answers are put into a table which has the same format as the table for the type instances in section 2. If we just specify a dot for the required answer, the query is treated as an "is_there ...?" query, and therefore the answer is "yes/no". The answers for other types of queries, like "insert", "delete", etc. are just a "yes" if the manipulation is successful; "no" otherwise.

The implementation of a GQL interpreter requires support from advanced graphics facilities. The system should provide efficient graphics mechanisms for pattern graph drawings. The system should also have facilities to support operations on pattern images and data model images, like enlarging, shrinking, copying, highlighting, etc. Icons, windows and menus can support a good language interface environment for users. A supporting system with those facilities will definitely make the implementation less expensive and the language more comprehensive.

An implementation of a GQL interpreter is now under development using the window facility of a Sun-3 workstation at University of Essex. The interpreter consists of three major processing modules, interface, constructor, and evaluator. The interface is responsible for providing a GQL environment. The environment is a frame which has a canvas window for displaying the database schema and making queries, and a control panel for calling GQL operators and functions. The GQL operators and functions are listed in menus and associated with control buttons. Users draw a pattern image by moving the mouse device and clicking its buttons. The other responsibility of the interface module is to translate the pattern images of a pattern graph into logical terms. Because there is a one-to-one mapping from an image to a term, the interface then translates a set of images into a sequence of terms. This module is implemented in C using the Sun3/Suntool window package.

The constructor then puts the sequence of logical terms together to get an expression according to a GQL grammar. This module is implemented in Quintus PROLOG which allows bidirectional calling between a C function and a PROLOG procedure. We build the interpreter on top of PROLOG, and therefore the evaluator translates the logical expression into a set of PROLOG clauses and evaluates the clauses.

6.2. Relations to Other Languages

Although GQL is a graphical query language, its syntax and semantics are close to natural language. All pattern graphs have the meaning "specify something such that ...". If we take into account the schema images that the graphs are drawn against, the meaning of the pattern graphs is "specify something from someplace such that ..." which is the main structure of SQL. SQL and GQL are both defined on relational calculus, but GQL provides a more natural way of qualifying/quantifying objects.

GQL Q/Q windows and built-in command "name_var(x)" make it possible to describe a query on the menu forms of objects and relationships within Q/Q windows. After a minor modification to syntactical forms of comparison operators (to change operational arcs into operational windows), a user can then assemble QBE's way of making a query. This makes GQL a more acceptable query tool for users.

6.3. Ease-of-use Feature of GQL

Reisner [Reis 81] has summarised the factors which may affect the "ease-of-use" feature of a query language. These are "syntactic form", "procedurality" and "underlying data model".

GQL syntax is a two dimensional syntax. Pattern graphs are drawn against visible model images of the database schema. They can be considered as a meaningful track of the traversal of the schema.

GQL's pattern images and rules of constructing pattern graphs are simple and comprehensive.

GQL is a non-procedural descriptive language which is believed to be a suitable language for users in general, but GQL may have some drawbacks for complicated queries that require a procedural description. The underlying data model of GQL defines an objective and explicit view of the world. Objects are also organised into an IS_A hierarchy and can be organised into "aggregation_of" hierarchy as well ([Du 87]). These two hierarchies and the network view of the schema help users to understand the database structure and also provide a sensible basis for graphically describing queries.

7. Conclusion

We have described several aspects of GQL. The distinguishing features of the language are summarised as follows. First, it is a non-procedural descriptive formal query language based on a subset of first order predicate calculus. Secondly, GQL provides a set of basic pattern images in which pattern graphs are constructed. Users draw pattern graphs for queries by using the pattern images and following the construction rules of pattern graphs. Thirdly, GQL defines a graphical mechanism for operations (functions) in general and thus is capable of representing complicated operations.

GQL can be extended to handle database schemas and to represent transitive closures. We can introduce new built-in commands and graphical methods to perform operations handling schemas, like traversing, duplicating, merging, extracting and defining database schemas. Transitive closure is a high order operation, but is quite useful for query specifications. To extend GQL for this purpose, we need to find a way to represent a recursive operation in graphical form.

An important extension to GQL is the definition of new operations. By embedding a procedural or a descriptive textual language, GQL would then allow users to define a particular operation on a particular type of objects or relationships. One approach is to define a "operation-def" window in which users input a piece of program code in the embedded language. Then the users are required to specify the parameters of the procedure and its operational image form (an operational window or an arc). GQL will accept the procedure as its built-in operator with the user defined graphical format.

References

[Abit 84] Abiteboul, S. and Hull, R. "IFO: A Formal Semantic Database Model", Proc. ACM SIGACT-SIGMOD Symp. on Principles of Database Systems, April, 1984, pp119-132.

[Azmo 84] Azmoodeh, M. Lavington, S. H. and Standring, M."The Semantic Binary Relationship Data Model of Information", 3rd BCS and ACM Symposium on Research and Development in Information Retrieval, Cambridge, July, 1984.

[Azmo 85] Azmoodeh, M. "BRMQ: A Data Base Interface Facility Based on Graph Traversals and Extended Relationship and Groups of Entities" CSM-78, Sept. 1985

[Bryc 86] Bryce, D. and Hull, R. "SNAP: A Graphics-based Schema Manager (Extended Abstract)", IEEE International Conf. on Data Eng., Los Angeles, Feb. 1986.

[Chen 76] Chen, P. "The Entity-Relationship Model - Toward a unified view of Data", ACM Trans. Database Syst. Vol.1, No.1, Mar. 1976, pp9-36.

[Du 87] Du, H. and Azmoodeh, M. "A Graphical Query Facility for MFD Interface System" Internal Report, CSM-107, Oct. 1987.

[Jian 85] Jiang, Y. J. and Lavington, S. H. "The Qualified Binary Relationship Model of Information", BNCOD-4, July, 1985.

[Kenn 85] Kenneth, J. G. et al. "ISIS: Interface for a Semantic Information System", ACM, SIGMOD, Vol.14, No.4, Dec. 1985.

[Reis 81] Reisner, P. "Human Factors Studies of Database Query Languages: A Survey and Assessment", Computing Surveys, Vol.13, No.1, March 1981, pp13-31.

[Wong 82] Wong, H. and Kuo, I. "Graphical User Interface for Database Exploration", Proc. of 8th Inter. Conf. on VLDB, Mexico City, Sept. 1982.

A conceptual model for the representation of statistical data in geographical information systems

G.Gambosi [+], E.Nardelli [++], M.Talamo [*]

[+] Istituto di Analisi dei Sistemi ed Informatica-C.N.R., Roma, Italy

[++] Istituto di Studi sulla Ricerca e Documentazione Scientifica-C.N.R., Roma, Italy

[*] Dipartimento di Informatica ed Applicazioni, Università di Salerno, Salerno, Italy

Abstract: In this paper we are concerned with the problem of the representation of statistical-scientific data in the field of geographical information systems. Our approach allows the modeling of both the descriptive and the spatial-geometric aspects of information. This ensures: (1) an easier formulation of statistical queries cointaining references to geographical attributes of data, (2) a way for an easy linking between statistical data and their geographical reference. We discuss the problem of modeling geographical data from a conceptual point of view: in this framework we show that it is possible to consider only the topological aspect of geographical information. We show that this allows to deal in a natural way with queries involving metrics via a suitable reduction to topological queries.

1. Introduction

A quite general application of the techniques typical of Statistical Database Management is relative to the representation of statistical data with a geographical counterpart, such as, for example, the spatial distribution of variables related to natural and socio-economic phenomena [S].

Usually, such data have to be accessed at different levels of spatial aggregation, where the basis of such aggregation can be expressed in terms of some (usually administrative) predefinite hierarchy or by the user by means of a direct (geometric) specification. For example, given a variable 'Avg_income per county', it should be possible to derive a variable 'Avg_income per state' (where the hierarchy County-State is represented in the system) and to derive a variable 'Avg_income per geographical region', where a specific partition of the set of counties in the area of interest has been defined (for example graphically) by the user or is the result of a previous query.

The first type of aggregation can be resolved in present Statistical DBMS by means of a 'County_in_State' association only if such association is present in the scheme. For what concerns the second type of operation, note that it requires that the user explicitly specifies all counties which belong to a same partition region.

Using an efficient and powerful model of representation of geographical information and geographical relations between data, it is possible to specify geographical aggregations by means of suitable geographical queries involving spatial-geometric relations.

1.1 Description of the problem and usual solutions

In geographical information systems entities are often organized in an administrative hierarchy which usually corresponds to spatial containment relations. The derivation of statistical data is accomplished by means of queries which strictly follow the administrative hierarchy.

An alternative and more flexible way is to use queries that involve also the manipulation of spatial-geometric relations. This gives the possibility to proceed following the spatial containment hierarchy and, moreover, makes it possible to pursue associative searching paths which have not an administrative counterpart.

In such a framework a central issue is to use a representation of geometric and descriptive aspects of geographical data that allows to efficiently manipulate such spatial-geometric relations.

This involves each abstraction level of the representation.

At the conceptual level it needs a model for describing in a complete and synthetic way the dual nature of geographical data.

At the logical and physical level it needs solutions for an efficient management of geographical data, which in particular makes it possible to efficiently answer queries involving both the geometry and the semantics of data [AGNT], [GGNST].

Various approaches have been developed to the problem.

They can be summarized as follows:

(1) geometric information regarding geographical data is represented away from descriptive information (and privileging the geometric paths of access to data) and cross references are added to allow a linking between the two aspects [CBGM]. A recent advance in this direction is the object oriented model presented in [H], which merges the two aspects of geographical information by assuming that geographical data are instances of suitable classes, able to represent both spatial-geometric and descriptive relations.

(2) geometric information is represented, like the descriptive information, in terms of the relational model: this usually leads to a low efficiency from the point of view of treatment of geometric-spatial queries [CRM], [C], [CF]. Variations on this approach include the modeling of spatial data as continuous entity, which are represented as an infinite set of tuples. Such entities are then 'embedded' in a discrete DBMS which takes care of discretization issues [N].

The rest of this paper is organized as follows. In section 2 a conceptual metascheme for geographical data is presented and in section 3 an example of the application of such model at scheme level is given. In section 4 some conclusions and open problems are given.

2. A conceptual metascheme for geographical data

Geographical data may be classified at two levels of abstraction. At the lower level we have the abstraction that groups instances of geographical data into <u>classes</u> whose elements have common properties (e.g. the class 'Town' as abstraction of Rome, Milan, Naples, ...). At the next, higher, level we have the abstraction that groups classes with common properties into <u>metaclasses</u> (e.g. the metaclass Geographical Entity as abstraction of classes 'Town', 'County', 'River').

At the metascheme level, geographical data may be classified, in a rough way, by means of three metaclasses, which take into account the descriptive nature of the information, the spatial-geometric nature of the information, and the relationships between these two aspects.

In a more precise way we can identify three metaclasses:

Geographical Entity : this metaclass represents the abstraction of all possible ways of classifying geographical data according to their descriptive nature;

Geometric Reference : this metaclass represents the abstraction of all possible spatial-geometric references which can be associated to geographical data;

Geographical Theme : this metaclass represents the abstraction of all possible relationships that associate, for a given geographical datum, its descriptive and geometric aspects.

In figure 1 a representation of such entities and the associations between them is given using the graphic formalism of ER diagrams.

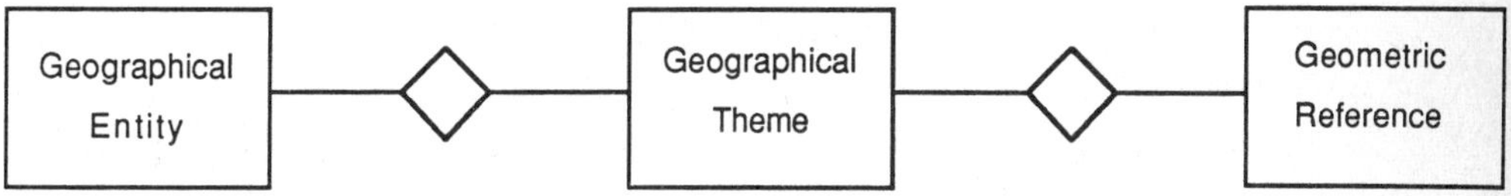

Figure 1: The initial diagram of the metascheme for geographical data

In the following paragraphs we shall examine with greater detail the metaclasses we have just defined.

2.1 Geographical Entity

This metaclass is a very general one and a deeper analysis leads to partition it in two parts. The key for such partition is the possibility of identifying instances of geographical data by their own name or not.

The abstraction of all classes which represent instances of geographical data identifiable by their own name will be called *Geographical Element* . An instance of this metaclass is, for example, the class 'Town', with values 'Rome', 'Milan', 'Florence', etc.

The second metaclass, instead, is the abstraction of all classes which represent instances of geographical data *not* identifiable by their own name. Such geographical data corresponds to values of a quantity distributed over the land. We call this metaclass *Geographical Property*. An instance of this metaclass is, for example, the class 'Land Height', with values '0-100', '101-200', '201-300', etc.

Notice that the partition between Geographical Element and Geographical Property is necessary if one wishes to represent a quantity distributed over the land, without necessarily considering the association between points of the land and values of the quantity, but taking into account only the set of values assumed by such quantity.

2.1.1 Geographical Element

As said above, this metaclass is the abstraction of all classes whose instances are objects identifiable by means of their name. Clearly, each class groups objects of homogeneous nature from a descriptive point of view. Some of these classes may refer to types of administrative corporations (e.g. 'County', 'State', 'Military Region'), some to types of physical entities (e.g. 'River', 'Lake', 'Mountain').

In each scheme of geographical databases, it is very likely that, between classes of this metaclass, there are associations of various kinds. An example is the association 'Lake-County' which relates each lake with the counties it touches. Moreover, a special type of associations, which is worth to be considered alone, is represented by aggregation associations: they relate a class belonging to a certain level in an administrative or physical hierarchy with subordinate classes and viceversa. Examples are the association 'County-State' which relates a state with all the counties it includes, and the association 'River-Hydrographic Bacin' which relates a hydrographic bacin with all its rivers.

As a consequence, in the metascheme it is necessary to introduce two more metaclasses, namely *Geographical Element Association* and *Geographical Element Aggregation* , to represent the abstraction of the two types of associations described above.

Moreover, as usual, each class is characterized by a certain set of attributes, one or more of which are keys for identifying the instances of the class. Therefore it needs to add a metaclass *Attribute* , which represents the abstraction of all possible attributes of classes, and a metaclass *Key Attribute* , in subset hierarchy with Attribute, which models those attributes which are keys.

2.1.2 Geographical Property

This metaclass is the abstraction of all classes whose instances are values (usually collected in some manner over an elementary predefined partition of the land) of a quantity. Obviously, each class groups values which refer to the same quantity. Some classes may refer to types of quantities which are distributed over the land in a continuous way (e.g. 'Land Height', 'Land Slope', 'Land Use'), others to types of quantities which are, instead, distributed in a discrete way (e.g. 'Population', 'Housing').

Some of the classes of this metaclass may be present in a scheme also as attributes of classes of Geographical Element, depending on how their instances have been collected. Consider, for example, the property which represents the average land height of certain regions. If the data about the land height have been collected on a state-by-state base, 'Land Height' will be represented as an attribute of the 'State' class (because in this case the height is considered as a propriety of states). Conversely, if the data about the land height have been obtained, directly or by means of some computation, from direct surveys, 'Land Height' will be a class by itself (because in this case the height is considered as a propriety of points of the land). Clearly, if the same concept is present in the scheme under the two structures, consistency issues may raise, which have to be controlled by the database administrator

For what regards associations between classes of this metaclass, they are not of an aggregative kind. Therefore, it is necessary to introduce in the metascheme only the metaclass *Geographical Property Association* , which represents the abstraction of all the associations between instances of the metaclass Geographic Property.

It is worth noting that the instances of the classes in Geographical Property have no attribute and are identified by means of themselves. Therefore, it is not necessary to introduce further metaclasses in the metascheme as it is already present the metaclass Key Attribute.

2.2 Geometric Reference

We defined this metaclass as the abstraction of all possible ways to give a spatial-geometric reference to geographical data. In principle this is an ambitious definition, but in practice, if we restrict ourselves, as usual in cartography, to the representation of information with a 2-dimensional geometric counterpart, the instances of this metaclass turn out to be represented by the usual concepts of Point, Curve, and Region. Therefore, the only classes in this metaclass will be 'Point', 'Curve' and 'Region'.

Notice that our aim is the modeling at the conceptual level only of the topological relations between geographical entities. The most important consequence of this approach is that one is not constrained to manage the metric aspects of geographic information at the conceptual level.

The central issue of our modeling of spatial-geometric reference is that, due to the non countability of the set R of real numbers, classes 'Point', 'Curve' and 'Region' are uncountable themselves. Thus, instances of such classes cannot have an associated identifier, defined as a finite sequence of characters from some finite alphabet.

We relax the usual condition that each entity must have an associated identifier, which makes it possible to uniquely identify such entity, by simply assuming that, given two *references* to entities of classes 'Point', 'Curve' and 'Region', it is possible to determine whether they refer to the same entity.

Thus, we are not interested in the direct management of entities of classes 'Point', 'Curve' and 'Region' but, instead, to the management of *references* to them introduced by the geometric descriptions of geographical information.

The key for identifying an instance of classes 'Point', 'Curve' and 'Region' is the instance itself, in the sense that we shall assume the existence of a predicate Identity(A,B), which, given two *references* A and B to instances of the same class, is true if and only if A and B refer to the same instance. Notice that all possible instances of classes 'Point', 'Curve' and 'Region' are implicitly present in each scheme. This assumption is well founded, in our case, as in a query is possible to specify in an arbitrary way the geometric description of an object. Therefore, like in classic data model the instantiation of an object is made by specifying its "name" (that is its key), in our model all we need is to specify the *reference* to the desired object.

The specification of conditions dependent on metrics in query formulation is expressed, at conceptual level, by means of the selection of suitable instances of classes 'Point', 'Curve' and 'Region' and by the reduction of such conditions to a set of topological conditions, represented inside the scheme by associations between classes 'Point', 'Curve' and 'Region'. In order to perform the selection of a geometric entity, it has to be possible to specify the reference to the desired entity either by means of a suitable formalism, as for example analytical functions, or by direct graphical specifications. At logical level, this can be realized by a function Define(F,R) which, given the description F (either by analytical function or by example) of a geometric entity, returns a reference R to it.

We introduce then a metaclass *Topological Association*, which represents the abstraction of all possible topological relations between classes 'Point', 'Curve' and 'Region'. It turns out that the only topological relations which are necessary to express queries dependent on metrics are the intersection relations: hence, it is assumed that each scheme includes the relations 'Point-Line-Int', 'Point-Region-Int', 'Line-Region-Int' and the reflexive relations 'Point-Point-Int', 'Line-Line-Int', 'Region-Region-Int'.

At logical level, such associations will not be implemented as relational tables, but by means of suitable functions, corresponding to the possible accesses to such associations. Such function can be easily implemented by means of computational geometry data structures and algorithms. This makes our approach interesting also from a practical point of view, since it makes it possible to base the physical implementation on efficient and standard techniques. We discuss extensively such issue in [GNT].

Since in the introduced model it is assumed that all instances of classes 'Points', 'Curves' and 'Regions' are represented, it is possible to easily reduce each query involving metrics in the selection of a suitable geometric entity followed by an equivalent query involving only topology.

Consider, for example, the following query: "*return all cities with a population greater than 50.000 at a distance not greater than 200 km. from Rome*". Such a query can be reduced to a first selection of the circular region with center in the instance of the 'Point' class associated to the instance 'Rome'of the 'City' class and with radius equal to 200 km.. This can be accomplished by means of the function Define previously introduced, which returns the reference C to the specified region. Let us denote as R(C) the region referenced by C: the query is then internally reduced to the following: "*return all cities with a population greater than 50.000 contained in region R(C)*". Note that such a query, together with the selection of R(C), can be represented directly as a path on the conceptual scheme.

2.3 Geographical Theme

The partition of the metaclass Geographical Entity induces on the metaclass Geographical Theme an analogous partition. *Geographical Element Theme* is the abstraction of all possible relationships which associate instances of geographical data, identifiable by name, to their spatial-geometric reference. *Geographical Property Theme* is the abstraction of all possible relationships which associate instances of geographical data, corresponding to values of a quantity, to their spatial-geometric reference.

Each relationship is made up of a set of tuples (R_i, V_i) such that:

- R_i is a reference to an instance of the Point or Curve or Region, which gives the spatial-geometric reference to V_i;
- V_i is an instance of the key attribute of a class belonging to Geographical Element or Geographical Property.

At this point it is possible to draw a complete diagram of the metascheme, using the graphic conventions of the Entity-Relationship model. Such a diagram is shown in figure 2.

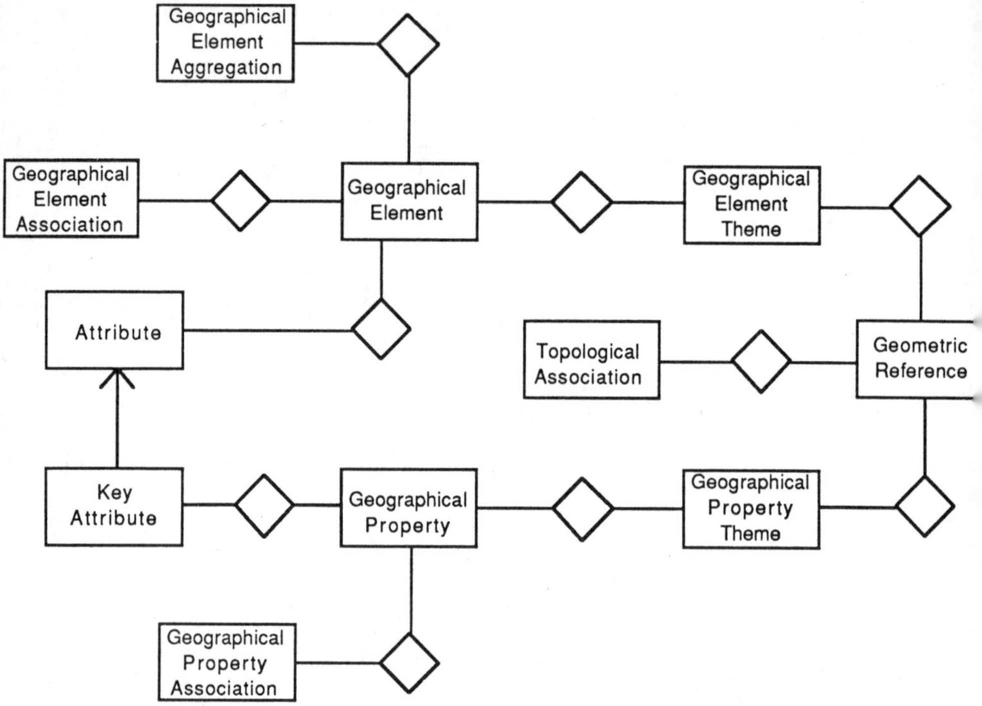

Figure 2: The complete diagram of the metascheme for geographical data

3. An example

In this paragraph we shall give an example of a scheme corresponding to the metascheme previously defined. In such a scheme classes 'River', 'Town', 'County' and 'State' represent different instances of metaclass Geographical Element, while classes 'Population', 'Land_Height' and 'Land_Use' are instances of the Geographical Property metaclass (see figure 3).

It is assumed that the geographical properties have been collected with reference to a predefined partition of the land in 1 mile x 1 mile squares. Class 'Population' represents the distribution of population on the territory in terms of number of inhabitants on each square, class 'Land_Height' represents land altitude, averaged on each square, while class 'Land_Use' represents the predominant use of land in each square.

For each class, a set of attributes is specified in the scheme in figure 3. Moreover, two instances of metaclass Geographical Element Aggregation ('Town-County' and 'County-State') are represented, together with a single instance 'Governor Residence' of metaclass Geographical Element Association.

Let us now consider the following query:"Return the percentage of counties in California in which grain is cultivated". Such query can be represented on the given scheme by the path of figure 4. Informally, the query can be answered by:

1. collecting and counting all pairs 'County in California' - 'Value "Grain" of Land Use' whose geometric references intersect;
2. collecting and counting all counties in California state;
3. returning the ratio between the values returned in 1 and 2.

4. Conclusion and open problems

In this paper, an original conceptual model has been presented for the integrated representation of geometric and descriptive aspects of geographical information. The introduction of such a model required the relaxation of some of the basic definitions of the relational data model, since the concept of key had to be modified to allow non countable sets as key values domains.

The main open problem of such approach is the definition of a powerful selection method to be used in the specification of values on such uncountable sets. It seems that, since such sets are represented by geometric entities such as points, curves and regions in a plane, the specification of an instance from such sets can be performed by a function, "by example", i.e. by pointing or drawing such an entity on a display or by a combination of such techniques. The real power of such approaches has anyway to be investigated, together with alternative selection methods.

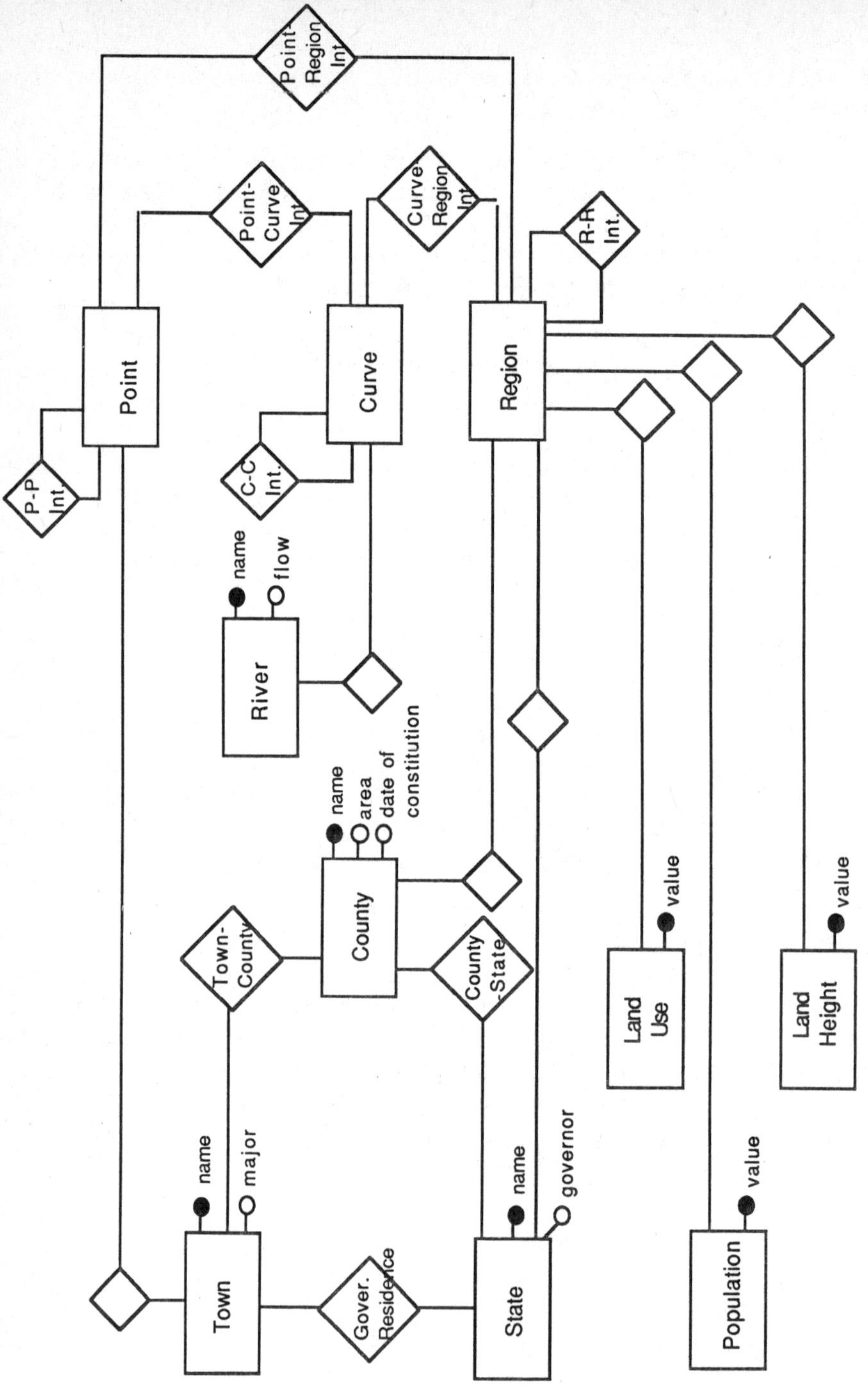

Figure 3: An example of application of the metascheme for geographical data

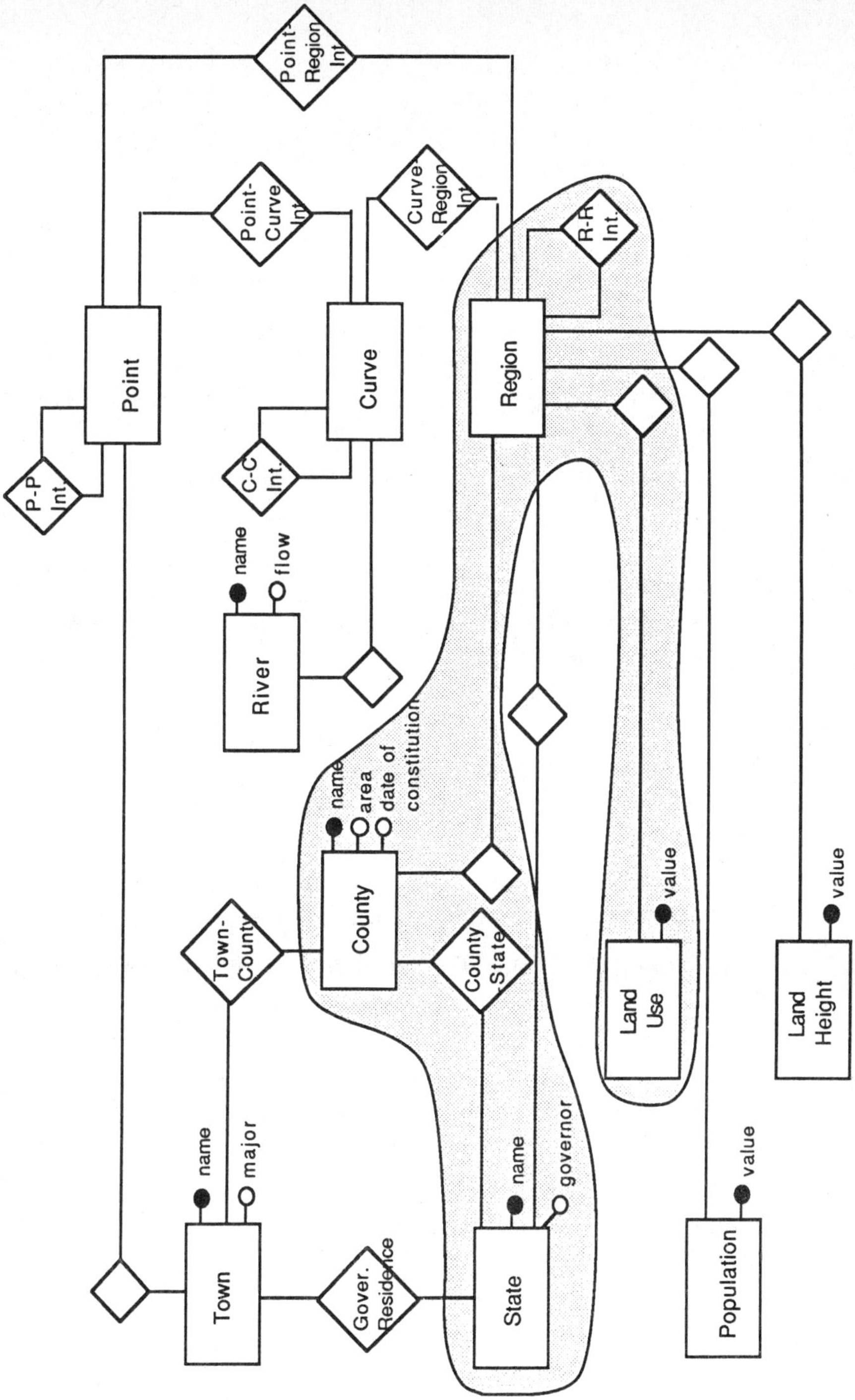

Figure 4: The path for answering the query presented in the text

References

[AGNT] G.Ausiello, G.Gambosi, E.Nardelli, M.Talamo: "GEODEQ: Uno strumento interattivo per la gestione e l'interrogazione di dati ed immagini cartografiche integrati", Proc. AICA National Conference, Palermo, 1986.

[C] N.S.Chang: "An integrated image analysis and image database management system", Ph.D. Thesis, Dept. of Electrical Engineering, Purdue University, 1980.

[CBGM] E.D.Carlson, J.L.Bennet, G.M.Giddings, P.E.Mantey: "The design and evaluation of an interactive geo-data analysis and display system", Proc. IFIP Congress 1974, North Holland Publishing Company, Amsterdam, 1974.

[CF] N.S.Chang, K.S.Fu: "A relational database system for images", TR-EE 79-28, Purdue University, May 1979.

[CRM] S.K.Chang, J.Reuss, M.H.McCormick: "An integrated relational database system for images", Proc. 1977 IEEE Workshop on Picture Data Description and Management, Chicago, Illinois, 1977.

[GGNST] G.Gambosi, C.Gaibisso, E.Nardelli, G.Soccodato, M.Talamo: "A proposal for the efficient representation and management of geometric entities in a geographic information system", Tech. Rep. 177, IASI, Roma, Dec. 1986.

[GNT] G.Gambosi, E.Nardelli, M.Talamo, "A model for the efficient manipulation of geographical data", in progress

[H] J.R.Herring: "TIGRIS: Object system implementation of a spatial geographic information system", INTERGRAPH Corp. report, 1986.

[N] G.Nagy: "Toward a high-level integrated image database system", Proc. AICA Workshop on 'Immagini e Territorio', Milano, 1987.

[S] A.Shoshani, "Statistical databases: characteristics, problems, and some solutions", Proc. of the 8th VLDB, Mexico, Sept. 82.

SPATIAL DATA BASE QUERIES:

RELATIONAL ALGEBRA VERSUS COMPUTATIONAL GEOMETRY

Robert LAURINI, Françoise MILLERET
Laboratoire Informatique Appliquée
Institut National des Sciences Appliquées de Lyon
F-69621 Villeurbanne Cedex

ABSTRACT

Conventional queries against relational databases can be expressed in relational algebra. But, when dealing with geometric and spatial queries, one also needs to use computational geometry algorithms.

Starting from examples taken in urban planning and in CADCAM, using different types of geometric modelling, we show what kinds of queries can be solved by relational algebra and computational geometry respectively. Among spatial queries, we essentially focus on:

- point-in-polygon queries,
- region queries,
- vacant places within a window.

Among spatial models, we present the conventional segment-oriented (wireframe) model requiring computational geometry algorithms for query evaluation, the Peano relation model allowing the using of algebra and a mixed model requiring both geometry and algebra. So, we show that the representation based on linear quadtrees together with Peano relations allows the solving of a spatial query subclass simply by using an extension of relational algebra called Peano Tuple Algebra.

We conclude this paper by pointing out the necessity to design spatial DBMS's in conjunction with the geometric representation and to extend query languages to deal with spatial queries.

KEY WORDS: Data Base Query, Relational Algebra, Computational Geometry, Geometric Modelling, CADCAM, Urban Planning, Peano Tuple Algebra.

I - INTRODUCTION

Spatial database management systems address to applications dealing with geometric and topological data such as geography, geology, CADCAM and robotics. One important issue in their design is how to handle queries against geometric information. Some off-the-shelf systems propose two kinds of data:

- attribute data to which relational algebra is applied,
- geometric (or graphic) data for which computational geometry is used to solve spatial queries.

By spatial queries, we mean questions regarding space occupancy and topological information such as:

- point-in-polygon queries,
- region queries,
- vacant places within a window,
- etc.

An example taken from the field of urban planning will be used for illustration.

The aim of this paper is be to define the limits between relational algebra and computational geometry when processing spatial queries in conjunction with different geometric models.

Let us take a preliminary example. Suppose we are interested in forestry. We have predefined zones represented by their boundaries (segment and points). Moreover, for each zones we have the number of trees. In the relational model, the following relationships must hold (see **Figure 1**):

R_1 (#zone, #segment)

R_2 (#segment, #point1, #point2)

R_3 (#point, x, y)

R_4 (#zone, tree-number).

Should we ask a query to retrieve the number of trees in a zone or a lot of zones, no problem will arise and the answer is obtained by relational algebra.

But, if we are interested in the number of trees in a region, defined by its boundary, and not corresponding to the previous zones, we need to use computational geometry and perhaps statistics.

Computational geometry will be necessary to determine zones-and-region intersections and statistics for some tree distribution (for instance uniform distribution).

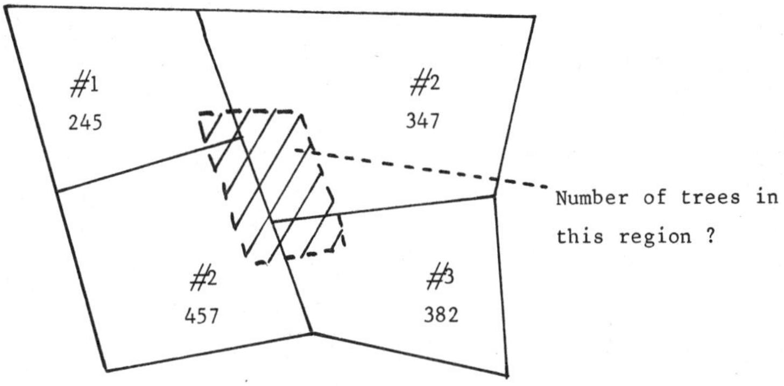

Figure 1: Example of query not solvable by relational algebra: What is the number of trees in the hatched region?

First, we will present a typology of spatial queries, secondly we give an town planning example to which we will apply different geometric models and give the algorithms to answer these queries. Last, by using a CADCAM example, we will give more details about the confrontation of relational algebra and computational geometry.

II - TYPOLOGY OF SPATIAL QUERIES

By spatial queries, we mean queries about space occupancy. Let us present some applications:

a) in geology, knowing a point (x_0, y_0, z_0), retrieve the layer including it;
b) when designing an engine in mechanical CADCAM, find all vacant spaces in order to fix a new part;
c) in urban planning, when delimiting a special area, find all landowners affected by such a decision;

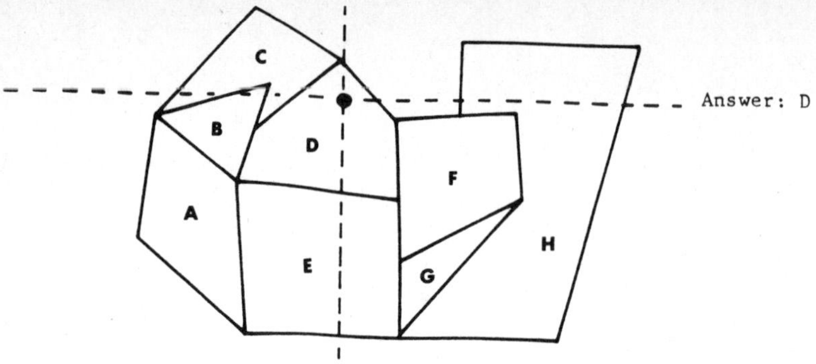

Figure 2: Point-in-polygon query within a tesselation

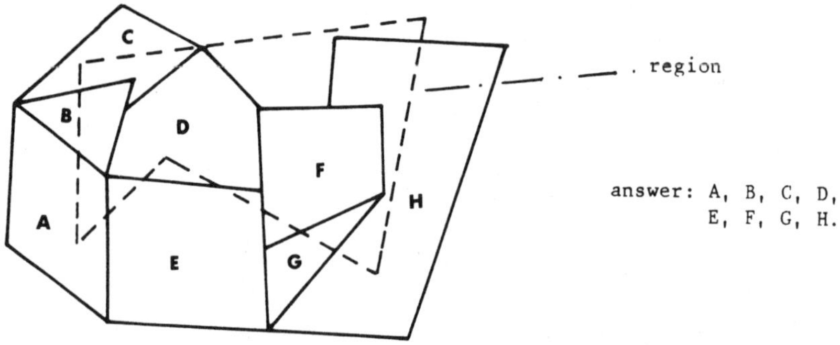

Figure 3: Region query within a tesselation

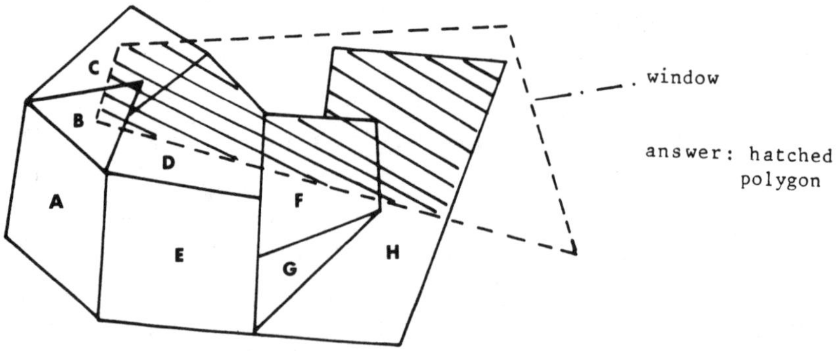

Figure 4: List of vacant places within a window

d) etc.

These queries or similar ones in other spatial applications can be represented by three major query types:

- **point-in-polygon query** i.e. considering a tesselation (defined as a polygonal space splitting) and an x_o, y_o point, retrieve what polygon it belongs to (see **Figure 2**);

- **region query** i.e. considering a tesselation and an area, retrieve all polygons including or intersecting it (possibly this area can be non-connected with separate portions and holes) (see **Figure 3**);

- **vacant spaces** within a window i.e. delimiting an empty polygon (possibly non connected) (see **Figure 4**).

III - QUERYING STRATEGIES ON AN EXAMPLE IN URBAN PLANNING

In this section, we will presentes associated to different geometric models.

But, before presenting the example, let us point out that spatial objects have the characteristics of incorporating an infinity of inner points. See **Figure 5**. In database theory, it is impossible to store such an infinity of items if we want to keep the extensional view of spatial objects. A possibility is to propose intensional models in connection with a mechanism for deriving their extensional expression.

Figure 5: The correct association, impossible to implement

3.1 Example description

To illustrate this problem, we will take an example from town planning. Let us suppose we have a cadaster and we want to retrieve all landowners affected by a new highway construction. See **Figure 6**.

To describe this situation, we need:

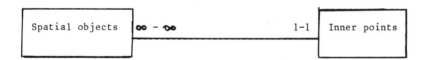

- a relation to describe ownership R_0(#landowner, #plot);
- one or several relations to describe the plots of land geometry (these relations will vary according to different kinds of geometric modelling)
- one or several relations to describe the future road geometry.

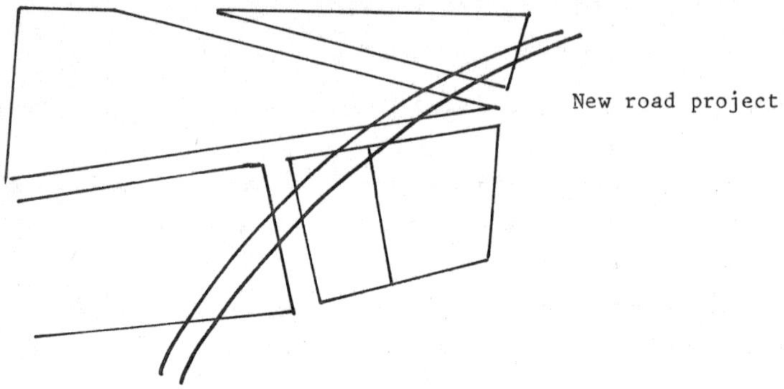

Figure 6: Example of a cadaster and a new road project

In this simple example, we will assume that the plot and road geometric representations are the same. Successively, we will examine:

- **a segment-oriented model,**
- **a surface-oriented model based on Peano relations,**
- **and a model mixing segment and surface oriented representation.**

As already mentonned, three kinds of query will be used:

- **point-in-polygon** in order to retrieve the landowner of a special point,
- **region query** to retrieve all the landowners affected by an area,
- **vacant places** in order to retrieve an empty polygon.

3.2 Segment-oriented model (or wireframe model)

3.2.1 Presentation of this model

In this geometric model, each plot is described by its boundary or more exactly by the list of bordering segments. The Entity-Relationship (E.R.) (CHEN, 1976) model is given in **Figure 7** .

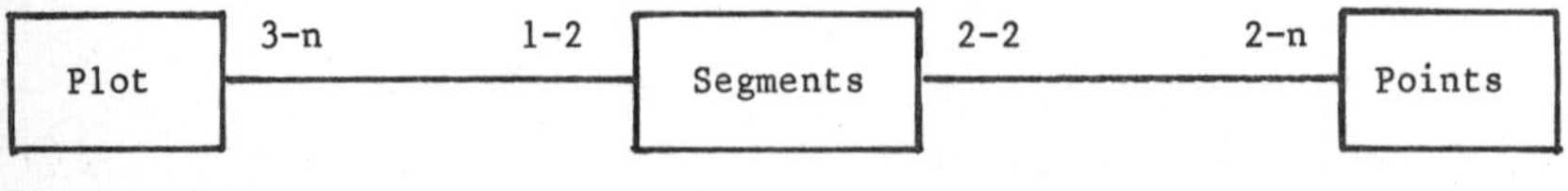

Figure 7: Conceptual model of segment-oriented representation.

Thus, in the relational model, we get:

R_0(#landowner, #plot)
R_{11}(#plot, #segment)
R_{12}(#segment, #point1, #point2)
R_{13}(#point, x, y).

3.2.2 Spatial queries

3.2.2.1 **Point-in polygon query.**

Considering an x_0, y_0 point, retrieve the landowner it belongs to (the highway is seen to be reduced simply to one point). Let us build a relation

Q_{11}(x, y)

with only one tuple $Q_{11}(x_0, y_0)$. To solve this problem by relational algebra, the only thing we can do is a join between R_{13} and Q_{11}. But, since a priori the x_0, y_0 point is not stored in the data base, we have no solution to this query.

However, by computational geometry, an answer can be found by considering the well-known half-line algorithm (PREPARATA-SHAMOS, 1986). Let us mention that starting from this point, we consider a half line L and we count the intersections between L and the polygon segments. It is established that when this intersection number is odd, the point x_0, y_0 is inside this polygon.

So, the algorithm is:

i) half-line algorithm on the R_{11} relation to obtain A_{111}(#plot);
ii) join between A_{111} and R_{12} to obtain A_{112}(#plot,#landowner);
iii) projection of A_{112} to get the answer A_{113}(#landowner).

3.2.2.2 **region query**

In this case, the query is composed of several relationships since the region is modeled by its segments (this the actual example problem):

Q_{121}(#segment, #point1, #point2)
Q_{122}(#point, x, y).

Similarly, starting with relational algebra, we get nothing interesting. By computational geometry, we can determine a polygon (possibly non connected) representing the geometric intersection with the tesselation. As a result we get:

i) A_{121}(#plot) obtained by computational geometry (intersection)
ii) A_{122}(#plot, #landowner) obtained by a join between A_{121} and R_0
iii) A_{123}(#landowner) obtained by projecting A_{122}.

3.2.2.3 **Vacant places**

In this case the query is composed by a relationship determining the window:

Q_{131}(#segment, #point1, #point2)
Q_{132}(#point, x, y).

The vacant places are given by a geometric difference (See PREPARATA and SHAMOS, 1986). The answering algorithm simply incorporates a difference. So we get:

A_{131}(#segment, #point1, #point2) and A_{132}(#point, x, y) are obtained by computational geometry (difference). Let us mention that A_{132} incorporates points of Q_{132}, R_{13} and new points derived from computational geometry such as intersections of segments of Q_{131} and R_{12}.

3.2.3 Conclusion

We have shown that with a segment-oriented model which described only the object boundary, spatial queries cannot be answered by using simply relational algebra, since we need to utilize values which are not stored in the database. To retrieve this information, computational geometry is needed.

In relational algebra, the result is always included in the database and we never produce new valued tuples. The only possibility is to store the surface itself. Let us examine a **surface-oriented model** based on Peano relations.

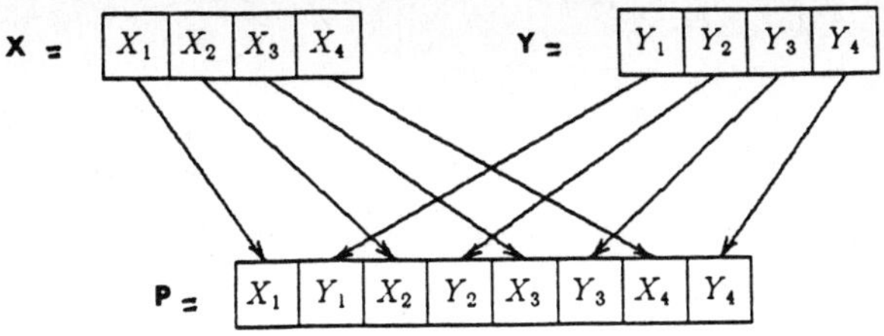

Figure 8a: Obtaining Peano keys by bit interleaving

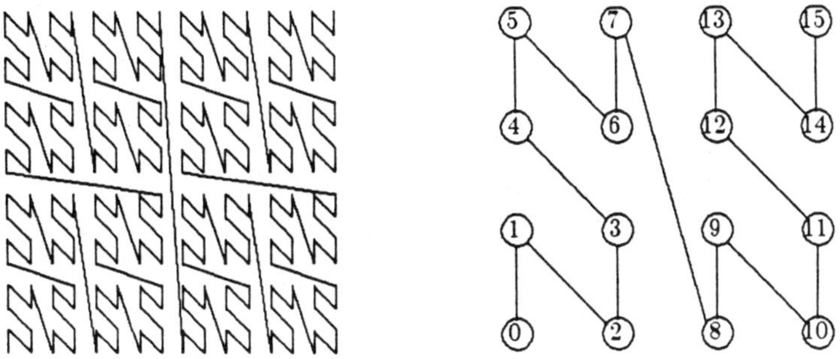

Figure 8b: Excerpts of the Space-filling Peano N-curve

Peano Key	size	color
0	2	black
4	1	black
5	1	white
6	1	white
7	1	white
8	1	white
9	1	white
10	1	black
11	1	white
12	2	white

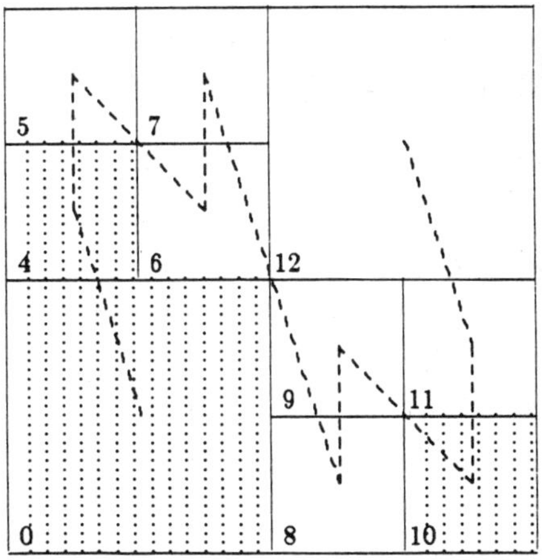

Figure 9: Examples of a quadtree and its description

3.3 Peano relation model

3.3.1 Model presentation

In several papers (LAURINI, 85, 87 and LAURINI-MILLERET, 86a, 86b, 87), we have defined a spatial database model whose characteristics are:

- **surface orientation avoiding the infinite number of tuples**
- **based on Peano space-filling curves**
- **based on quadtrees and octrees**
- **tuple algebra to solve spatial queries.**

In this model, a squared 2D space is described by a recursive splitting into homogenous quadrants and all these quadrants are sorted by their Peano key. In the 3D space, we deal with octants and cubes. Peano keys p derived from fractal space-filling curves and the more practical way to obtain them is by the bit interleaving of the x anx y coordinates. See **Figures 8a 8b and 9.** It has also been established than some spatial queries can be solved by **Peano Tuple Algebra** (LAURINI, 87). About quadtrees, refer to SAMET (1984). Peano key based quadtrees are also named linear quadtrees (GARGANTINI, 1983) and Peano keys are also called Morton sequence (MORTON 1966) and z-value by ORENSTEIN (1986).

Concerning the town planning example, the conceptual model is (see **Figure 10**):

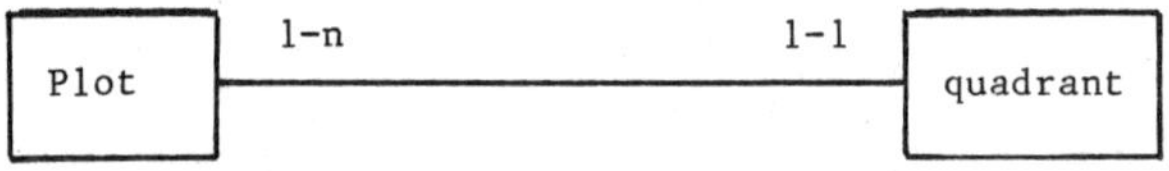

Figure 10: E.R. model of a surface-oriented model based on quadtrees

In the relational model, we have (Peano relations):

R_{21}(#plot, Peanokey, size)

In the quadtree model, a quadrant can be disaggregated into four other squares (respectively eight for an octree). Thus , a R_{21} tuple can be disaggregated into four squares (respectively eight cubes). This is one of the Peano relation peculiarities i.e. one tuple can be equivalent to four other tuples. In order to handle consistent and compact objects, several conformance levels are necessary (See LAURINI 87 and LAURINI-MILLERET 87).

Let us mention that **Peano Tuple Algebra (PTA)** is a relational algebra in which some operations need tuple aggregation/disaggregation provided that certain constraints hold.

In **Figure 11**, an example is given showing union and intersection of two quadtrees (**Figure 11a and 11b**). In this example, we can see how the Peano tuple algebra differs from the relational algebra. To evaluate the union (**Figure 11c**), since (3,1) is overlapped by (0,2), the tuple (3,1) must be deleted. Moreover R(8,1), R(9,1), R(10,1) and R(11,1) must be aggregated to give R(8,2). For the intersection (**Figure 11d**), R(0,2) must be disaggregated and only R(3,1) must be kept.

3.3.2 **Point-in-polygon query**

Considering an x_0, y_0 point, we determine its corresponding Peano key p_o by bit interleaving and we form a Peano relation:

Q_{21}(Peanokey, size)

with only one tuple $Q_{21}(p_0,$ e). This size is arbitrarily taken equal to e corresponding to the minimum resolution square (point), since in fractal geometry a point can be defined by a square (or a cube) as small as possible.

By a Peano tuple join between Q_{21}(Peanokey, size) and R_2(#plot, Peanokey, size), we obtain A_{211}(#plot, Peanokey, size) with only one tuple. Let us mention that a Peano join (see LAURINI, 87) is a usual equi-join in which some disaggregation possibly needs to be performed. Then a natural join with R_0 and a projection onto #landowner, produces the A_{213} result. Let us resume the steps of this algorithm:

i) By a Peano join between R_2, and Q_{21}, we get A_{211}(#plot, Peanokey, size)

ii) By a natural join between A_{211} and R_0, we get
A_{212} (#plot, Peanokey, size, #landowner)

iii) By a projection of A_{212}, we get the result A_{213}(#landowner).

3.3.3 **region query**

In this case the algorithm is exactly the same except that the query

Peano Keys	size	color
0	1	white
1	1	white
2	1	white
3	1	black
4	2	white
8	1	black
9	1	black
10	1	white
11	1	black
12	2	black

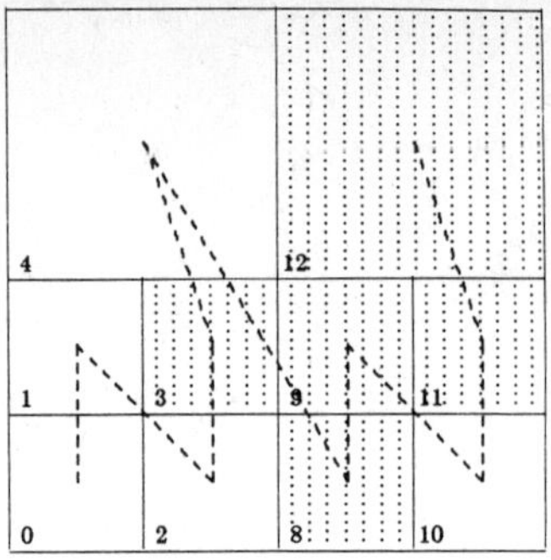

Figure 11a: First object described by a quadtree

Peano Keys	size	color
0	2	black
4	1	black
5	1	white
6	1	white
7	1	white
8	1	white
9	1	white
10	1	black
11	1	white
12	2	white

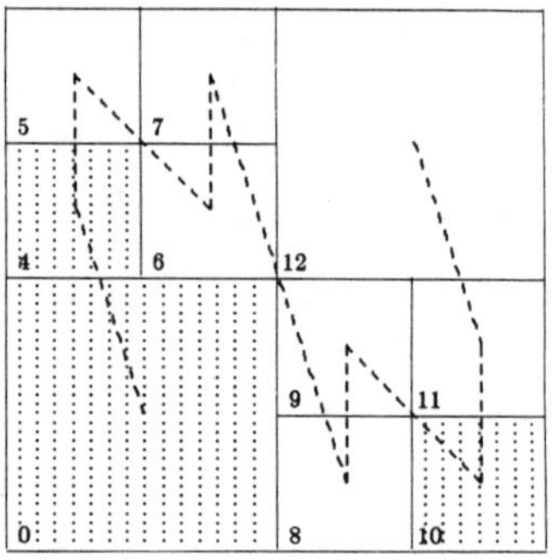

Figure 11b: Second object described by a quadtree

Peano Keys	size	color
0	2	black
4	1	black
5	1	white
6	1	white
7	1	white
8	2	black
12	2	black

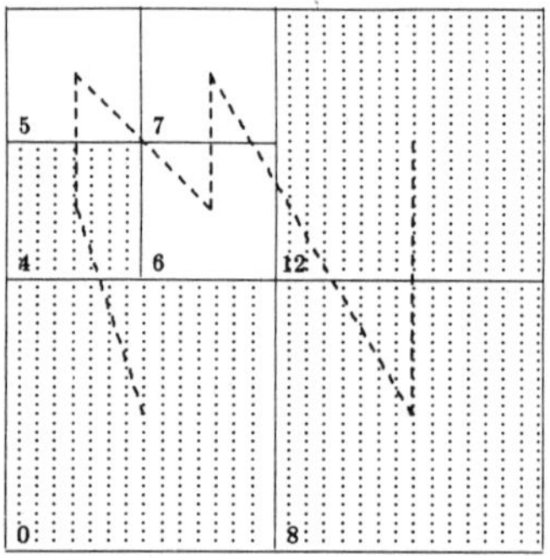

Figure 11c: Union of quadtrees #1 and #2

Peano Keys	size	color
0	1	white
1	1	white
2	1	white
3	1	black
4	2	white
8	2	white
12	2	white

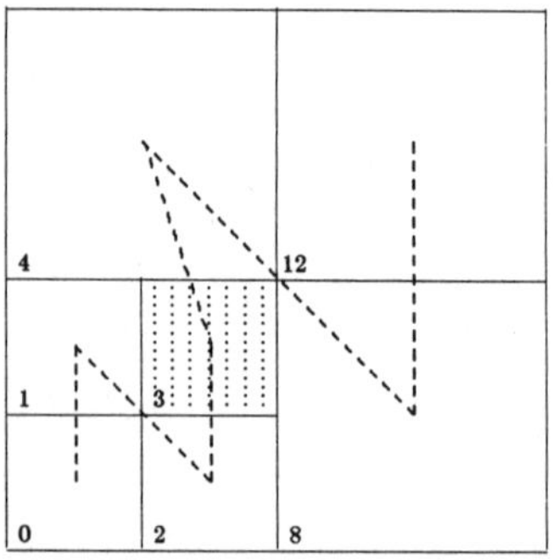

Figure 11d: Intersection of quadtrees #1 and #2

Q_{22}(Peanokey, size)

contains several tuples, leading to several tuples in A_{221}, A_{222} and A_{223}.

i) By a Peano join between R_2, and Q_{22}, we get A_{221}(#plot, Peanokey, size)

ii) By a natural join between A_{221} and R_{21} we get A_{222}(#plot, Peanokey, size, #landowner)

iii) By a projection of A_{222}, we get the result A_{223}(#landowner).

3.3.4 **Vacant places**

Here, the query:

Q_{23}(Peanokey, size)

contains several tuples and we have to perform a Peano difference (See LAURINI, 1987). The answer represents the internal area of the vacant places.

i) Projection of R_{21}(#plot, Peanokey, size) to get A_{231}(Peanokey, size)

ii) Apply a Peano difference A_{231} minus Q_{23} and we obtain the result A_{232}(Peanokey, size).

3.3.5 **Conclusions**

By using Peano relations, spatial queries can be solved by a tuple algebra whose only modification is the existence of aggregation/disaggregation procedures. It is necessary to mention that we do not deal with an exact representation since some parcel boundaries are approximated by "staircases". With an infinite number of tuples, we have an exact model and queries like region queries can be performed by conventional relational algebra. But how to perform a join with an infinite number of tuples? Of course, Peano relations are an approximate model of spatial object, but this approximation can be associated with a given resolution level. Now, let us examine an **exact surface-oriented representation** based on Peano relations for internal parts and segments for the boundaries (mixed representation).

3.4 MIXED MODEL

3.4.1 Presentation

A stimulating way to deal with an exact **surface-oriented** representation is to mix a Peano relation model and a **segment-oriented** one.

In this case, some quadrants can be intersected by segments. See AYALA et al, 1985 for more details (called extended quadtrees) and **Figure 12** for an example. The corresponding conceptual model is given in **Figure 13**.

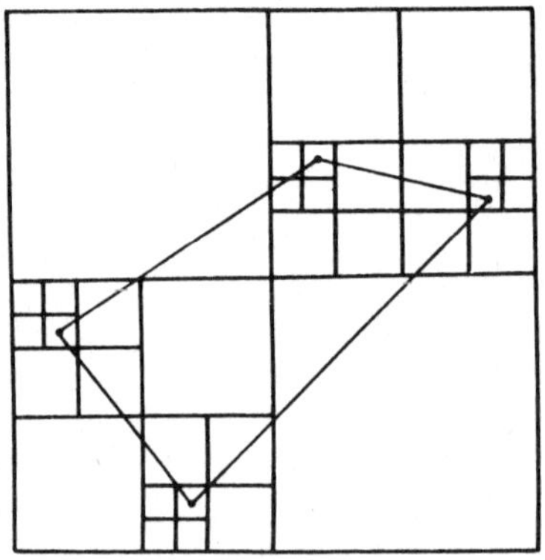

Figure 12: Example of an object described by extended quadtrees.

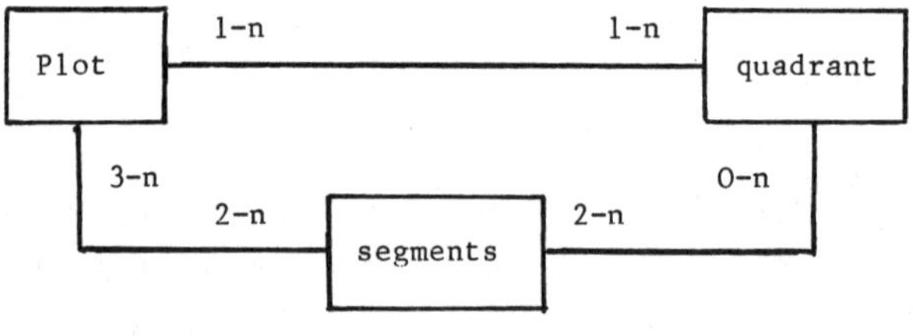

Figure 13: Conceptual model of a mixed representation

Since only a limited number of quadrants is intersected by segments, we choose to store this number as an attribute (inter-number). Moreover, #point is replaced by Peano keys (p_1 and p_2) representing both the point identifier and its coordinates.

R_{31}(#plot, Peanokey, size, inter-number)

R_{32}(#plot, #segment)

R_{33}(#segment, p_1, p_2)

3.4.2 **Point-in-polygon query**

As previously explained, we deal with a query Q_3(Peanokey, size) with only one tuple: $Q_3(p_0, e)$. We perform a Peano join with R_{31} giving A_{311}(#plot, Peanokey, size, inter-number). Now, we have several cases.

If we deal with only one tuple, this is the answer and we continue by a relational join as explained in §3.3.2. But, if A_{311} consists of several tuples, we have the situation given in **Figure 14** and **Table 1.**

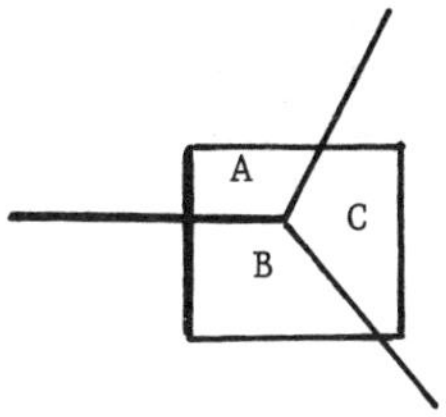

Figure 14: Square cut by several segments described in Table 1

#plot	PeanoKey	size	inter-number
A	p_0	e	3
B	p_0	e	3
C	p_0	e	3

Table 1: Description of a square cut by several segments

So, by Peano tuple algebra, we have a first guess that we need to ameliorate by computational geometry in order to find only one tuple. The algorithm is:

i) A_{311}(#plot, Peanokey, size, inter-number) determined by a Peano join on R_{31} and Q_3.

ii) Should we have several tuples, amelioration by computational geometry is needed to get only one tuple using R_{32} and R_{33}

iii) By a natural join between A_{311} and R_{31} we get A_{312}(#plot, Peanokey, size, inter-number, #landowner)

iv) By a projection of A_{312}, we get the result A_{313}(#landowner).

3.4.3 region query

Here the query is composed of several relations: Q_{321} as a Peano relation for surface storing and Q_{322} for the borderline:

Q_{321}(Peanokey, size)
Q_{322}(#segment, P_1, P_2)

In this case, by Peano tuple algebra, a Peano join between Q_{321} and R_{31} will give a relation

A_{321}(#plot, Peanokey, size, inter-number)

All tuples with an inter-number equal to zero will be kept. Then by computational geometry, considering all A_{321} tuples with an inter-number different from zero, Q_{321} and R_{33}, we will reduce the number of A_{321} tuples. Then, the algorithm is similar to the previous one:

i) A_{321}(#plot, Peanokey, size, inter-number) determined by a Peano join between Q_{321} and R_{31}.

ii) Should we have several tuples, amelioration by computational geometry is needed to reduce the tuple number.

iii) By a natural join between A_{321} and R_{21} we get A_{322}(#plot, Peanokey, size, inter-number, #landowner)

iv By a projection of A_{322}, we get the result A_{323}(#landowner).

3.4.4 **Vacant places**

The goal is to determine an empty polygon within a window given by its two representations:

Q_{331}(Peanokey, size, inter-number)
Q_{332}(#segment, p_1, p_2)

In this case we need to determine the boundaries and the internal area. First we compute the internal part by Peano Tuple Algebra and the boundary by computational geometry. The algorithm is:

i) Projection of R_{31}(#plot, Peanokey, size) to get A_{331}(Peanokey, size) and of Q_{331} to get A_{332}(Peanokey, size).

ii) Apply a Peano difference A_{331} minus A_{332} to obtain A_{333}(Peanokey, size) representing the internal area.

iii) Join R_{31} and A_{333} to obtain A_{334}(Peanokey, size, inter-number).

iv) On all A_{334} tuples with inter-number different from zero, use a computational geometry algorithm to determine A_{335}(#segment, p_1, p_2)

IV - QUERY EVALUATION VIS-A-VIS GEOMETRIC MODELLING

If we compare the previous algorithms used to answer our three queries (see **Table 2**), we see that only the spatial object representation based on Peano relations allows the use of algebra to answer spatial queries.

However, there exist other types of spatial queries. Let us mention also the **geometric projection** and **cross-section**. We will examine these queries using a CADCAM example.

A solid can be described by its **wireframe structure**, its **Peano relation representation** and a **mixed representation**. Here, the point-in-polygon problem becomes a point-in-polyhedron one. Similarly, wireframe representation requires computational geometry and Peano relations, Peano Tuple Algebra. For the mixed representation, we need both.

The first query will be a cross-section defined by a plane and we want to retrieve all cross-cut objects. The second query will be a geometric projection to determine the projected section into the flow on the xOz plane (the y coordinate will be deleted).

4.1 Wireframe representation

The ER model of a wireframe-represented solid is given in **Figure 15**:

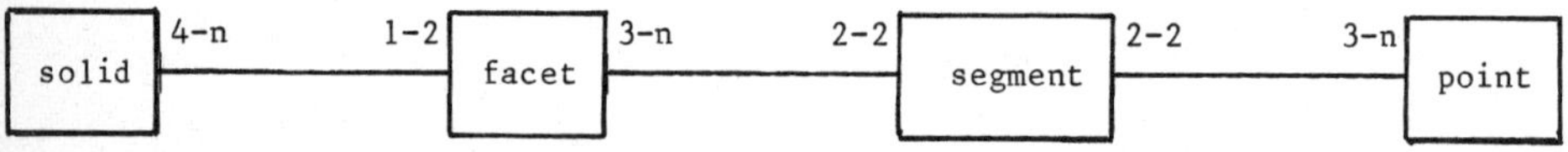

Figure 15: Wireframe representation of a solid

	segment-oriented model	Peano relations	mixed model	
nature of space representation	segment-oriented exact 1D model only boundaries	surface-oriented approximated 2D model only internal parts	surface and segment exact- 2D model both parts	
database type	intensional	both intensional and extensional	both intensional and extensional	
data base query	result not yet stored in the DB. needs to be reconstituted	result yet stored in the DB	result yet stored	
point-in-polygon query	CG + RA	PTA	PTA	CG + PTA
region query	CG + RA	PTA	PTA	CG + PTA
vacant places in a window	CG	PTA	PTA	CG + PTA

Table 2: Some spatial query algorithm comparison

In the relational model, we can write:

R_{40}(#solid, type, material)
R_{41}(#solid, #facet)
R_{42}(#facet, #segment)
R_{43}(#segment,#point1, #point2)
R_{44}(#point, x, y, z).

4.1.1 **Cross-section**

We select a cross-section plane

$$F(x, y, z) = Ax + By + Cz + D = 0$$

and we compute the intersections with all solids by computational geometry so as to obtain a set of relationships:

A_{411}(#solid, #facet),
A_{412}(#facet, #segment),
A_{413}(#segment, #point1, #point2)
A_{414}(#point, x, y, z).

To obtain the result, we have to perform a relational projection to get A_{415}(#solid).

4.1.2 **Projection**

To determine the geometric projection, we project all 3D segments into the projection plane (giving 2D segments) and we have to compute the planar convex hull by computational geometry (PREPARATA and SHAMOS, 1986). The result is a set of relations A_{421}(#segment, #point1, #point2) and A_{422}(#point, x, z).

4.2 Solids described by Peano relations

Here the solids are represented by a set of octrees. In other words they are described by only one relationship (remember that we deal with 3D Peano keys by interleaving x, y and z). The E.R. model is given in **Figure 16**. Each tuple represents a cube.

Figure 16: E.R. model of an octree-represented solid

R_{50}(#solid, type, material)
R_{51}(#solid, Peanokey, size).

4.2.1 **Cross-section**

To determine a cross-section, we start from the F(x, y, z) plane and we approximate it with an octree: so we get a 3D representation of this plane limited by two staircases described by (a complementary window can be useful):

Q_5(Peanokey, size).

To compute cross-cut objects is very simple with Peano Tuple Algebra:

i) We perform a Peano join between R_{51} and Q_5 to give A_{511}(#solid, Peanokey, size)

ii) We perform a relational projection to get A_{512}(#solid).

4.2.2 **Projection**

To perform a geometric projection, we use the Peano geometric projection operator (LAURINI, 87): by projecting an octree onto a coordinate plane we get a quadtree. The crude relations obtained are in the First Conformance Form (1CF) and it is easy to get them in 3CF by deleting overlapings and compacting the results (see Peano Tuple Algebra). In this example, we have to delete the y coordinates. Let us denote Peanokey[x,z], the previous 3D Peano key reduced to 2D. The answer is:

A_{52}(Peanokey[x,z], size)

4.3 Mixed representation

In this case, solids are doubly defined. See **Figure 16**.

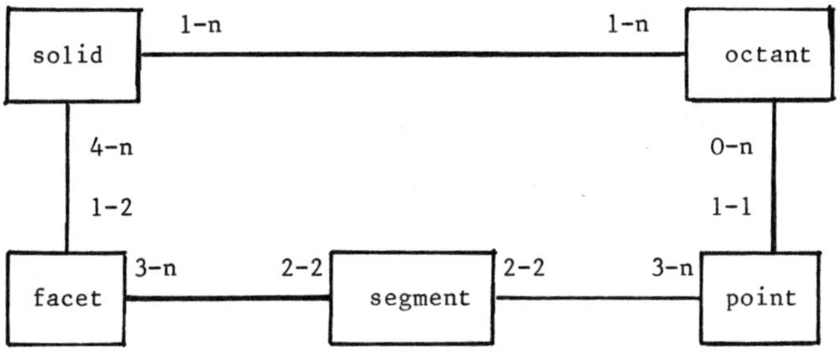

Figure 16: E.R. model of a mixed representation of solids

So we have 5 relations:

R_{60}(#solid, type, material)

R_{61}(#solid, #facet)

R_{62}(#facet, #segment)

R_{63}(#segment, p1, p2)

R_{64}(#point, x, y, z)

R_{65}(#solid, Peanokey, size, inter-number)

4.3.1 **Cross-section**

To determine a cross-section, we can determine an approximate result by Peano Tuple Algebra, as previously explained to get A_6(#solid). Remember that in this result, we have all intersected solids plus some solids neighboring the cross-section plane. Afterwards, we can ameliorate this result with computational geometry by deleting outside solids.

4.3.2 **Projection**

By Peano Tuple Algebra, we obtain a quadtree. In order to avoid staircased approximate limits, we can ameliorate this result by computational geometry. The result is:

A_{61}(Peanokey[x,z], size, inter-number)
A_{62}(#segment, p_1, p_2)

V - CONCLUSIONS

In this paper, we have shown the implications for spatial queries of mixing mix computational geometry (CG) and relational algebra (RA) or Peano Tuple Algebra (PTA). Our conclusions are:

- with a **wireframe model**, we deal with an intensional model whose extension is impossible to derive completely. It seems necessary to use a computational geometry package to answer spatial queries and combine it with a SEQUEL-like language; its advantage is an exact 1D representation and its drawback is that the object's internal parts are not described and the results are not yet stored in the data base.

- with a surface model based on linear quadtrees/octrees, we deal with an extensional/intensional model named **Peano Relations**. Moreover, we have an approximate 2D/3D geometric representation. An important set of spatial queries can be answered by Peano Tuple Algebra which is a very limited extension of relational algebra (only tuple aggregation/disaggregation needs to be implemented); the advantage is the description of the internal parts and the disadvantage the absence of boundary description;

- with a **mixed model**, we have a double representation for objects, an exact 1D (for borderlines) and an approximate 2D/3D (for internal parts); to answer spatial queries, we need combined algorithms (Peano Tuple Algebra and Computational Geometry).

Finally, only the description of spatial objects based on Peano Relations allows the simple use of algebra (Peano Tuple Algebra) for processing an important subclass of spatial queries. The other models always need amalgamations of computational geometry and algebra.

So, concerning spatial DBMS design, it is necessary to know what kind of geometric models will be used in order to incorporate some computational geometry algorithms. There are also important implications for the design of spatial query languages. A possibility is to impose the geometric model.

In this text, we have taken examples issued from town planning and CADCAM, but similar problems arise in other fields such as geology, geography, robotics, and computer vision.

REFERENCES

AYALA D, BRUNET P, JUAN R, NAVAZO I (1985) Object representation by means of non minimal division quadtrees and octrees. ACM Transaction on Graphics. Vol 4, 1 June 85 pp 41-59

CHEN PP (1976) The Entity-Relationship Model: Toward a Unified View of Data. ACM Transactions on Database Systems 1, 1, pp 9-35.

GARGANTINI I (1983) Translation, Rotation and Superposition of Linear Quadtrees. International Journal of Man-Machine Studies. Vol 18, 3, March 1983, pp 253-263.

LAURINI R (1985) Graphics Data Bases Built on Peano Space-Filling Curves. EUROGRAPHICS'85, Nice, September 8-13/1985. pp 327-338. Edited by CE VANDONI, NHPC.

LAURINI R (1987) Manipulation of Spatial Objects with a Peano Tuple Algebra. University of Maryland, Center for Automation Research Technical Report. CAR TR 311.

LAURINI R, MILLERET F (1986a) Les Clefs de Peano: un nouveau modèle pour les bases de données multidimensionnelles et les bases d'images. 2èmes journées BASES DE DONNEES AVANCEES, Giens, France, April 22-25/1986 pp 211-230. Published by INRIA, Rocquencourt, France.

LAURINI R, MILLERET F (1986b) Les clefs de Peano en Synthèse d'images: Modélisation et Opérateurs. Semaine Internationale de l'Image Electronique, Nice, April 21-25/1986 pp 659-675. Published by CESTA, Paris.

LAURINI R, MILLERET F (1987) Peano Relations in CADCAM Databases. International Conference IEEE "Data and Knowledge Systems for Manufacturing and Engineering", Hartford, Connecticut October 19-20/1987.

MORTON GM (1966) A Computer Oriented Geodetic Database and a New Technique in File Sequencing. IBM Canada-Ontario report, March 1966.

ORENSTEIN J (1986) Spatial Query Processing in an Object-Oriented Database System. Proceedings of ACM/SIGMOD'86, Washington DC. pp 326-336. Edited by C. ZANIOLO.

PREPARATA F, SHAMOS M (1986) Computational Geometry: an Introduction. Springer-Verlag

SAMET H (1984) The Quadtree and Related Hierarchical Data Structures. Computing Surveys. Vol 16, June 1984, pp 187-260.

DESIGN OF THE USER-INTERFACE FOR AN OBJECT-ORIENTED STATISTICAL DATA-BASE

Erik Malmborg
Statistics Sweden, U/ADB
S-11581 STOCKHOLM Sweden

Abstract

The concept of an Object-Oriented Statistical Database (OOSD) is rather new. The approach taken in this paper unifies two different traditions leading to OOSD. The first tradition is from Object-Oriented Programming Systems and languages in the Smalltalk tradition. The other tradition is from database systems able to handle more complex objects than records (or segments). Examples of such systems can be found in e g CAD/CAM and Office Information Systems.

The emphasis of the paper is on the user interface. The development towards modern Window-Icon-Mouse interfaces is described. The specific interfaces to be designed are a graphical meta-data browser, a graphic table-design language and the interaction between these.

The paper includes an analysis of different approaches for the interaction between query (or table-design) languages and meta-data handling, leading to the proposed "Macintosh-style" interaction.

0. Introduction and structure of paper

This paper describes parts of the design of a planned software system named TBE-2. TBE stands for Table By Example and alludes to the famous IBM language Query By Example (Described in e.g. Date-85). At Statistics Sweden there has been a great deal of discussion on how to extend the ideas from QBE into a Table Design Language. TBE(-1) (Nilsson-84) is a result from these discussions. TBE-1 is an algebraic language working on boxes (matrices). It is theoretically based on the box-theory of Sundgren-73. A modern interactive user-interface was planned but never materialized. ABE (Klug-82) has a user interface similar to that planned for TBE-1. QBE, TBE-1 and ABE are based on text terminals with full-screen handling. TBE-2 is based on the technology of bit-mapped graphic screens and will have a "Macintosh-style" interface.

The paper is divided into six sections. The first 3 give background material on Object-Oriented Statistical Databases, Macintosh style user interfaces and on the interaction between statistical query languages and meta-data.

Sections 4 - 6 describe the design of the TBE-2 user interface. Section 4 describes the graphical meta-data browser. Section 5 describes the table design language. Section 6 contains further material on the interaction between table design, meta-data and data. Some indications of further plans are given at the end of section 6.

1. The meaning and goal of an Object-Oriented Statistical Database

There seem to be two traditions leading to a possible concept of an Object Oriented Statistical Database (OOSD):

- Traditional databases handle records (segments, tuples . . .), i.e. simple concatenations of terms. In many areas there has been a demand for databases handling more complex "objects". Such areas are e.g. VLSI-design, CAD/CAM and picture analysis. The storage/retrieval of statistical tables (matrices, boxes . . .) is a similar problem. The emphasis is on data base (on secondary storage), and the demand is from the users. This is a problem in search of a solution. Scientific papers in this tradition normally have references to papers on "data structures for VLSI" and/or "CAD databases". An example from VLDB-85 is "An Extensible Object-oriented Approach to Databases for VLSI/CAD" by Afsarmanesh et al. (Afsarmanesh-85)

- The concept of Object oriented systems (not specifically "databases") has evolved from programming languages (SIMULA-67, Smalltalk . . .) and User Interfaces (Smalltalk, MacIntosh . . .). The "object" in such systems are further developments of "procedures" and "classes" from earlier languages. A basic idea is that an object (class) should handle all operations on the instances of that class. This gives highly modular systems. The storage/retrieval on secondary storage of the instances could be examples of such operations, but the emphasis is on "object-oriented". The "object-oriented" approach in this sense is almost a solution in search of problems. Papers in this tradition typically refer to e.g. the Rank Xerox Smalltalk project (Goldberg, Kay, Ingals . . .). An example from VLDB-85 is "An Object-Oriented Environment For OIS Applications" by Niestrasz/Tsichritzis (Niestrasz-85).

To me it seems that these are two traditions, possibly leading to research on "Object Oriented Databases" in general and "Object Oriented Statistical DBMS" in particular.

A simple solution to the problem of the unclear concept is to propose a definition that is pertinent to both traditions above. I think this is basically what is done in the report from the working group on Object Oriented Statistical DBMS from the third conference in this series (Luxembourg July-86). The report can be found in Statistical Software Newsletter 1,1987.

The system to be designed is somewhere inbetween a Data Base Management System and a tool for System Development. Typically the requirement on a DBMS is to give the user an answer within seconds. In the case of complex questions or reports minutes might be acceptable. System Development, on the other hand, is normally considered a complex process with users and analysts together building an application giving the required functionality. This process takes weeks, months or even years.

In the tool to be designed users and/or analysts should be able to create applications giving the desired information within hours or perhaps days. Technically the system is not to include its own database format, but generate programs able to use existing file formats. The technical aspects of program generation are not the topic of this paper, but were discussed in Malmborg/Chowdhury-84.

2. The development towards Macintosh-style user interfaces

The Macintosh computer from Apple has become the symbol of a new style of man-machine interface. This is somewhat unfair to XEROX, as the major research and development behind this type of "WIMP"-interface was made at the XEROX PARC laboratories. The Smalltalk system from XEROX (Kay-77, BYTE-81, Goldberg-83,-84, Krasner-83) is a major development. As will be described later in the paper Smalltalk is the language of choice for my project. STAR (Smith-82) is another important XEROX-system on the way to modern "WIMP"-interfaces. Smith-82 is strongly recommended reading for those interested in the goals and strategies for WIMP-interfaces. WIMP stands for:

- Windows, i.e. the dividing of the CRT-screen into several areas, and the flexible methods for moving, opening, closing, enlarging.... the windows.
- Icons are symbols on the screen, denoting the different "objects" hidden inside the computer (programs, data...).
- Mouse is the device for pointing on the screen.
- Pull-down or Pop-up menues are flexible ways for giving commands to the computer (or to the objects chosen by pointing to an icon and "selecting" it before pulling down or popping up the menu).

There are further aspects of this type of interface than can be seen on the WIMP-surface. The interface should e.g.

- be mode-less, i.e. the user should not be locked into modes as in a hierarchical menu-system.
- use a familiar conceptual model (as e.g. the desktop metaphor of the Macintosh).
- be What You See Is What You Get (WYSIWYG)

The developments in this area are far from complete. IBM's Systems Application Architecture (SAA) will probably in a few years give this type of interface to the users of IBM mainframes, office computers and personal systems.

There is a rather strong coupling between WIMP-interfaces and Object Oriented Programming (OOP). In OOP, programming is by sending messages to objects rather than by invoking procedures. In a WIMP-environment objects are selected by pointing to the icon representing them. The appropriate command is then chosen as an option in a menu. Cox-86 gives a good discussion on the operator/operand model of traditional languages (programming and command) versus the message/object model.

3. The interaction between statistical query languages and meta data

The project described in this paper is heavily influenced by two rather different papers. Both however share a common understanding of the problems with traditional query or table design languages against statistical data bases. This problem view is also similar to that of e.g. Shoshani-82, Chan-80, McCarthy-82.

The sheer volume and complexity of metadata needed for finding appropriate data and formulating a query against a large statistical database puts extreme demands on the user. If there is no integrated meta-data base, the documentation of the contents in a census, environmental, world trade or other large statistical database consists of hundreds of pages.

One solution to the problem is to create an integrated meta-database used by the Data Base Management System to drive a menu-oriented interface for the user. This relieves the user from much search in the written database documentation. Solutions in this category can be found in the SUBJECT-system (Shoshani-82, Chan-80) and in the AXIS-system (Nordbäck-82,-83).

Another slightly different solution is to have an interactive meta-database where the user can search for relevant data, and then specify his questions or table needs. The query language and the meta-datasystem can be integrated. The system designed in this paper emphasizes such integration.

One of two main influences mentioned above is Sato-86, which describes an implemented statistical database for the Japanese National Land Agency. When retrieving information from this database the user is helped by the Land Information Dictionary, i.e. the meta-database. This meta-database is structured by using a semantic data model based on the extended relational data model RM/T (Codd-79, Date-85). In the Land Information Dictionary the user can search ("browse") the data on two levels.

The first level is the "conceptual" level where all conceptually obtainable data can be found. This corresponds to all possible questions that can be formulated using the concepts introduced.

The second level clarifies which parts of the conceptual data correspond to data actually stored in the Data Base. One reason for the distinction is that often no complete cross-section table exists, but merely a number of subtables. This is due to the demands of statistical accuracy or the convenience of survey routines (Sato-86).

One important aspect of Sato-86 is the handling of aggregate data on the semantic level. The key idea is not to use any specific constructs such as summary sets (Johnson-79). A similar approach was suggested in Malmborg-86.

In Sato-86 the potential explosion on the number of different objects (or entities) is handled by only modelling the most detailed cross products (of the category attributes) on the conceptual level. This anyhow allows questions and data for the physical database to work on all subsets of the set of category attributes in the conceptual file. The figure below (fig 6 from Sato-86) illustrates this.

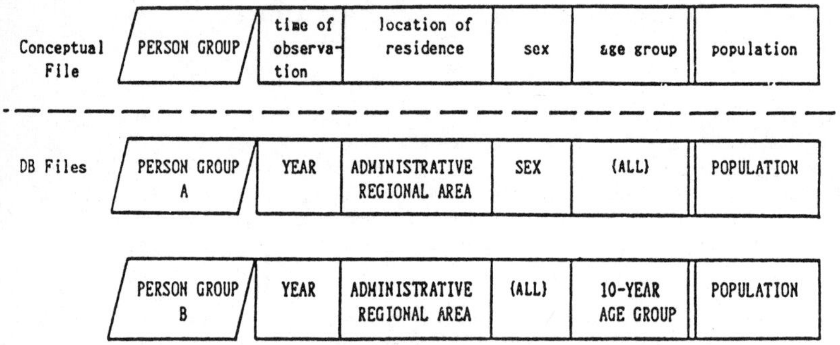

Fig.6 A Partial DB schema of Sample Statistical DB

The second paper that has had a major influence on my project is Wong-82. The system described in that paper is a Graphical User Interface for Database Exploration (GUIDE). The system is an Interactive Metadata Browser, where the user can interactively explore a data model presented on a graphical display. The data model is based on a version of the Entity-Relationship model (ER-model, Chen-76, Teoray-86). The user has a lot of options for building his own graphical view of the for him relevant part of the database. He/she can set schema detail, change focus, change radius, hide, zoom and move schema (see Wong-82 for details). When the view is ready it can be used for queries against the data base.

The GUIDE schema is a "normal" ER-model. There is no provision for summary sets and aggregate functions. The paper indicates plans for introducing summary sets referring to Johnson-79.

4. The graphical metadata browser

This and the next two sections describe the planned software system. The project is in its early phases and only simple "mock-ups" exist. Where the project will end is a question of financial funds. The planned software system in this first phase will be used as a demonstration tool to create an interest in later phases. The prototype will be fully runnable, and will show some main ideas of the project. As the system is programmed in Smalltalk-80 with the superior facilities of this language for structuring and reusing code there is a good hope of reusing code from the prototype in the final system. The prototype will work within the Smalltalk environment, with its drawback in the form of rather low speed. As the power of PC's and workstations grow this might not be a problem. The final system will however for the reason of portability generate code in the C-language rather than work in the Smalltalk environment. The tool and the metadata will probybably still be in Smalltalk.

The first part of the system to be described is the Interactive Graphical Metadata Browser. At Statistics Sweden we have an established System Development Model(the SCB-model, described in Sundgren-84, Malmborg-82,-83,-84). The SCB-model is similar to the ER-approach (Chen-76), but has its own roots (Sundgren-73). The graphic notation is somewhat different from the ER-approach. The SCB-model has gradually evolved towards a mix of an Extended Entity Relationship model (as in Teorey-86) and Jackson structured Design (as in Jackson-83). The models introduced in the first version of TBE-2 (the name of the tool, cf Nilsson-84 where TBE(-1) is described), will correspond to the simple ER-models.

The SCB model contains a special notation for specifying statistical tables. The user of TBE-2 will not use this "alpha/beta/gamma-notation", but it will be used in the explanation of the tool. The following extract from Malmborg-84 contains a short presentation of a simple example:

> In the most simple situation we have an information base containing static information on a population of entities. The entities are grouped into disjoint classes. The input is obtained by observation (including questionnaires). There are no dynamics to be consided. All input and output are by problem definition "snapshots". This type of problems exists in practice, and can be handled by e.g. a simple object-oriented approach.
>
> The SCB System Development Model is seen to be fully adequate for this class of situations. As an example I will present a (simplified) model used for a survey of fishing and fishing equipment.

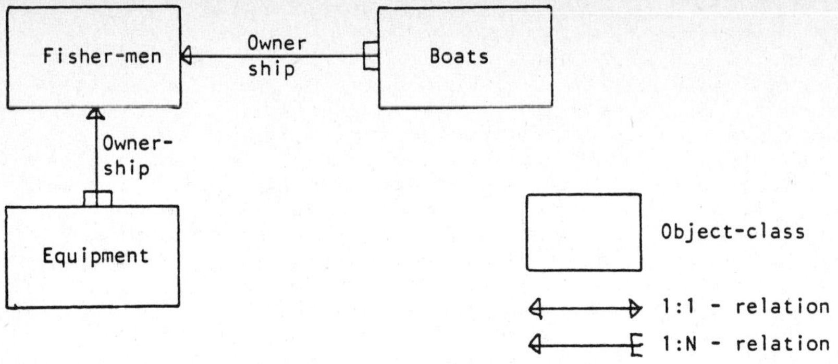

Note that this simple "infological" model contains a lot of agreement on semantics:

- It is assumed that each boat has a single owner. This is for the purpose of registration. Of course it might not be true in an economic sense.
- Boats and Equipment are seen to be totally disjoint classes of entities. Of course this is only an agreement of concepts made for this survey. In another situation they might be considered as different entities from a common class.

The output from the survey will be statistical tables. The content of these can be expressed in e.g. the alpha/beta/gamma-notation (Sundgren-73) which is a part of the Systems Development Model. An example:

α-part (selected population) : fisher-men in fresh-water (lakes)

β-part (selected variables) : number of equipment

γ-part (distributed by) : equipment type and district

	seine	trawl	fyke net	...
Lake Vänern	100	...	...	
Lake Vättern	...	...	...	
Lake Mälaren	...	...	...	
...				

TBE-2 will contain a graphic editor for creating the infological models. When using the browser, objects (entities) are selected by moving the pointer onto the symbol and clicking the mouse-button. The result will be a pop-up menu where different options will be given:

- Present a verbal description of the object class.

- Present a list of files (or relations) from the database representing extension of the object class. Each file is described in several aspects (population, level of detail ..)

- Present a list of variables (attributes) for the object.

By clicking on a variable in the last list, a list of categorizations for the variable pops up on the screen. In our example geographic location might be categorized by Lake District (as in the table above) or Region code. The use of generalization hierarchies for grouping of category attributes is discussed in Sato-86.

The categorizations are presented as lists of values (the lake names in our example). These lists correspond to the values (texts) in the table's column- and row labels (stub and heading). This visual aspect will be used in the next section.

5. The table design language

The idea for the table design part of the program is based on an observation of the similarity between a "tiled" window in Smalltalk or on the Macintosh and a statistical table. Some common aspects:

- both can be seen as a rectangular area subdivided into a number of rectangular parts. In this paper the outer rectangle is called a window, and the inner rectangles are called panes.

- The contents of the panes within a window are interdependent. On the Macintosh we might typically have two panes, one with a grapical and one with a textual representation of the same model (or document). When we move around in the model both views (panes) are updated. The same is true for a statistical table. If we change the chosen values for the stub, the central pane of the table should be updated.

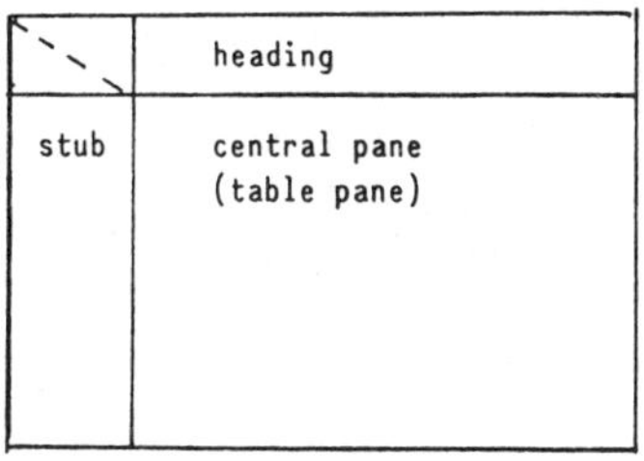

Fig 5.1

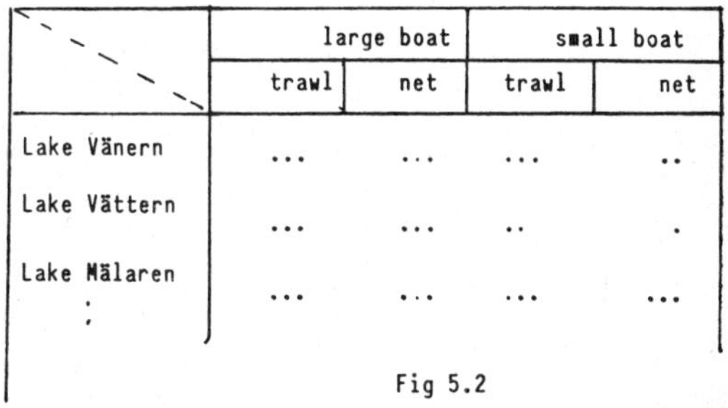

large boat
small boat
trawl
net
trawl
net
Lake Vänern
Lake Vättern
Lake Mälaren

Fig 5.2

The way for the user to work is to pop up an empty

table-window with panes for stub, heading and a center pane (for the figures to be computed). Additional panes may be introduced for table title, footings ... .

To specify the contents and structure of the table the user browses around in the metadata-window (cf section 4). When the appropriate set of labels for stub or heading is found, it is selected in the metadata window. Selections are by default indicated by text inversion both in Smalltalk and on the MacIntosh. Moving this selection into the table window is accomplished either by cut & paste (standard MacIntosh technique) or by dragging a symbol between the windows (another standard technique). Both alternatives will be tested to find out which is most natural to the user. The process is repeated for each gamma-variable (category attribute). If more than two gamma-variables are used, the stub and/or the heading is strucured into the appropriate number of levels (cf fig 5.2 where the heading is 2-level). The number of levels is chosen from the pane's pop-up menu.

The appropriate function for computing the cells (count, sum, mean ...) is chosen from the pop-up menu of the central pane in the table-window. If the beta-part is based on a variable this is chosen in the meta-data window (e.g. age in a table with mean age of equipment).

If the table is more complex and contains several concatenated subtables and/or totals the approach can be extended by more options on the table-pane's menu.

The goal is to create a very intuitive and natural interface for the user. A key point is that the "visual" finding of the needed labels for the table (stub or heading) in the meta data makes the user feel confident in his design.

To describe this type of interaction in a text is awkward. The idea is that the user should need only minimal documentation, but find the way of specifying the information needs (the table) obvious. The problem of describing user interfaces of this kind is well known (Goldberg-84, Schmucker-82). The specification of the "intuitive" MacIntosh user interface standard in Schmucker-86 is a hefty 50 page appendix. The process of developing this kind of interface from the early XEROX-research into the present MacIntosh standard represents more than 10 years of intensive research and development.

In other words, it is surprisingly difficult to design a really simple and intuitive user interface. Further experience shows (Smith-82) that the designer and the user often come to different opinions on what is simple and intuitive. Of course the user's opinion (by definition) is correct, but it shows the need for prototyping and experiments.

6. The interaction between table design, meta data and data.

In the preceeding two sections I have presented a simple interaction between the meta data (the browser) and the table design. Actually I think this is only a first level of interaction. As was discussed in section 3, Sato-86 and Malmborg-86 make the proposal that aggregate data (summary sets in Johnson-81) should be represented on the conceptual level as ordinary entities, rather than as special modelling constructs (such as summary sets).

Both on logical and physical level aggregated data can be looked upon in two different ways. Logically one alternative is as entities (objects) or relations (Codd-meaning). The other logical alternative is as matrices or boxes (Sundgren-73). In the first alternative category attributes are explicit, in the second implicit (i.e. part of meta-data rather than data). The same two choices exist on the physical level, giving four possible combinations.

What is done in the table window (section 5) can be seen as the specification of an aggregate data set. This specification can be taken back to the meta-data window to constitute an (aggregate) object. This object can either correspond to a physically created summary file or only be a specification without extensions. The inheritance mechanisms is Smalltalk makes it possible to handle such entities as "abstract objects". We can use the specification to generate the program to create the summary file if needed. It is possible to have several such specifications for a single entity type, giving several ways to create a specific summary file.

The idea of "abstract objects" can be extended to cater for the handling of event-oriented databases. We can specifiy abstract objects representing the states of objects at any specific time point as a function of the events. Malmborg-82, -83, -84 present the modelling issues behind this potential extension of TBE-2. Such extensions would make the tool into a general purpose system development tool, rather than a table design tool. The ideas in this last paragraph will be elaborated in another paper.

REFERENCES

Afsarmanesh, H., McLeod, D., Knapp, D., Parker, A.
An Extensible Object-Oriented Approach To Databases For VLSI/CAD
Proc 11th Int Conf on Very Large Data Bases, Stockholm 1985

BYTE (-81)
Special Issue of the journal BYTE presenting Smalltalk.
Several articles by the creators of the language.
BYTE Aug-81

Chan, P., Shoshani, A. (-80)
SUBJECT: A directory driven system for organizing and accessing large statistical databases.
Proc 6th Int Conf on Very Large Data Bases, Montreal 1980

Chen, P.P-S: (-76)
The Entity-Relationship Model. Toward a Unified View of Data.
ACM Transactions on Database Systems, Vol. 1, No 1, March 1976

Codd, E.F. (-79)
Extending the Database Relational Model to Capture More Meaning.
ACM Transactions on Database Systems, Vol. 4, No 4, Dec 1979

Copeland, G., Maier D. (-84)
Making Smalltalk a Database System
ACM Sigmod Record, June 1984

Cox, B.
Object-oriented Programming: An Evolutionary Approach
Addison-Wesley 1986

Date, C. (-85)
An Introduction to Database Systems, 4th ed., Vol 1 and 2.
Addison-Wesley, 1985

Goldberg, A. (-84)
Smalltalk-80 - The Interactive Programming Environment.
Addison-Wesley, 1984

Goldberg, A.,Robson, D. (-83)
Smalltalk-80 - The Language and its Implementation.
Addison-Wesley, 1983

Jackson, M.A. (-83)
System development
Prentice Hall, 1983

Johnson, R.R. (-81)
Modelling Summary Data
Proc ACM SIGMOD Int Conf on Management of Data, 1981

Kay, A.C. (-77)
Microelectronics and the Personal Computer
Scientific American, September 1977

King R.
A Database Management System Based on an Object-Oriented Model.
Proc conf on Expert Database Systems 1986. Benjamin/Cummins 1986.

Klug, A.
Abe, A Query Language for Constructing Aggregates-by-example.
Proceedings of the First LBL Workshop on Statistical Database Management. Menlo Park, California 1982.

Krasner, G. (-83)
Smalltalk-80 - Bits of History, Words of Advice.
Addison-Wesley, 1983

Malmborg, E. (-82)
The OPREM-Approach - An extension of an OPR-approach to include dynamics and classifcation.
Statistiska Centralbyrån S/SYS-E12

Malmborg, E. (-83)
An analysis of systems design methodologies using the ISO-framework.
Second Scandinavian research seminar on information modelling and data base management (Tampere 1983), also as SCB S/SYS-E14.

Malmborg E. (-84)
Stepwise formalization of information systems specifications by extending a simple object-oriented approach.
Seventh Scandinavian research Seminar on Systemeering (Helsinki-84), also as SCB P/ADB-E20

Malmborg, E., Chowdhury, S. (-84)
Program generation as a method for the production of portable software - an experiment and a proposal for a methodolgy. Paper in Swedish SCB, ADB-METOD 1984:04

Malmborg, E. (-86)
On the Semantics of Aggregated Data.
SCB, U/ADB - E25, also in Proc 3rd Int Workshop on Stat Database Man, Luxembourg 1986

McCarthy, J.L. (-82)
Metadata Management for Large Statistical Databases
Proc 8th Int Conf on Very Large Data Bases, Mexico City 1982

Niestrasz, O., Tsichritzis, D.
An Object-Oriented Environment For OIS Applications
Proc 11th Int Conf on Very large Data Bases, Stockholm 1985

Nilsson, G. (-84)
Table by Example - TBE
SCB, Feb 1984

Nordbäck, L., Widlund, A. (-82)
AXIS - The Manager of Very Large Statistical Databases.
Proc COMPSTAT 1982 (5th Symposium at Toulouse)
Physica-Verlag, Wien 1982

Nordbäck, L. (-83)
Problems, Plans and Activities concerning the Economic Databases at Statistics Sweden.
Proc 2nd Int Workshop on Stat Database Man,1983 (Los Altos)

Olsson, L. (-82)
The role of databases in the dissemination of statistics
SCB, S/SYS - E13, also in Proc ISIS-82 seminar

Sato, H., Nakano, T., Fukazawa, Y.,Hotaka, R. (-86)
Conceptual Schema for a Wide-Scope Statistical Database and Its Applications
Proc 3rd Int Workshop on Stat Database Man, Luxembourg 1986

Schmucker, K.J.
Object-Oriented Programming for the Macintosh.
Hayden Book Company, New Yersey 1986

Shoshani, A.
Statistical Databases: Characteristics, Problems, and some Solutions.
Proc 8th Int Conf on Very Large Data Bases, Mexico City 1982

Smith, D.C., Harslem, E.
Designing the Star User Interface
BYTE, April 1982

Sundgren, B. (-73)
An Infological Approach to Data Bases
SCB, URVAL nr 7

Sundgren, B. (-84a)
Conceptual design of data bases and information systems
SCB P/ADB-E19

Sundgren, B. (-81)
Statistical data processing systems - Architectures and Design Methodologies

SCB S/SYS - E11

Sundgren, B. (-85)
Outline of an Algebra of Base Operators for Production of Statistics.
SCB, P/ADB-E21

Teorey, J.T., Yang, D., Fry, J.P.
A Logical Design Methodology for Relational Databases Using the Extended Entity-Relationship Model.
ACM Computing Surveys, Vol 18, No 2, June 1986

Teitel, R.
Statistical Databases and Database Systems.
Proc Bureau of the Census First Annual Research Conference
Reston, Virginia 1985

Wong, H.K.T., Kuo, I. (-82)
GUIDE: Graphical User Interface for Database Exploration.
Proc 8th Int Conf on Very Large Data Bases, Mexico City 1982

KNOWLEDGE BROWSING - FRONT ENDS TO STATISTICAL DATABASES

Geoffrey A. Stephenson
LOGICA SA/NV, Place Stephanie 20 - Bte 2
B - 1050 Bruxelles

Abstract:

A problem facing users of statistical databanks is obtaining secondary information relating to the statistics. Such information includes methods of collection, problems with individual observations, precise definitions and the conceptual framework underlying the data. Recent advances in knowledge engineering provide the methodology for making such data available online. A demonstration support system, using hypertext in the first instance, has been developed covering a small well defined domain in social security statistics.

Introduction

This paper describes the first phase in a study of the use of knowledge engineering techniques to improve access to statistical data. The access problems that the study confronts are those where the potential user lacks detailed knowledge of the available data, either technical knowledge related to the production of the statistics or an accurate conceptual model of the knowledge domain. The aim of the work is to provide a front-end to a statistical database, that will make available to the user the information (secondary data or meta-data) required to choose relevant data from the database and interpret it correctly. The database is a collection of social statistics that has been created by the Luxembourg Income Study. It consists of derived micro-data sets, based on income survey data from more than twelve countries. In the first phase, an online hypertext database of information related to the social security systems of the countries concerned has been produced and will be used by post-graduate participants in a research summer school. The paper sets out the framework in which the study is taking place, the methodology used and the future steps.

Background

One of the principal problems facing users of statistical data banks or databases, is that of finding the correct data. This problem has a number of dimensions:

- the user may not know what data is available, with extensive official data banks there may be thousands of data series or other sources, and in general these are poorly indexed.

- the user may be faced with a variety of sources from which to choose, without sufficient descriptive material to make that choice. For example, he may have to choose between different levels of aggregation (SITC, Nimexe), or different definitions (of unemployment), with limited documentation.

- the user may not know enough about the subject material to choose the correct source. Statisticians may make subtle distinctions of definition to correctly catch important economic aspects of the system they are measuring, but these may be of importance in only a limited number of cases or for specific purposes; the user may be unaware of the consequences. For example, in trade data it is important for certain products to distinguish between real imports (for consumption) and transitory imports for either re-export (Europort) or for partial processing/use and return (diamond cutting, oil rigs).

- the user may require access to subsidiary information to correctly interpret the primary data he is processing. The nature of this subsidiary data may not be at all evident. For example, the existence of a dock strike or strike of data collectors, may have a significant impact on data series, which is not immediately visible if the user is not aware of the problem. Often this type of problem is only noted in footnotes to the published data.

Whilst the information to help the user may be available in publications of various sorts, at present little of this secondary information is available on-line with the data itself. The organisation of this type of data and its presentation to the user at the appropriate time is an area where little progress has so far been made. This is due partly to the costs involved, both in storage and data production, and partly to a lack of suitable techniques.

The Luxembourg Income Study (LIS) [1] is an international research programme that has been running, since 1984, at the Centre for Studies in Population, Poverty and Socio-economic Policy, in Luxembourg. Its aim has been to provide a database of income survey statistics for international comparison studies eg [2]. Since the initiation of LIS, a number of conferences have been held to discuss studies using the international data produced (eg Sloan Foundation Conference on "The changing well-being of the aged and children in the United States: intertemporal and international perspectives." Luxembourg May 1987). It became evident at the Sloan Conference in 1987, that the research workers using the data were having difficulty with understanding the contexts undelying the international data. Each expert had a model of social security based on the institutions in his country of origin. These models did not map directly from one country to another as the institutional frameworks were different. Although extensive documentation of the data sets was available it was clearly insufficient to provide the necessary background data for fully understanding the situation in individual countries. Some work on improving the situation was begun by Prof.L.Rainwater using LOTUS 1-2-3. At about the same time the Statistical Office of the European Communities asked for papers for a Seminar on Development of Statistical Expert systems (DOSES, held in December 1987 in Luxembourg), as a first stage in initiating a new research programme. A paper [3] was prepared for this seminar, and a short demonstration of hypertext for social statistics given. The present work is the extension of that demonstration to cover sufficient information to form a base for evaluation of the techniques used.

The paper contains a section outlining the concept of knowledge browsing, a description of the browser devised for the evaluation and some of the problems encountered.

Knowledge browsing

A considerable amount of research has been directed towards improving the human-computer interface, and to modelling computer based information access (e.g. [4], [5]). Fig. 1 shows the relationship of different access activities to the user's certainty of what information is wanted and to his knowledge of how to access that information. The diagram is not intended to be taken literally as many of the activities will overlap in specific instances.

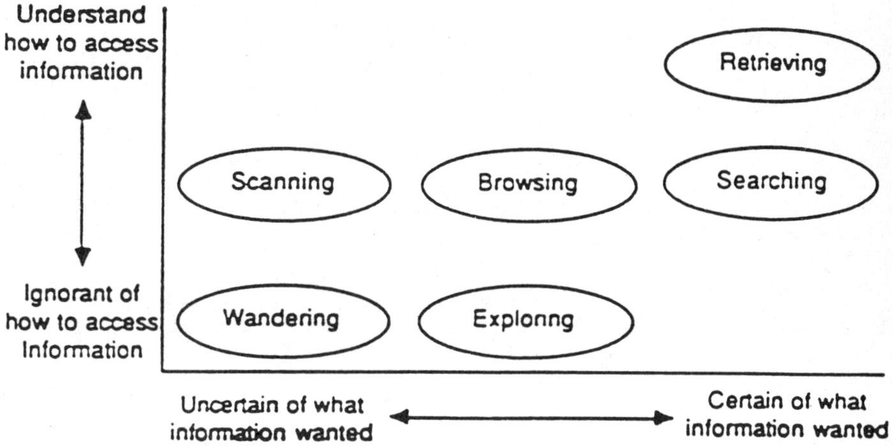

Figure 1

Definition of activities

Retrieving - if the user knows what information is wanted, knows that it is in the system and how to get at it. An example would be a statistician accessing a time series in a database of industrial production series.

Searching - if the user knows what information is wanted, believes the system can provide it but doesn't know exactly how to get at it. An example would be the statistician consulting a database via an on-line index, or using a help facility.

Browsing - if the user's goal is ill defined and weakly held, and he is unsure where to look. This essentially means that at each decision point a user's momentary interests select the path to follow. An example,for numeric databases, would be a statistician looking for leading indicators for a given series.

Scanning - This is a type of browsing where users aren't looking for a particular piece of information but are scanning for things that might be of interest. An example would be running some exploratory data analysis on all the variables in a survey.

Exploring - if the user is trying to determine the limits, constraints and capabilities of the system. This behavior is characterised by no particular information access goal being followed but effort is applied to understanding the structure and content of the system. An example would be reading the user manual for an online database system such as COMEXT (SOEC foreign trade database).

Wandering - This is totally unstructured and usually occurs when the user has no understanding of the domain that is being accessed.

At present, systems do not provide support for the full range of such activities. In many cases they support only one type of activity. In the case of economic and social statistics, the systems available generally support access to a collection of numeric data sets and facilities to manipulate selected data and display it in a mixture of numeric and graphic forms. In some cases it is possible to store intermediate results for later use. The capability to store textual information alongside the numeric data is very limited, and there is no support for complex structured text to help in the retrieval of related items.

Treating browsing (and to a more limited extent searching and scanning) as defined above, as the main activities that we wish to add to the current environment, we can set out more detailed requirements. There are four main capabilities that we wish to have whilst the user is interacting with the system:

- first, the structure of the knowledge domain should be observable, so that the user can form an accurate model of the topic. The user will probably start with an incomplete or partially inaccurate model of the topic. The user can also form an idea of the scope of the system if he can observe it.

- second, this observable information space should be easy for the user to move around in and he should be able to control this movement.

- third, the user should not be forced to make decisions that he does not understand or wish to make. Browsing should allow the user to look at alternatives, make tentative decisions and if he wishes undo them, until they arrive at a decision they are happy with. The user should be able gradually to refine his model of the topic or the decision he has to make (e.g what to retrieve).

- fourth, the system should support recognition. Rather than having to recall information from memory the user should be able to move around the information space and recognise the required items. This capability is usually provided by menu lists, as opposed to command interfaces.

A number of techniques exist to support the type of environment identified above, and they can be classified into a number of different areas:

- navigation is concerned with the ability of the user to understand where he is in the information space and how his position relates to a more general view. Signposts can provide an indication of where one is and routes to items of interest.

Information maps can provide a graphic display of the locality where the user is currently and linked objects that are close by. Such maps are also useful to show the user the path he has used and where he has been.

- movement is concerned with techniques to help the user move around the information space. Audit trails, bookmarks and been-here indicators are all possible aids, as are pre-set trails to follow and auto-retrace facilities.

- different types of links can be set in the information space to provide a hypertext like facility. These would include, example links, definition/explanation links, analogy links, source links and bibliographic links.

- filters allow the user to remove certain items from the context. They may be used to constrain browsing to certain paths through the information space or to limit the volume of information displayed.

- access mechanisms such as indices, thesauri, query systems and keywords may be used to enable the user to search the information space.

- adaptation allows the user to modify the information space to correspond more closely to his needs and interests. Creating his own links, marking points of interest and collecting items into a scrapbook are examples of the techniques available.

Hypertext

Hypertext [5] (also called non-linear text), covers a number of mechanisms which allow text, graphics etc. to be cross-referenced in such a way that they can be accessed on a computer via direct display of the cross reference links. In concept hypertext is fairly simple: windows on the screen are associated with objects in a database (for example small chunks of text), and links are provided between these objects, both visually (via labelled tokens such as highlighted words) and stored in the database (as pointers). A number of hypertext products have recently come on to the market with differing characteristics. The LIS group use IBM-compatible PCs (plus mainframes) and so the aim was to use a package running on that range of machines. For the DOSES seminar the GUIDE package from OWL International Inc. was used. There were some difficulties in porting this between machines, as it runs as a MICROSOFT Windows application and this has to be reconfigured for different types of mouse etc. (which created problems where the rodents used were non-standard). Two products are available free for limited use: TextPro from Knowledge Garden and PC-Hypertext from MaxThink. An earlier version of TextPro called Hype is also freely available. The free systems are downloadable from COMPUSERVE and BIX.

The study decided to make the data available using both GUIDE and TextPro, as they are significantly different and an evaluation of some of the alternatives is thus possible. At one level the products are similar. Text is stored in files and passages are marked as independent blocks. These blocks are threaded together by linking words or phrases in one block to another block. When a link word is activated (via mouse or cursor movement) the associated block is added to the screen as a new window. A simple method for backtracking is available. The principal differences are in input method, degree of windowing (GUIDE allows links within documents and thus jumps without opening a new window), backtrace information and visual environment.

TextPro is part of a larger system that allows rules and other advanced features to be used via a language compiler. At the TextPro level it works on simple ASCII files, blocks are delineated by /(label) and /end, label being a word or phrase marked elsewhere in the text by enclosing between #m and #m (eg #mBelgium#m). (In Hype the / was .. and #m was \, an example of some input is shown in Figure 2). When a label is activated, the text file is searched for the related block, and this creates response time delays for large files (significant on an IBM PC XT like machine, when the file exceeds 50000 characters). Windows are fixed size, and carry the label at the top in a manner that allows all the labels to be visible when the windows are active. Thus the user can see the chain of labels he has activated. Backtracking is by successively closing the windows in the order they were activated. The system can be used with or without a mouse. In the earlier Hype version, window positions and sizes were under control of the writer, and a limited capability to link independent files was available.

```
j) housing
k) miscellaneous
..end(social)
..bo
..BELGIUM O
..window(belgium o) ,,,,72,
ORGANISATION IN BELGIUM

Ministry of Social Security
For unemployment:  Ministry  of Employment and Labour

NATIONAL SOCIAL SECURITY OFFICE (collection and distribution of
contributions except for industrial injuries)

Organisation is classified into:
\bo1\- sickness, maternity and invalidity
\bo2\- old-age and survivors
\bo3\- employment injuries and occupational diseases
\bo4\- family allowances
\bo5\- unemployment
..end(belgium o)
..end(bo)
..bo1
..window(bo1) ,,,,72,14
SICKNESS, MATERNITY AND INVALIDITY (B ORG)

NATIONAL SOCIAL SECURITY OFFICE (collection and distribution of
contributions except for industrial injuries)
```

Figure 2. Example of Hype input, showing block separators and link markers.

PC-Hypertext is also part of a larger system, that is oriented to tools for breaking up and linking text. The text version has the capability to add into the text (standard ASCII file) activation strings (delineated by ‹ ›), that can be either the names of other text files or DOS commands (including therefore calls to other programs). When the strings are activated the current file is replaced with the chosen file or the command is executed. It is possible to backtrack through earlier windows. The disadvantage of the system is that only one activation string per line is allowed, so that the highly dense cross-reference matrices used in the database (described below) would not have been possible. The system does however provide the means to produce information about the cross-links and lists of files independently of the main access method (maps of links primarily intended for checking validity).

GUIDE is a more flexible hypertext package for both text and graphics, but has no additional facilities. Text is entered to the system using the GUIDE editor, or via the standard Windows clip-board from any word processor that supports clip-board. Links are added separately and can exist between independent GUIDE files. Links are activated with a mouse. There are three types of links:

- cross-reference links, when activated, either open a new window or jump to a new location, depending on whether the link is to a word or phrase in a different or the same document (file).

- note links, open a new temporary window (while the mouse button is pressed), and are intended for short explanations eg source of information.

- replacement links, substitute new text in the passage, and are used to elaborate items. They allow text to be artificially shortened, maximising window contents, and then expanded as required by the user.

Links are added independently of the text production, and can be set up between different files. An advantage of this is that the links are independent of the text format, unlike the other systems, where the marked link text has to be unique as it is the label for the associated text block (see Figure 2). Backtracking for note links is immediate, for replacement links is by reactivation (the substitution is reversible) and for cross-reference links there is a back-track button in the window that enables the jumps to be reversed sequentially.

Database description

The database set up with the hypertext systems consists of information from two CEC publications [6] [7] plus general information on the structure of social security statistics. The information in [6] consists of details of different aspects of social security eg health care, old-age support, and covers financing, organisation, legislation, benefits, etc. for each of the EEC countries. This is arranged in the publication in tabular form and consists mainly of text with some numeric data eg contribution rates. The basic blocks of information are text fragments describing a sub-topic for a specific country (in principle there should also be a time dimension). Users need to be able to look at data for a country, for a sub-topic, or for a topic. The hypertext is structured so that from the top level pages, which contain information on the system itself (how to move around) and the structure of the data (lists of topics and countries), the user can move to progressively more detailed information via summary pages (description of topic, sub-topic, country system, country-topic structure) choosing his route from embedded cross-references, or can go directly to pages of jumps for fast access to data. These jump pages consist of one or more levels of cross-tabulations of topics/sub-topics by country and activating a cell marker jumps the use directly to the relevant lower page. Figure 3 shows an example of some GUIDE input, with links visualised, which includes a jump table.

For expert access button on ⇨ *EXPERT.* ⇨

The document covers a number of aspects of social security schemes :

⇨ *ORGANISATION* ⇨, ⇨ *FINANCING* ⇨, ⇨ *HEALTH CARE* ⇨, SICKNESS / CASH BENEFITS, MATERNITY

for the countries :

⇨ *BELGIUM* ⇨, ⇨ *DENMARK* ⇨, ⇨ *GERMANY* ⇨, ⇨ *GREECE* ⇨, ⇨ *FRANCE* ⇨, ⇨ *IRELAND* ⇨, ⇨ *ITALY* ⇨, ⇨ *LUXEMBOURG* ⇨, ⇨ *NETHERLANDS* ⇨. ⇨ *UNITED KINGDOM* ⇨

QUICK ACCESS TABLE:

	FIN	ORG	HC
⇨B⇨	⇨ *B* ⇨	⇨ *B* ⇨	⇨ *B* ⇨
DK	⇨ *DK* ⇨	⇨ *DK* ⇨	⇨ *DK* ⇨
D	⇨ *D* ⇨	⇨ *D* ⇨	⇨ *D* ⇨
H	⇨ *H* ⇨	⇨ *H* ⇨	⇨ *H* ⇨
F	⇨ *F* ⇨	⇨ *F* ⇨	⇨ *F* ⇨
IRL	⇨ *IRL* ⇨	⇨ *IRL* ⇨	I⇨ *RL* ⇨
I	⇨ *I* ⇨	⇨ *I* ⇨	⇨ *I* ⇨
LUX	⇨ *LUX* ⇨	⇨ *LUX* ⇨	⇨ *LUX* ⇨
NL	⇨ *NL* ⇨	⇨ *NL* ⇨	⇨ *NL* ⇨
UK	⇨ *UK* ⇨	⇨ *UK* ⇨	⇨ *UK* ⇨

Figure 3. Example of GUIDE input, showing links and jump table.

In addition to the ability to move down the page hierarchy directly, either slowly or quickly, a number of threads have been built in to enable the user to move horizontally. For example having moved from the top level to a topic summary and then to a specific country topic structure description, it is possible to move to the 'next' country at that level along a horizontal thread. It is of course possible to move through any pre-chosen sequence by using the jump pages, backtracking to them between data page choices.

In relation to our list of techniques for helping the user, most are available in one or other of the packages:

- navigation: in both systems the data structure includes summary pages that describe the context and menu pages that give signposts to the next lower level. The jump table pages enable more expert users to overview the structure of the database. In the TextPro system the window titles enable the user to see where he has come from, which is a local context information map. Neither system provides a map of the connections, that can be viewed independently.

- movement: both systems support pre-set trails and auto-retrace. The database has been designed with numerous movement possibilities for retrieval, searching and scanning, through the menu and jump table pages. This recognises that the database is not only of interest for browsing.

- links: GUIDE provides more than one type of link, the note links are used for explanations and short expansions that will be read but are not of major importance. The different types of links are more conceptual than physical.

- filters: are not supported directly. However, the degree of explicit movement supported allows the user to control his path closely.

- access mechanisms: GUIDE allows the user to specify words for a text search in a document independently of the path facilities. This is a simple free text search and does not support logical operators. The access is from a pull-down menu. There are no other facilities.

- adaptation: GUIDE allows the user to add new links, although within the experimental context this will probably not be allowed. Both systems have online editors that can be used to cut and paste text to another document (user scrapbook). In practice this is much easier in the Windows environment of GUIDE as the second document can coexist in another window and so this facility will only be offered from the GUIDE version. Although GUIDE has the ability for the user to change the screen layout using the normal Windows capabilites, this is a difficult facility for novices to use.

There are a number of facilities missing that would improve the environment:

- a facility for the users to control the windowing so that they can decide which parts of windows are simultaneously visible. This needs to be more structured than is currently possible with Windows.

- the capability to move from this application to another and vice-versa. This would enable the user to call up the hypertext as an online conceptual help facility, or to call a database access as the last stage in a path in the hypertext. In the LIS context this is not possible as the main numeric databases are not on the same systems as the hypertext.

- additional tools to support information maps, so that the user can view the connections.

Experience with setting up the system has shown that the construction of hypertext adds about 100% to the time for inputting the data. There is a lack of tools for constructing the links. There were problems with presentation of data due to inflexibility of the window environment; where automatic line wrapping occurs and the window size is varied at runtime by the system, tabular presentation fails; conversely if line wrapping is not supported the user may lose ends of lines or have to keep moving the window relative to the text. More control over the environment is required both for the designer and the user. For example, if the designer could control the window size, as in Hype, and the user could easily move it on the screen, the result would be better.

The text was not available in machine readable form and so was entered using a word processor package. This created layout problems as the structure defined using the word processor eg indentations, was sometimes destroyed by the windowing in the hypertext system. This can only be overcome by training the secretarial staff to lay out hypertext documents, or by drastically simplifying the text layout, and is an example of some of the unforeseen stylistic problems in creating hypertext.

Future directions

The current version of the hypertext database is only a partial solution to the general LIS requirement for a front end to the main databases. It does not cover all the information required and is not sufficiently integrated with the data retrieval system. However, it should provide a basis for evaluating the techniques proposed. The next stage of the project is to present the system to the users at the LIS Summer School and obtain their responses to using it. The students will have access to both systems and to the original text. There will be some observation of use by the students, facilities for them to leave comments that may enable modifications to the systems to be made during the two week period, and structured interviews at the end on their experiences. If the response to the hypertext solution is positive, the next phase will be to define the facilities required from the interface more precisely and to develop a system to supply them. This should be possible either by using a more flexible system (for example the full system of which TextPro is a sub-set) or by developing independent tools to supply the missing elements (eg information maps). The enhanced system can then be used by the main LIS research group to expand the hypertext knowledge base.

References.

1. T.Smeeding, G.Schmauss and S.Allegreza. "An introduction to LIS." LIS-CEPS Working paper No.1. 1985.

2. T.Smeeding, R.Hauser, L.Rainwater, M.Rein and G.Schaber. "Poverty in major industrialised countries." LIS-CEPS Working paper No. 2. 1985.

3. G.Stephenson and I. Clowes, "Knowledge interrogation system for social and economic statistics." DOSES Proceedings 1987 (to be published).

4. D.A.Norman and S.W.Draper. Eds. "User centred system design - New perspectives on human-computer interaction." published by Lawrence Erlbaum. 1986.

5. I.Clowes. "Study of browsing techniques." ESPRIT EMG ASSISTANT project, internal report. LOGICA Cambridge. 1987.

6. J.Conklin. "Hypertext: an introduction and survey." IEEE Computer Sept.1987. pp.17-41.

7. "Comparative tables of the social security schemes in the member states of the European Communities." 13th ed. Commission of the European COmmunities. Luxembourg. 1985

8. "European system of integrated social protection statistics (ESSPROS) Methodology - Part 1." Eurostat. Luxembourg. 1981.

STATISTICAL RELATIONAL MODEL

Sakti P. Ghosh
Computer Science
IBM Almaden Research Center
San Jose, California.

Abstract

We will outline some of the important problems of extending the relational model for processing statistical data bases. We achieve this by augmenting to relational algebra few more numerical operations without altering the relational algebra. This algebra is referred to as statistical relational algebra. Applications of this extended extended relational model for solving some common statistical problems are discussed in the paper.

1. Introduction

In the last seven years many researchers in databases and statistics have been exploring and contributing in the area of statistical database management. The biannual workshop on statistical and scientific databases have contributed to this advancement also (Wong (1982a), Hammond & McCarthy (1983). Many of the research have concentrated on inventing new data models and data languages for processing statistical databases (Sue (1983), Shoshani (1982), Wong (1982), Ghosh (1984). Recently attempts have also been made to invent new algebra for statistical databases (Fortunato, Rafanelli, Ricci & Sebastio (1986)). Some work has also been done in the area of statistical metadata (Ghosh (1985), but no attempt has been made upto now to extend the relational model to handle statistical processing also. The relational data base model was invented by Codd (1970) to provide a simple representation of logical structured data using relational algebra. Many attempts have been made to extend the relational algebra for other data models (including applications like: office data, distributed data). These have heavily emphasized only the logical/communication aspects of data. Most of the data models have claimed that they are suitable for handling statistical data, which is true but they are very inefficient.

The goal of this paper is to extend the Codd's relational algebra by augmenting it with commands which perform statistical processing, without altering the relational structure of the data. Thus the extended relational model will have the capability of simultaneously performing logical structural and statistical manipulation of data when it is executing a command. The syntax of a query language for performing basic statistical processing within the relational DBMS environment and some proposed extensions of SQL (1983) are discussed. This language extensions can also be included in INGRES (Stonebraker (1986). This extension shifts many of the tasks of the statistical application programmers to the DBMS, and thus increasing efficiency of statistical applications.

One important feature of relational algebra is that a relational operation can be executed on a table (or two tables for joins) and the result is also a table. On a relational table it is possible to define separate types of algebra on columns (domains) and rows (tuples). We shall define the operations of relational algebra on the domains (attributes) and numerical algebra on the rows or tuples of the relation. This type of algebra is most advantageous for many statistical database management problems. Thus it is possible to define the relational operations of projection, selection, ordering, union, intersection, joins, outer joins on domains $D_0, D_1, \ldots\ldots, D_n$ of the relation R. We can also define numerical operations (see Ghosh (1987)) of sum, subtraction, multiplication, division, vector dot products, restricted dot products, vector products, restricted vector products on the rows of the relation. In order to utilize the full capability of the extended algebra for statistical database processing, we shall also add some non-relational operations to the domains of the relational table, such as: vector product, restricted vector product, etc. These operations on domains will preserve the closure property of the algebra (on the domains) for the relation.

Let us denote the relational model as

$$R(ID, D_1, D_2, \ldots\ldots, D_n)$$

where the domain ID is the identifier domain of the tuples. $D_1, D_2, \ldots\ldots, D_n$ are the attribute domains, some of which are numerical domains. Unless otherwise required, it is assumed that the domains $D_1, D_2, \ldots\ldots, D_n$ are numerical domains and are used for multivariate statistical analysis. The tuples of $R(ID, D_1, D_2, \ldots\ldots, D_n)$ correspond to data observed on individuals and are assumed to be statistically independent. The tuples of this relation are denoted by $r_1, r_2, \ldots\ldots, r_m$.

Let us denote by $\mathcal{A}(R)$ the category numeric algebra defined over the relation R. The structure of $\mathcal{A}(R)$ has been outlined and the operations will achieve the following:

(i) An extension of the relational algebra on the columns of R.

(ii) An extension of numerical algebra on the rows of R.

(iii) A set of operations between multiple tables.

2. Some new category numeric operations.

We shall discuss some operations which are combination of relational and statistical operations, thus enriching the structure of $\mathscr{A}(R)$.

Select-power(x)-aggregate: Some times it is necessary to select some of the tuples on a subset of domains of the relation based on some predicates and aggregate on a subset of the numerical domains. Suppose, we have a relation $R(D_1, D_2, ..., D_n)$. We want to select the tuples on the domains $D = (D_{i_1}, D_{i_2}, ..., D_{i_k})$ based on the criteria C_1 and then create aggregates on the domains $(D_{j_1}, D_{j_2}, ..., D_{j_l}) \subseteq (D_{i_1}, D_{i_2}, ..., D_{i_k})$. Sometimes it may be necessary to calculate squares or p^{th} power (p = 2, 3,.....) of the items in the domains $(D_{j_1}, D_{j_2}, ..., D_{j_l})$ before aggregation. We shall introduce one generic operation to perform this category-numeric operation on a relation, and it will be denoted as: SELECT-POWER(p)-AGGREGATE. A syntax for this operation, as an extension of SQL, may be:

SELECT-POWER(p)-AGGREGATE

$\equiv$ SELECT $D_{i_1}, D_{i_2}, ..., D_{i_k}$

FROM R

POWER(p)-AGGREGATE on $D_{j_1}, D_{j_2}, ..., D_{j_l}$

INTO R_1

WHERE C_1; (2.1)

Example 2.1: Consider the following relational table:

R(COUNTY	CITY	POPULATION	BUDGET)
Dixon	Dodge	250	35.5
Dixon	Orange	250	45.0
Dixon	Clara	500	37.2
Marin	Uba	350	45.2
Marin	Sun	350	32.0
Alameda	David	750	86.0

The result R_1 of the operation

SELECT COUNTY, POPULATION, BUDGET
FROM R
POWER(1)-AGGREGATE on BUDGET
INTO R_1
WHERE POPULATION < 400 ;

is given by the relation

R_1(COUNTY	POPULATION	BUDGET)
Dixon	250	80.5
Marin	350	77.2

Select-product-shift(x)-aggregate: This operation enables us to select certain set of tuples from a relation based on a criteria C_1, and then calculate the product aggregates between items (numerical values) of different domains, which have a column position difference of X (in the ordering of the domains of the relation). The parameter X can have positive or negative sign. Positive sign will indicate right shift and the negative sign will indicate left shift. Thus the operation SELECT-PRODUCT-SHIFT(+1)-AGGREGATE on $R(D_1, D_2, \ldots\ldots, D_n)$ will calculate the product aggregates of the domain pairs $D_1D_2, D_2D_3, \ldots\ldots, D_{n-1}D_n$. The same operation with X = -2, i.e. SELECT-PRODUCT-SHIFT(-2)-AGGREGATE will calculate the product aggregates for the following pairs of domains: $D_1D_3, D_2D_4, \ldots\ldots\ldots, D_{n-2}D_n$. The shift does not rap around like a circle. This non-cyclic property enables the computation of all bivariate correlation coefficients of a multivariate statistical distribution by choosing X = +1, +2,......, +(n-1). A formal syntax for this operation may be:

SELECT-PRODUCT-SHIFT(+/-p)-AGGREGATE
$\equiv$ SELECT $D_{i_1}, D_{i_2}, \ldots\ldots, D_{i_k}$
FROM R
PRODUCT-SHIFT(+/-p)-AGGREGATE on $D_{j_1}, D_{j_2}, \ldots\ldots, D_{j_l}$
INTO R_1

WHERE C_1; (2.2)

Example 2.2: Consider the data of the relation R (COUNTY, CITY, POPULATION, BUDGET) given next:

R(COUNTY	CITY	POPULATION	BUDGET)
Dixon	Dodge	250	35.5
Dixon	Dodge	300	45.0
Dixon	Clara	500	37.2
Marin	Uba	350	45.2
Marin	Sun	400	32.0
Alameda	David	750	86.0

The result R_1 of the operation

```
SELECT COUNTY, POPULATION, BUDGET
  FROM R
  PRODUCT-SHIFT(+1)-AGGREGATE on POPULATION, BUDGET
  INTO R1 (COUNTY, PR-AGGR)
  WHERE POPULATION < 400 ;
```

is given by the relation

R_1(COUNTY	PR-AGGR)
Dixon	22375.0
Marin	28620.0

Union-power(x)-aggregate:
Union-product-shift(x)-aggregate:
Intersection-power(x)-aggregate:
Intersection-product-shift(x)-aggregate:
Difference-power(x)-aggregate:
Difference-product-shift(x)-aggregate:

These six operations are similar to the two operations (based on SELECT) discussed before, only the component dealing with relational operation in each is different. The relational operations: UNION, INTERSECTION and DIFFERENCE have been defined by Codd(1970), so they are not repeated here. The structure and syntax of these six operations are similar to their counterpart operation with SELECT component, so we will not discuss in detail all the six operations.

Aggregate-distribution: Most of the statistical data is based on multiple attributes, some of which are category and some numerical. One of the simplest and most important step in statistical summarization of data, is to create multivariate frequency distributions. These multivariate frequency transformations (tables, distributions, etc.) are created by counting the number (frequency) of records, which have the same value set for a given subset of attributes, say, D_{i_1}, D_{i_2}, ..., D_{i_k}. The D_{i_u}s can be either numeric or category attributes. This operation will be denoted as AGGREGATE-DISTRIBUTION. The syntax for this operation is:

AGGREGATE-DISTRIBUTION ON $D_{i_1}, D_{i_2}, \ldots\ldots, D_{i_k}$
FROM R
INTO $R_1(D_{i_1}, D_{i_2}, \ldots\ldots, D_{i_k}, f)$

WHERE C_1; (2.3)

R_1 has one additional domain f, whose items are the frequency counts for the value vectors specified by the attributes $D_{i_1}, D_{i_2}, \ldots\ldots\ldots, D_{i_k}$ in a tuple. The values of f are obtained by first projecting the relation R on the domains $D_{i_1}, D_{i_2}, \ldots\ldots\ldots, D_{i_k}$ (without eliminating duplicate tuples). Then adding the items of the frequency domain for a value vector of $D_{i_1}, D_{i_2}, \ldots\ldots\ldots, D_{i_k}$ (if no frequency specified then every tuple has frequency 1) and assigning the sum to f. Then eliminating the duplicate tuples.

Example 2.3: Consider the following relation R.

R(CITY	AGE-GROUP	HEIGHT-GROUP	FREQUENCY)
San Jose	10	60	253
San Jose	20	60	302
San Jose	30	60	412
San Jose	10	65	237
San Jose	20	65	242
San Jose	30	65	205
Cupertino	10	65	323
Cupertino	20	65	222

Suppose we execute the operation:

```
AGGREGATE-DISTRIBUTION ON   AGE-GROUP, HEIGHT-GROUP
    FROM R
    INTO R1 (AGE, HEIGHT, FREQUENCY )
    WHERE CITY = San Jose or Cupertino;
```

Then the relation R_1 is given by

R_1(AGE	HEIGHT	FREQUENCY)
10	60	253
20	60	302
30	60	412
10	65	560
20	65	464
30	65	205

The result of the following operation

```
AGGREGATE-DISTRIBUTION ON   AGE
    FROM R1
    INTO R2 (AGE, FREQUENCY );
```

is given by

R_2(AGE	FREQUENCY)
10	813
20	766
30	617

We shall now introduce two new operations based on combination of relational joins with power-aggregates and power-shift-aggregates. These operations are discussed next.

Join-power(x)-aggregate: The simplest structure of this operation would be to join two relations based on a joining attribute equation and then performing aggregates on certain numerical domains of the joined relation. The syntax for such an operation may be expressed as:

$$\begin{aligned}
&\text{JOIN-POWER(p)-AGGREGATE} \\
&\quad \equiv \text{SELECT } D_{i_1}, D_{i_2}, \ldots\ldots, D_{i_k} \\
&\quad\quad \text{FROM } R_1, R_2 \\
&\quad\quad \text{WHERE } R_1.D_{u_1} = R_2.D_{u_2} \text{ AND } C_1 \\
&\quad\quad \text{POWER(p)-AGGREGATE on } D_{j_1}, D_{j_2}, \ldots\ldots, D_{j_l} \\
&\quad\quad \text{INTO } R_3;
\end{aligned} \tag{2.4}$$

Note: The domains $R_1.D_{u_1}$ and $R_2.D_{u_2}$ are domain compatible and $D_{j_1}, D_{j_2}, \ldots\ldots, D_{j_l}$ are numerical domains.

Example 2.4: Consider the following relations:

PARENT (P-NAME P-AGE)

P-NAME	P-AGE
John	25
Mary	23
David	33

CHILD(C-NAME P-NAME C-AGE C-HEIGHT)

C-NAME	P-NAME	C-AGE	C-HEIGHT
Sue	Mary	3	30
Dick	John	2	25
Nancy	John	4	32
Robert	Mary	5	33
Tracy	Mary	7	38

Result of the following operation

```
SELECT P-NAME, C-AGE
   FROM PARENT, CHILD
   WHERE PARENT.P-NAME = CHILD.P-NAME, AND C-HEIGHT > 20
   POWER(2)-AGGREGATE on C-AGE
   INTO PAR-AGE ( P-NAME, SQ-C-AGE );
```

is given by

PAR-AGE (P-NAME	SQ-C-AGE)
John	20
Mary	83

Relational-set-recursive-function: In many time series data analysis it is important to compute certain recursive function with respect to time, e.g. (i) in Markov process it is necessary to compute the transition probabilities between consecutive time states, (ii) in statistical process control based on CUSUM, it is necessary to compute recursively cumulative sums or run lengths as a sequence in time, etc. In a statistical database, in general, there are needs to compute such recursive functions on one or more attributes for certain category classes on selected tuples which are sequenced on the values of a particular attribute (which can be time stamp attribute). The general syntax for such category numeric operations is given by:

RELATION-SET $D_{i_1}, D_{i_2}, \ldots\ldots\ldots, D_{i_k}$
FROM R [, R_2].........[, R_n]
WHERE C_1
RECURSIVE on D_j([*Ascending*|*Descending*])
FUNCTION f

INTO R; (2.5)

The different subcommands covered by the RELATION-SET have been discussed before. The function f is a recursive function. The index for recursion is determined by the ascending/descending values of the domain D_j in the tuples selected by the other three suboperations of this operation. We shall now discuss some special cases of this operation.

Special cases.

Case (i): Run lengths on a subset of domains of a relation. The syntax for such a command is given by:

SELECT $D_{i_1}, D_{i_2}, \ldots\ldots, D_{i_k}$
FROM R
WHERE C_1
RECURSIVE on D_j (Ascending)
FUNCTION RUN-LENGTHs on $D_{j_1}(k_1^+), D_{j_2}(k_2^+), \ldots\ldots, D_{j_l}(k_l^+)$

INTO R_1; (2.6)

Note: The run-lengths on the items of the domain D_{j_u} are computed based on the threshold value k_u^+ for $u = 1,2,\ldots,l$. Suppose the values of D_{j_u} (sequenced w.r.t. values of D_j) are: 3.5, 2.3, 6.7, 1.2, 3.4; and $k_u^+ = 2.0$. Then the run-lengths for the values of D_{j_u} are 1, 2, 3, 0, 1.

Note: Statistical functions like: mean, median, quantiles, etc. of run lengths can be included in the statement as a component of FUNCTION.

Example 2.5: Consider the following relation:

DEPOSIT(	BATCH	STATION	TIME	THICH1	THICH2)
	B1	S1	7:00	2.3	3.4
	B2	S2	7:00	2.9	3.0
	B1	S1	8:00	2.4	3.5
	B2	S2	8:00	2.8	3.0
	B1	S1	7:30	2.9	3.2
	B2	S2	7:30	3.0	3.2
	B3	S1	8:30	2.8	2.9
	B3	S1	9:00	2.3	2.8
	B4	S3	7:30	4.6	2.9

The result of the following operation

```
SELECT BATCH, STATION, TIME, THICK1, THICK2
   FROM  DEPOSIT
   WHERE BATCH ≠ B4
   RECURSIVE on TIME (Ascending)
   FUNCTION  RUN-LENGTHs on THICK1 (2.80+), THICK2 (3.00+)
   INTO DEP-RUN ;
```

is given by:

DEP-RUN(BATCH STATION TIME RUN1 RUN2)

BATCH	STATION	TIME	RUN1	RUN2
B1	S1	7:00	0	0
B1	S1	7:30	0	1
B1	S1	8:00	0	2
B2	S2	7:00	0	0
B2	S2	7:30	1	1
B2	S2	8:00	2	2
B3	S1	8:30	0	0
B3	S1	9:00	0	0

Case (ii): Cumulative Sum (CUSUM) is extensively used in statistical process control systems based on sequential testing (van Dobben de Bruyn (1968), Yashchin (1984)). It involves computing the following cumulative sum S_i from a sequence of observations x_i, $i = 1,2,\ldots\ldots;$

$$S_i = \max \{ S_{i-1} + (x_i - k),\ 0\}$$

where $S_0 = 0$, and k is a threshold.

The syntax for an operation for creating CUSUMs for automatic statistical process control is given by:

SELECT $D_{i_1}, D_{i_2}, \ldots\ldots\ldots, D_{i_k}$
FROM R
WHERE C_1
RECURSIVE on D_j (Ascending)
FUNCTION CUSUMs on $D_{j_1}(k_1), D_{j_2}(k_2), \ldots\ldots, D_{j_l}(k_l)$
INTO R_1; (2.7)

Example 2.6: Let us consider the relational table DEPOSIT given in example 4.9. The result of the following operation

```
SELECT BATCH, TIME, THICK1, THICK2
   FROM  DEPOSIT
   WHERE STATION ≠ S3
   RECURSIVE on TIME (Ascending)
   FUNCTION  CUSUMs on THICK1(2.80), THICK2(3.00)
   INTO CUSUM (BATCH, TIME, CUSUM1, CUSUM2);
```

is given by:

CUSUM (

BATCH	TIME	CUSUM1	CUSUM2)
B1	7:00	0.0	0.4
B1	7:30	0.1	0.6
B1	8:00	0.0	1.1
B2	7:00	0.1	0.0
B2	7:30	0.3	0.2
B2	8:00	0.3	0.2
B3	8:30	0.0	0.0
B3	9:00	0.0	0.0

3. Conclusions

The main theme of this paper was to invent a set of commands, which can manipulate statistical data (which contains logical and numerical structured data) and still preserve the relational representation. Thus we have extended relational algebra to handle statistical databases. We have also developed a syntax to handle these commands. The syntax is an extension of SQL. In this extension of SQL, we have added new subcommands in the SQL statements to handle the statistical manipulation of the data. It is well known, that database users have been performing statistical analysis in the application program environment but our feeling is that, if the database management system has the capability to take over some of the basic statistical manipulation of the statistical information, then not only it would simplify application programming but also will increase the performance of statistical analysis programs.

One of the questions raised is that: What is the completeness of this extended algebra? The question is difficult to answer, but we will discuss some of the classes of statistical analysis problems that are covered by this extended algebra.

(i) Relation-set-power(x)-aggregate operations will provide logical database creation solutions for the class of univariate statistical moments computations in a statistical database management (SDBM) environment. It can also be used to provide solutions to creation of category classification tables, which are used in computing contingency χ^2 for measuring associations.

(ii) Relation-set-product-shift(x)-aggregate operations provide simple and powerful statistical database management solutions for the class of total correlation coefficient computations in multivariate SDBM environments. These total correlation coefficients are used in computing multiple correlation coefficient, multiple regression analysis, covariance matrix analysis, partial correlation analysis, etc.

(iii) Join-power(x)-aggregate operation will solve the statistical problems indicated in (i) and (ii) for distributed SDBM systems.

(v) Relational-set-recursive-function operations have been designed to solve SDBM problems associated with stochastic process analysis. These have many applications in automated manufacturing process control and statistical quality control using CUSUMs.

4. Acknowledgement

The author would like to thank Dr. Kwan Wong of IBM Almaden Research Center for his suggestion, support and encouragement of this work.

5. REFERENCES

1. van Dobben de Bruyn D. S. (1968). *Cumulative Sum Tests Theory and Practice,* Published by Hafner Publishing Co., New York.

2. Ted (E. F.) Codd (1970). *A Relational Model of Data for Large Shared Data Banks.* Comm. ACM, Vol. 13, No. 6, pp. 377-387.

3. E. Fortunato, M. Rafanell, F. L. Ricci, A. Sebastio (1986). *An Algebra for Statistical Data.* Proc. of the Third Int. Workshop on Stat. & Scientific DB Magt., pp. 122-134.

4. Sakti P. Ghosh (1984). *Statistical Relational Tables for Statistical Database Management.* IBM Research Report No. RJ 4394; IEEE Trans. of Software Engineering vol. SE-12, No. 12 pp. 1106-1116, Dec. 1986.

5. Sakti P. Ghosh (1985). *Statistics Metadata: Linear Regression Analysis.* Proc. of Int. Conf. on Foundations of Data Organization, Kyoto, Japan, May 21-24, pp. 3-12. Also published as IBM Research Report (1984) No. RJ. 4444.

6. Sakti P. Ghosh (1987). *Numerical Operations on Relational Database.* IBM Research Report No. 5605. Accepted for publication in IEEE Trans. of Software Engineering.

7. R. Hammond and John L. McCarthy, (Edited 1983). *Proceedings of the second International Workshop on Statistical Database Management.* Los Altos, California, September 27-29, 1983.

8. Shoshani A. (1982). *Statistical Databases: Characteristics, Problems, and Some Solutions.* Proc. of the 8th Int. Conf. on Very Large Data Bases (VLDB), pp. 208-222.

9. SQL/Data System Application Programming (1983). IBM Program Product. SH24-5018-0.

10. Stonebraker M (edited 1986). *Ingres Papers.* published by Addision-Wesley Publication, New York, NY, USA.

11. Stanley S. Y. W. Su (1983). *SAM*: A Semantic Association Model for Corporate and Scientific-Statistical Databases.* J. of Information Sc., vol. 29, pp. 151-199.

12. Harry K. T. Wong (edited 1982). *A LBL Perspective on Statistical Database Management.* published by Lawrence Berkeley Laboratory, University of California, Berkeley, CA-94720.

13. Harry K. T. Wong (edited 1982a). *Proceedings of the First LBL Workshop on Statistical Database Management.* Melno Park, California,

14. Emmanuel Yashchin (1984). *CUSUM Control Schemes: Method, Analysis, Software.* IBM Research Report, RC 10718.

15. Proceedings of the Third International Workshop on Statistical and Scientific Database Management. Luxembourg, July 22-24, 1986.

A MODEL OF SUMMARY DATA AND ITS APPLICATIONS IN STATISTICAL DATABASES

Meng Chang Chen, Lawrence McNamee, Michel Melkanoff
Computer Science Department, UCLA

Abstract. The summary (statistics) data model described herein is an extension of the relational model. The concept of category (type or class) and the additivity property of some statistical functions form the basis of this model. In this approach category shields details of a database instance from users, and plays an important role in deriving new statistics data. *Statistics data* is a trinary tuple consisting of <statistical function, category, summary>. The additivity property allows new *statistics data* to be generated without having to access the original database. *Statistics data* is meta-knowledge summarized by statistical functions of the detailed information typically stored in a conventional database. Unfortunately, deciding whether a category is derivable from a set of categories, in general, is NP-hard. The proposed generating category set can resolve the intractability problem of the category derivation. The derivation of new *statistics data* within a relation or on multi-relations is investigated, and the efficiency and correctness of the stored *statistics data* are guaranteed when the original database is updated or when new *statistics data* is obtained. Finally potential applications and security concerns applying to this model are also discussed.

1. INTRODUCTION

The characteristics and problems of statistical databases have been introduced in [1, 22, 6, 12, 23]. The major difference between statistical database management systems(SDBMS) and conventional database management systems(DBMS) is that conventional databases manage individual records, whereas statistical databases deal with summaries of raw databases and tend to ignore the details of individuals. They also differ in the type of queries they can handle. In effect a conventional DBMS cannot store statistical data nor process statistical applications efficiently [6]. Thus another approach to statistical database management systems needs to be taken.

The summary data model proposed herein aims at prompt responses to statistical queries by modeling and processing stored *statistics data*. This data model also serves as a bridge between database management systems and sophisticated statistical functions. A category (called type or class in other papers) is a set of tuples(or objects) satisfying the definition of the database. While the definition of a category is static, the tuples of an instance of this category vary as the database is updated. A statistical function, which is a set function, inputs a set of tuples and generates a real number result, called a summary. An additive statistical function is homomorphic from the set disjoint union to an additive operator on real numbers. *Statistics data* contains a category, an additive statistical function, and a summary which is a result of the statistical function of an instance of the category. If *statistics data* is either precomputed and stored or derivable from other *statistics data*, the usage of *statistics data* in answering statistical queries can reduce the search time tremendously since there is no need to scan through the original database.

The amount of *statistics data*, which could be very huge, is determined by the granularity of a category and the number of additive statistical functions required, and is not proportional to the number of tuples in the database. The proposed summary data model provides a technique for storing the necessary *statistics data* and a method to derive new *statistics data*; thus efficient storage and management of *statistics data*, and prompt responses to statistical queries is possible. In the relational model, join operations are used to obtain information on related relations. As in this model similar operations are proposed to derive *statistics data* on related relations from precomputed *statistics data*. When database updates create an inconsistency problem between *statistics data* and the original database, this model provides an efficient algorithm to reconcile the difference. Besides providing fast response time to statistical queries, *statistics data* can also serve as the mechanism by which aggregation-based integrity maintenance and query optimization are supported. In addition, this model can be applied to statistical query processing on distributed databases.

2. RELATED WORK

Smith et al [24] discuss the concept of aggregation and generalization in database design. Following their classification, statistics data in our model consists of the "non-inheritable attributes" which "are often formed by applying 'summarizing' functions to instances of a category". Walker [27] looks into the homomorphism property of relational algebra and proposes a measurement of an inexact answer. Sato [21] gives a primitive definition of category and its derivability property. Ikeda and Kobayashi [16] use a data dictionary/directory to store information by classification of category attributes and statistical procedures, and provide some algebraic operations for interactive statistical analysis. Koening and Page [18] discuss update problems for derived relations. In their work, all the derived relations are realized, and a chain rule is used to update the derived data. Klug [17] gives a formal treatment to aggregate functions in relational algebra and calculus without the use of the notation 'duplicates.' Nwokogba and Rowan in [19] describe the derivability problem of the non-disjoint category set and in [20] propose a statistical parameterization model whereby summaries are stored for further analyses. Ghosh in [12] advocates storing summaries(metadata) for subsequent usage, and introduces extension versions of SQL and QBE for summary tables. In [13, 11] Ghosh also investigates the problem of using metadata in real time linear regression analysis, and discusses in depth the applications and properties of metadata in statistics. Hebrail [15] describes the homomorphism property of statistical functions, properties of summary data, and their applications, based on the partition model in the environment of a single relation or universal relation. Fortunato et al [7] give an algebraic interpretation of operations on summaries, whereas Ghosh [14] introduces category numerical operations into relational algebra for supporting statistical analyses. In [10, 25], access methods are proposed for storing partial results for statistical processings, and in [8, 28] the computing complexity of answering queries from partial sums is discussed.

From our survey of the above papers, the following features of our work have not been previously described.

1. The introduction of category which shields details of a raw database from users and the operations on category for *statistics data* derivation.
2. The categorization of statistical functions into additive and computed functions.
3. A formalism and a proof of the time complexity for the derivability problem.
4. An introduction of the generating category set and a proposal for its generation and consistency

maintenance algorithm.

5. A discussion and proposal of operations for the derivation of *statistics data* of multi-relations using the relational model.

3. SUMMARY DATA MODEL

Many applications can be described naturally by entity types, relationships, and attributes [4, 3]. In a relational database, entity types and relationships are represented as relations [26]. A relation scheme, R, is a collection of attributes. Each attribute, α, has a domain, denoted as Domain(α). Let the cross product of the domain of each attribute of a relational scheme R be denoted as Domain(R). A relation (table) instance is any subset of Domain(R). A database scheme(schema) is a set of relation schemes. In statistical databases, an attribute may be a category attribute or a numeric attribute or both. The values of category attributes describe the qualitative characteristics of tuples, while numeric attributes contain the quantities that are summarized by statistical operations. The classification of category and numeric attributes might differ among statistical database users.

3.1 Category

The concept of category in this model is similar to type or class in many knowledge representation systems. A *category* can be defined either explicitly as a set of tuples(or objects), or implicitly as a sentence of relational algebra or calculus. An *orthogonal category* is a category which can be represented as a cross product of subsets of domains for each attribute. The subsets of domains are categorized into *interval* subsets, like { $15 \le \text{Age} \le 30$ }, and *non-interval* subsets, like {red, white, blue}. Hereafter, when there is no ambiguity, categories are assumed orthogonal.

Figure 1 is an example of a relation instance which will be used throughout this paper. The category 'male employees in CS department' is represented as:

$$N\times\{CS\}\times\{male\}\times\{1...100\}\times\{Manager,Engineer,Secretary\}\times\{15...100\}$$

where $\times$ denotes cross product. For simplicity, a category can also be represented in a shorthand which removes domains that are not restricted. Thus the above category can be abbreviated as $\{CS\}\times\{male\}$.

EMP#	Department	Sex	Age	Position	Income(k)
001	AD	male	42	Manager	75
006	AD	male	40	Manager	60
014	AD	female	55	Secretary	35
030	EE	female	42	Manager	60
034	EE	female	35	Engineer	52
057	EE	male	28	Engineer	45
089	EE	male	23	Secretary	25
095	EE	female	29	Engineer	40
121	CS	male	49	Manager	62
124	CS	male	40	Engineer	55
143	CS	female	27	Engineer	34
177	CS	male	31	Secretary	31

Figure 1 A relation instance of employee data

Categories are the objects of interests of statistical queries such that the details of the database are shielded from users, and are used as objects in the derivation of new *statistics data.* A *category instance* is a set of tuples resulting from the restriction of a relation instance by a given category. From the operational point of view, this restriction operation is precisely the set intersection of the relation instance and the category, or a selection operation in languages like SQL. Thus a category instance of category C of relation instance R is properly denoted as C∩R.

3.2 Statistical Functions and Statistics data

Hereafter *Real* denotes the set of real numbers, and *RS* denotes the set of all relation instances of a relation scheme, R.

Definition $S: RS \rightarrow Real \cup \{ \lambda \}$ is called a statistical function which summarizes a set of tuples against some *attribute* (*s*)‡. A statistical function is said to be of *degree n* if it summarizes against n attributes. The output of a statistical function is called the *summary* of the input relation instance using the statistical function. *Statistics data* is a 3-tuple consisting of the name of a statistical function, a category, and a summary. □

Example The category instance of male CS employees of relation instance as in Figure 1 is the set of tuples with EMP# {121, 124, 177}. The statistical function, "average income" of degree 1, generates the summary, 39.3. For another statistical function, "sum of products of age and income" of degree 2, the summary is 6199. The *statistics data* for the above examples are, respectively,

<average income, {CS}×{male}, 39.3>, and

<sum of products of age and income, {CS}×{male}, 6199>. □

Statistics data are precomputed and stored in this model. The advantage of using *statistics data* in answering statistical queries is the reduction of response time, as conventional database management systems need to scan through raw databases. However, the amount of all possible *statistics data* is of the order of the product of the sizes of all possible categories and the number of statistical functions. In the worst case, the number of all possible categories can be up to $\prod_{a_i \in R} 2^{|Domain(a_i)|}$. The huge amount of potential *statistics data* not only requires a large storage capacity and a lengthy processing time, but also complicates the integrity maintenance problem when the database is updated. Thus, it is not practical to store all possible *statistics data*. Instead, in our model only selected *statistics data* are stored which will then be used to derive new *statistics data* by applying inference mechanisms.

3.3 Additive and Computed Property of Statistical Functions

Definition A statistical function, S, is called *additive* if $\forall r_i, r_j \in$ RS, and $r_i \cap r_j = \emptyset$, s.t. $S(r_i \cup r_j) = S(r_i) +_S S(r_j)$, where $(Real \cup \{ \lambda \}, +_S)$ forms a commutative group. □

The *additivity* property allows additive statistical functions to generate the summary of a set of tuples directly from the summaries of a partition of the set. Ordinarily the operation $+_S$ is an arithmetic addition. The requirement of a group guarantees the existence of an inverse for elements with respect to $+_S$, which permits direct computation for updating. In fact, this additivity property is a homomorphism from a boolean lattice formed by disjoint sets to the set of real numbers and λ with addition. The following proposition shows that if a statistical function is additive, given a relation instance, and if the summaries (of this statistical function) of two disjoint category instances are known, then the summary

‡ λ is defined as: $\forall e \in Real \cup \{ \lambda \}, \lambda + e = e$. Apparently, $S(\phi)=\lambda$, where ϕ denotes empty set.

of the union of two disjoint category instances can be calculated directly from the known summaries. Thus, when deriving a new summary for an additive statistical function, there is no need to know the details of the relation instance from which the summary comes, but one only needs to know the categories and their associated summaries.

Proposition 1 Given an additive statistical function S, and relation instance R, for categories, C_i, C_j, C_k, and requiring that $C_i \cap C_j = \varnothing$ and $C_k = C_i \cup C_j$, then $S(R \cap C_k) = S(R \cap C_i) + S(R \cap C_j)$.

Proof Since $C_i \cap C_j = \varnothing$ and $C_k = C_i \cup C_j$, then, $\forall$ R, $(R \cap C_i) \cap (R \cap C_j) = \varnothing$ and $(R \cap C_k) = (R \cap C_i) \cup (R \cap C_j)$. Following the additivity property of S, we obtain the result $S(R \cap C_k) = S(R \cap C_i) + S(R \cap C_j)$. ■

Proposition 1 also states that $S(R \cap C_i) = S(R \cap C_k) + (S(R \cap C_j))^{-1}$. The additive functions include many functions like sum, square sum, cardinality, while some functions like maximum and minimum are not additive since *(Real*$\cup$*{ λ }, maximum)* and *(Real*$\cup$*{ λ }, minimum)* are not commutative groups. Though some functions, like average, are not additive, they can still be computed directly employing the summaries of other additive statistical functions or computed results of other 'nice' statistical functions.

Definition A statistical function S is called *computed* from the results of other additive or computed statistical functions $S_1, \ldots, S_n$, which are called the component functions of S, if S can be represented as:

$$\exists f,\ \forall r_i \in RS,\ S(r_i) = f(S_1(r_i), \cdots, S_n(r_i)),$$

where f: $(Real \cup \{ \lambda \})^n \rightarrow (Real \cup \{ \lambda \})$, and there is no cycle(i.e. quasi-order) in the definition of computed functions. □

The concept of *statistics data* can be applied to very complex applications. For example, Ghosh in [13] shows how *statistics data* can achieve real time linear regression analysis.

Example The *covariance* S_{xy} is a computed statistical function, since

$$S_{xy} = \frac{1}{n-1} \sum_{j=1}^{n} (x_j - \bar{x})(y_j - \bar{y})$$

$$= \frac{1}{n-1} \left[\sum_{j=1}^{n} x_j y_j - \frac{1}{n} \left(\sum_{i=1}^{n} x_i \right) \left(\sum_{i=1}^{n} y_i \right) \right]$$

S_{xy} can be computed directly from the results of additive functions, including n (cardinality), $\sum_{j=1}^{n} x_j y_j$ (sum of products), and $\sum_{i=1}^{n} x_i$ and $\sum_{i=1}^{n} y_i$ (sum). □

3.4 Derivability

Proposition 1 connotes an approach to derive new *statistics data.* In this section, the computing procedures of *statistics data* and the computational complexity of derivation process is discussed.

Definition Let A and B be categories. A *proper difference*, denoted by $\ominus$, is defined as $A \ominus B = A - B$ if $B \subseteq A$. A *disjoint union*, denoted by $\oplus$, is defined as $A \oplus B = A \cup B$ if $A \cap B = \emptyset$. A category is *derived* from a set of categories, if it can be expressed as a finite expression consisting of categories in the set as operands, $\oplus$ and $\ominus$ as operators, and matched pairs of parentheses. □

A category is equal to an expression if it is equal to the execution result of the expression, which can be considered as the operational semantics of the expression.

Proposition 2 For a relation instance R_1 and an additive statistical function S, if a category C_k can be derived from a set of categories C, the summary of the category instance, $S(C_k \cap R_1)$, can be also obtained from the summaries or inverses of the summaries of the associated *statistics data.*

Proof Given a set of categories C, and a category C_k which is derivable from C. Then C_k can be represented by a finite expression, EXP, such that

$$C_k = EXP(C_{i_1}, \cdots, C_{i_n}),$$

where $C_{i_j} \in C$ as operands, $\ominus$ and $\oplus$ as operators, and with matched parentheses.
Intersecting R_1 with both sides of the equations, we obtain

$$C_k \cap R_1 = EXP(C_{i_1} \cap R_1, \cdots, C_{i_n} \cap R_1).$$

This equation is still satisfied since $\cap$ is distributive over $\oplus$ and $\ominus$, and any subexpressions in $EXP(C_{i_1} \cap R_1, \cdots, C_{i_n} \cap R_1)$ is still defined under $\oplus$ and $\ominus$.
Then, applying the additive statistical function on both sides, we arrive at

$$S(C_k \cap R_1) = S(EXP(C_{i_1} \cap R_1, \cdots, C_{i_n} \cap R_1)).$$

Finally applying the following axioms iteratively, $S(A \oplus B) = S(A) + S(B)$ and $S(A \ominus B) = S(A) + (S(B))^{-1}$, to the expression, where A and B are subexpressions, the original expression eventually is replaced by an expression with $S(C_{i_j})$ or $S(C_{i_j})^{-1}$ as operands, addition as operations, and with matched parentheses. ■

The above proposition indicates an approach for obtaining a new summary of additive statistical functions from precomputed *statistics data*, i.e. finding categories which are needed to construct the target category first, then summing up their corresponding summaries or inverses of summaries. For computed statistical functions, we execute their component functions first, then execute the computed functions. However, proposition 3 below shows, in general, that this approach suffers from computational intractability.

Example This example shows how to calculate the covariance of Age and Income of all the employees in the category {Dept=AD}. Given a relation instance R_1 as represented Figure 1 and a set of categories, G, $\{g_1, g_2, \cdots, g_n\}$ where g_1 = {AD}×{male}, g_2 = {AD}×{female}, g_3 =

{EE}×{male}, g_4 = {EE}×{female}, g_5 = {CS}×{male, female}, and g_6, $\cdots$, g_n are arbitrary categories. The *statistics data* of categories, g_i, and additive functions, Cardinality, $\sum Age$, $\sum Income$, and $\sum Age \cdot Income$ are represented in tabular form as in Figure 2(a). The category of interest is

$$\{\text{Dept=AD}\} = g_1 \cup g_2.$$

For each additive function S, S({Dept=AD}$\cap R_1$) = S($g_1 \cap R_1$) + S($g_2 \cap R_1$).

The *statistics data* of the category {Dept=AD}, shown in Figure 2(b), can be obtained by summing up from some summaries in Figure 2(a). Since the function of covariance is a computed function with component functions, Cardinality, $\sum Age$, $\sum Income$, and $\sum Age \cdot Income$, the covariance can be obtained simply by using the *statistics data* of Figure 2(b). Figure 2(c) shows the calculation for covariance. □

Category	Cardinality	$\sum Age$	$\sum Income$	$\sum Age \cdot Income$
g_1	2	82	135	5550
g_2	1	55	35	1925
g_3	2	51	70	1835
g_4	3	106	152	5500
g_5	4	147	182	7117
...	...	...	...	...

(a) Statistics data of category set G

Category	Cardinality	$\sum Age$	$\sum Income$	$\sum Age \cdot Income$
{Dept=AD}	3	137	170	7475

(b) Statistics data of the category {Dept=AD}.

$$S_{Age,Income}(Domain(R) \cap R_1) = \frac{1}{n-1}\left[\sum_{j=1}^{n} Age_j \cdot Income_j - \frac{1}{n}\left(\sum_{i=1}^{n} Age_i\right)\left(\sum_{i=1}^{n} Income_i\right)\right]$$

$$= \frac{1}{2}\left[7475 - \frac{1}{2} \times 137 \times 170\right] = -2085$$

(c) Calculation of covariance

Figure 2. Calculate the Covariance of Age and Income

Proposition 3 To determine whether a category is derivable from a set of categories is an *NP-hard* problem.

Proof (in sketch) [2] We show that this problem is NP-hard by constructing a polynomial transformation from the well-known NP-complete problem, namely, the SAT problem [9], to a one-attribute derivability problem.

A *SAT* problem is defined as:

Given a set U={u_1, u_2, $\cdots$, u_n} of variables, a collection CC={c_1, c_2, $\cdots$, c_m} of clauses over U,

QUESTION: Is there a satisfying truth assignment for CC?

A set is constructed for each variable u_j or $\overline{u}_j$ in U of the SAT problem,

$$\{a_j, u_j, (c_{j_1}, u_j), \cdots, (c_{j_d}, u_j) \text{ where } u_j \in c_{j_i} \text{ in SAT}\}.$$

a_j prohibits that both sets for u_j and $\overline{u}_j$ be selected. Selection from the collection of sets for each u_j and $\overline{u}_j$ is equivalent to the truth assignments of the SAT problem. After the selections, all the (c_i, u_j) will be reduced to c_i. If all the c_i in CC are obtained, the SAT problem is satisfied. Finally, a garbage collection mechanism is needed to remove unnecessary elements. ■

3.5 Generating Category Set

The intractability problem results from the fact that categories overlap each other. Our solution to achieve a fast derivation is to decompose the overlapping category sets into disjoint category sets, which still preserve at least the same expressiveness, i.e. the decomposed disjoint category set can derive all the categories derivable from the former. For two orthogonal categories A and B, their intersection is also orthogonal, while A-B and B-A might not be orthogonal. The only approach to preserve the orthogonal property on A-B and B-A is to decompose A-B and B-A into a group of smaller orthogonal categories.

Proposition 4 For two orthogonal categories of n attributes, there exists a set of at most 2*n+1 disjoint orthogonal categories such that A, B, and A$\cap$B are derivable from the set.

Proof We will prove this proposition by induction on the number of attributes. Let A = $a_1 \times a_2 \times \cdots \times a_m$ and B = $b_1 \times b_2 \times \cdots \times b_m$.

For i=1.

Case 1. $b_1 \subseteq a_1$ & attribute 1 has interval subset.

The set $\{a'_1, b_1, a''_1\}$ meets the requirement, where a'_1, b_1, a''_1 are disjoint and their union is a_1.

Case 2. Otherwise.

The set $\{a_1 - b_1, b_1 - a_1, a_1 \cap b_1\}$ meets the requirement.

For i=k$\geq$1.

Assume the assertion holds for i=k.

For i=k+1.

Let A = $a_1 \times a_2 \times \cdots \times a_{k+1}$ and B = $b_1 \times b_2 \times \cdots \times b_{k+1}$,

Case 1. $b_{k+1} \subseteq a_{k+1}$ & attribute 1 has interval subset.

Construct five categories,

$c_1 = a_1 \times a_2 \times \cdots \times a'_{k+1}$,

$c_2 = a_1 \times a_2 \times \cdots \times a''_{k+1}$,

$c_3 = (a_1 \times a_2 \times \cdots \times a_k - b_1 \times b_2 \times \cdots \times b_k) \times b_{k+1}$,

$c_4 = (a_1 \times a_2 \times \cdots \times a_k \cap b_1 \times b_2 \times \cdots \times b_k) \times b_{k+1}$

$c_5 = (b_1 \times b_2 \times \cdots \times b_k - a_1 \times a_2 \times \cdots \times a_k) \times b_{k+1}$,

where a'_{k+1}, b_{k+1}, and a''_{k+i} are disjoint and their union is a_{k+1}.

From the assumption in step i=k, there exists a set CS with at most 2*k+1 disjoint orthogonal categories such that $(a_1 \times a_2 \times \cdots \times a_k - b_1 \times b_2 \times \cdots \times b_k)$, $(a_1 \times a_2 \times \cdots \times a_k \cap b_1 \times b_2 \times \cdots \times b_k)$, and $(b_1 \times b_2 \times \cdots \times b_k - a_1 \times a_2 \times \cdots \times a_k)$ are derivable.

Thus c_3, c_4, and c_5 are derivable from CS$\times b_{k+1}$.

Since A = $c_1 \oplus c_2 \oplus c_3 \oplus c_4$, B = $c_4 \oplus c_5$, A$\cap$B = c_3, A and B and A$\cap$B are derivable from CS$\times b_{k+1} \cup \{c_1\} \cup \{c_2\}$ which has at most 2*(k+1)+1 categories.

Case 2. Otherwise.

Redefine c_i as:

$c_1 = a_1 \times a_2 \times \cdots \times a_k \times (a_{k+1} - b_{k+1})$,

$c_2 = b_1 \times b_2 \times \cdots \times b_k \times (b_{k+1} - a_{k+1})$,

$c_3 = (a_1 \times a_2 \times \cdots \times a_k - b_1 \times b_2 \times \cdots \times b_k) \times (a_{k+1} \cap b_{k+1})$,

$c_4 = (a_1 \times a_2 \times \cdots \times a_k \cap b_1 \times b_2 \times \cdots \times b_k) \times (a_{k+1} \cap b_{k+1})$,

$c_5 = (b_1 \times b_2 \times \cdots \times b_k - a_1 \times a_2 \times \cdots \times a_k) \times (a_{k+1} \cap b_{k+1})$.

The reasoning is the same as case 1
except $A = c_1 \oplus c_3 \oplus c_4$, $B = c_2 \oplus c_4 \oplus c_5$, $A \cap B = c_3$.

This completes the proof. ■

In real applications, the original category set whose elementary category is frequently utilized by users is decided by the Data Base Administrator(DBA) during the database design phase. In order to solve the possibly computationally intractable problem, this category set has to be decomposed into a disjoint category set by decomposing any pair of intersecting categories following Proposition 4. The size of the resulting disjoint set will be no greater than $\prod_i s_i$, where s_i is the number of distinct symbols of attribute i in the original category set.

Definition A disjoint category set C is called a *generating category set* of a relational scheme R if $\forall$t, t $\in$ Domain(R), $\exists C_i \in C$, such that t$\in C_i$. A subset, MC, of a generating category set is called the *minimum cover* of a category C_d, if $C_d \subseteq \bigcup_{C_i \in MC} C_i$ and $\forall C_j \subseteq$ MC, $C_d \not\subseteq \bigcup_{\substack{C_i \in MC \\ C_i \neq C_j}} C_i$. □

In this proposed data model, each relation scheme has a corresponding generating category set. In addition, *statistics data* for each category in the generating category set are precomputed and stored for some selected additive statistical functions. When a summary of a category is queried, an unique minimum cover can be determined. Hence the queried summary can be obtained from the stored *statistics data*, or a partial solution plus an unsolvable partial query will result. If new *statistics data* becomes available and its associated category C_d meets the following condition,

$$\exists C_k, C_1, \cdots, C_m \in \text{GS, and } C_k \ominus (C_d \ominus C_1 \ominus \cdots \ominus C_m) \text{ is orthogonal,}$$

the *statistics data* of C_k will be replaced by the *statistics data* of $C_k \ominus (C_d \ominus C_1 \ominus \cdots \ominus C_m)$ and $(C_d \ominus C_1 \ominus \cdots \ominus C_m)$. If the above condition is not met, the *statistics data* of C_d might be stored temporarily. An access method, which is not covered in this paper. is provided to store, update, manage and access the *statistics data*.

Example Assume we have the *statistics data* as shown in Figure 2(a). A query,

"What's the total salary of female employees?",

has a minimum cover $\{g_2, g_4, g_5\}$. However this query cannot be answered directly from the *statistics data*. A partial answer results in \$187,000, which can be obtained from the *statistics data* of g_2 and g_4, and an unsolvable partial query "What's the total salary of female employees in CS department?". If this query is solved from the original database and the answer is \$34,000, then the *statistics data* of g_5 can be replaced by <$\sum$*Income*, {CS}×{female}, 34k> and <$\sum$*Income*, {CS}×{male}, 148k>. □

3.6 Statistics Data on Related Relations

In the relational model, a relation represents an entity set or a relationship in an Entity-Relationship(ER) model or its extensions and variations. In addition, a collection of related relations, which are linked via identical values of common attributes of the relations, represents entity sets and their relationships. In a normal situation, those related relations can be considered as a lossless decomposition of the pseudo universal schema. Hereafter, this lossless decomposition property is assumed. Thorey, et al [26] gives a clear survey of methods for transforming an extended ER conceptual model into a relational model.

A query about the summary on two related relations defines two categories, *summarized category* and *qualified category*. Assume numeric attributes appear in one relation scheme only. A summarized category is defined on the relation scheme upon which statistical function summarizes, and a qualified category is defined on the other relation scheme. Summary on the summarized category C_1 and the qualified category C_2 over two related relations R_1 and R_2, is represented as $S(\pi_{R2}((R_1 \otimes R_2) \cap (C_1 \otimes C_2)))$ or $S((R_1 \otimes R_2) \cap (C_1 \otimes C_2))$, where $\otimes$ denotes natural join. The former notation restricts the duplications of tuples in R_2 resulting from the join operation, while the latter corresponds to the normal join operation. Each representation meets the semantics of certain applications. As a rule of thumb, if the summarized category is on the '1' side of a relationship '1:N', the 'π_{R2}' is used.

Example Let R1={Student, Course, Grade} where a student may take many courses, and R2={Student, Working-Hours}. The query,

"What is the average working hours of students who have a grade C or below?",

defines the summarized category Domain(R2) on R2 and the qualified category {Grade ≤ C} on R1. This query should be interpreted as $S(\pi_{R2}((R_1 \otimes R_2) \cap (C_1 \otimes C_2)))$, where C_1={grade ≤ C}, C_2=Domain(R2), while the query,

"What is the average grades of students who work more than 10 hours a week?",

might be interpreted as $S((R_1 \otimes R_2) \cap (C'_1 \otimes C'_2))$, where the qualified category C'_1=Domain(R1), and the summarized category C_2={Working-Hours ≥ 10}. □

Since the number of possible categories on two related relation schemes could be as large as the product of the numbers of possible categories of each relation scheme, the *statistics data* over related relations are not stored in order to reduce a potentially large storage requirement. As in relational algebra where a (natural) join operation is required in order to obtain information from two related relations, a similar concept is exploited in this model. This data model provides operations (which are stricter than the homomorphic functions over join operations described in [27]) to obtain a summary of two related relations without joining the relation instances.

Definition Let A be the common attributes of two related relation schemes R1 and R2, let R_1 and R_2 be relation instances of R1 and R2 respectively, let C_1 be a category of R1, let $C=\{c_i\}$ be a disjoint category set of R2, let S be an additive statistical function on the numeric attributes in R2, and let $S(R_2,C)=\{\, S(R_2 \cap c_i) \mid c_i \in C \,\}$ be a collection of *statistics data*, and let π denote projection.
The function *S-join* is defined as

$$S\text{-}join((R_1 \cap C_1), S(R_2, C)) = \begin{cases} \displaystyle\sum_{\pi_A(c_i) \subseteq \pi_A(C_1 \cap R_1)} S(R_2 \cap c_i) & \text{if } \forall c_i \in C, \pi_A(c_i) \cap \pi_A(C_1 \cap R_1) = \emptyset \text{ or } \pi_A(c_i), \\ \textit{undefined} & \textit{otherwise}. \end{cases}$$

The function *S′-join* is defined as

$$S'\text{-}join(C_1, S(R_2, C)) = \begin{cases} \sum\limits_{\pi_A(c_i) \subseteq \pi_A(C_1)} S(R_2 \cap c_i) & \text{if } \forall c_i \in C, \pi_A(c_i) \cap \pi_A(C_1) = \emptyset \text{ or } \pi_A(c_i), \\ \text{undefined} & \text{otherwise.} \end{cases}$$

□

S-join joins a collection of *statistics data* with a category instance, and obtains *statistics data* over the two related relation schemes when S-join is defined; otherwise, S-join is undefined. S-join and S'-join are not defined when any c_i overlaps, but is not contained in $C_1 \cap R_1$ or C_1, respectively. The difference between S-join and S'-join is that the relation instance of R1, R_1, is not shown in the definition of S'-join, i.e. no details of R_1 is required in S'-join. The following proposition shows how S-join is applied.

Proposition 5 If S-join($(R_1 \cap C_1)$, S (R_2, C)) is defined, and numeric attributes appear in R2 only, then

$$S(\pi_{R2}((R_1 \otimes R_2) \cap (C_1 \otimes C_2))) = S\text{-}join((R_1 \cap C_1), S(R_2, C)),$$

where C is a partition of the summarized category C2, and R_1 and R_2 are relation instances of R1 and R2 respectively.

Proof Since $\cap$ is distributive over $\otimes$,

$$S(\pi_{R2}((R_1 \otimes R_2) \cap (C_1 \otimes C_2))) = S(\pi_{R2}((R_1 \cap C_1) \otimes (R_2 \cap C_2))).$$

Numeric attributes appear in R2 only, so the formula can be reduced to

$$S(\pi_{R2}((R_1 \cap C_1) \otimes (R_2 \cap C_2))) = S(R'_2 \cap C_2),$$

where $R'_2 = \{r_2 \mid r_2 \in R_2, \exists r_1 \in (R_1 \cap C_1), \text{ and } \pi_A r_2 = \pi_A r_1\}$.

Since C is a partition of C2,

$$S(R'_2 \cap C_2) = \sum_{c_i \in C} S(R'_2 \cap c_i).$$

Moreover,

$$S(R'_2 \cap c_i) = \begin{cases} S(R_2 \cap c_i), & \text{if } \pi_A(c_i) \subseteq \pi_A(C_1 \cap R_1), \\ 0 & \text{if } \pi_A(c_i) \cap \pi_A(C_1 \cap R_1) = \emptyset \\ \text{others.} & \end{cases}$$

Since S-join($(R_1 \cap C_1)$, S (R_2, C)) is defined, $\forall c_i \in$ C, $\pi_A(c_i) \cap \pi_A(C_1 \cap R_1)$ is equal to $\emptyset$ or $\pi_A(c_i)$, thus,

$$S(R'_2 \cap C_2) = \sum_{\pi_A(c_i) \subseteq \pi_A(C_1 \cap R_1)} S(R, c_i)$$

Following the definition of *S-join*,

$$S(\pi_{R2}((R_1 \otimes R_2),(C_1 \otimes C_2))) = S\text{-}join((R_1 \cap C_1),S(R_2,C)).\blacksquare$$

The above proposition shows that *statistics data* on related relations can be obtained by applying the S-join operation if the following conditions are satisfied: (1) C_2 is derivable; (2) the category instance $C_1 \cap R_1$ is known; (3) S-join is defined. The apparent advantage of applying S-join is to save the join operation on the original relations. In some situations, the *statistics data* cannot be obtained from S-join, although it can be obtained by joining the original relation instances.

It is more difficult to obtain the summary $S((R_1 \otimes R_2) \cap (C_1 \otimes C_2))$, because the duplications of tuples of R_2 generated from the join operation with R_1 is determined by the contents of R_1. In fact, it is not possible to obtain $S((R_1 \otimes R_2) \cap (C_1 \otimes C_2))$ without knowing the details of the original relations or without adding constraints onto the relations, except in the trivial case that every category instance, $c_i \cap R_2$ has only one tuple. The following proposition shows if the restriction that A be a key or superkey of R1 is added, then $S((R_1 \otimes R_2) \cap (C_1 \otimes C_2))$ may be obtained by applying S'-join.

Proposition 6 If A is a key or super key of R1, and S'-join is defined, then

$$S((R_1 \otimes R_2) \cap (C_1 \otimes C_2)) = S'\text{-}join(C_1,S(R_2,C))$$

Proof Since $\cap$ is distributive over $\otimes$,

$$S((R_1 \otimes R_2) \cap (C_1 \otimes C_2)) = S((R_1 \cap C_1) \otimes (R_2 \cap C_2)).$$

Since numeric attributes appear in R2 only, and the values of A in R_1 is unique, thus the formula can be reduced to

$$S((R_1 \cap C_1) \otimes (R_2 \cap C_2)) = S(R'_2 \cap C_2),$$

$$\text{where } R'_2 = \{\, r_2 \mid \exists r_1 \in (R_1 \cap C_1),\ \pi_A r_2 = \pi_A r_1, \text{ and } r_2 \in R_2 \,\}.$$

$$\{r_2 \mid \exists r_1 \in \cap C_1,\ \pi_A r_2 = \pi_A r_1, \text{ and } r_2 \in R_2\} =$$

$$\{r_2 \mid \exists r_1 \in C_1 \& r_1 \in R_1,\ \pi_A r_2 = \pi_A r_1, \text{ and } r_2 \in R_2\} =$$

$$\{r_2 \mid \exists r_1 \in (C_1 \cap R_1),\ \pi_A r_2 = \pi_A r_1, \text{ and } r_2 \in R_2\}.$$

The first equation is due to R1 and R2 having the property of a lossless join. The second equation is due to A being a key or superkey of R1. Thus, $R'_2 = \{r_2 \mid \exists r_1 \in C_1,\ \pi_A r_2 = \pi_A r_1, \text{ and } r_2 \in R_2\}$.
Since C is a partition of C2, obviously,

$$S(R'_2 \cap C_2) = \sum_{c_i \in C} S(R'_2 \cap c_i).$$

Moreover,

$$S(R'_2 \cap c_i) = \begin{cases} S(R_2 \cap c_i), & \text{if } \pi_A(c_i) \subseteq \pi_A(C_1), \\ 0 & \text{if } \pi_A(c_i) \cap \pi_A(C_1) = \emptyset \\ others & \end{cases}$$

Since S'-join((C_1), S(R_2 ,C)) is defined as a given condition, $\forall\ c_i \in$ C, $\pi_A(c_i) \cap \pi_A(C_1)$ is equal to $\emptyset$ or $\pi_A(c_i)$, thus,

$$S(R'_2 \cap C_2) = \sum_{\pi_A(c_i) \subseteq \pi_A(C_1)} S(R, c_i)$$

Following the definition of S'-*join*,

$$S((R_1 \otimes R_2),(C_1 \otimes C_2)) = S'-join((C_1), S(R_2, C)). \blacksquare$$

When A is a key or superkey of R1, Proposition 5 is applicable to $S((R_1 \otimes R_2) \cap (C_1 \otimes C_2))$. For the same reason, Proposition 6 is applicable to $S(\pi_{R2}((R_1 \otimes R_2) \cap (C_1 \otimes C_2)))$, if A is a key or superkey of R2. Summaries over more than two relations can be obtained accordingly. Without loss of generality, since join is commutative and associative, summary over n relations $S(\pi_{R2}(((R_1 \otimes \cdots \otimes R_{n-1}) \otimes R_n)) \cap ((C_1 \otimes \cdots \otimes C_{n-1}) \otimes C_n)))$ or $S(((R_1 \otimes \cdots \otimes R_{n-1}) \otimes R_n) \cap ((C_1 \otimes \cdots \otimes C_{n-1}) \otimes C_n))$ can be obtained by applying Proposition 5 or 6. However, our model fails to generate a summary whose numerical attributes scatter on two relation schemes without joining the relation instances. For the computed functions, they are always computed after all of their component functions are computed.

Example Consider again the relation instance of employee data shown in Figure 1.

Position	Degree Required	Other descriptions
Engineer	Yes	...
Manager	Yes	...
Secretary	No	...

Figure 3(a) Relation of position description

Dept	Position	Cardinality	$\sum$*Income*
AD	Manager	2	135
AD	Secretary	1	35
EE	Manager	1	60
EE	Secretary	1	25
EE	Engineer	3	92
CS	Manager	1	62
CS	Secretary	1	31
CS	Engineer	2	89

Figure 3(b) Statistics data of employee relation

Figure 3(a) is the relation of position description where attribute 'position' is a key and the only key. Figure 3(b) is a collection of *statistics data* of Figure 1. Here we show how to answer the query,

"What is the average salary of those positions that require degrees?".

Adopting the notation in Proposition 5, let R_1 be the employee relation, and R_2 be the position description relation. The query defines the qualified category C_1={Degree Required=Yes}; the summarized category C_2 is Domain(R2). Since the common attribute, 'Position', is the key of R2, Proposition 6 is applicable. Let's check the applicability of S'-join. First, C_1 is known. Second, C_2 is derivable. Lastly, if S' join is defined, the *statistics data* can be obtained. us S$((R_1 \otimes R_2) \cap (C_1 \otimes C_2))$ = S'-join$((R_1 \cap C_1)$,S$(R_2$,C)), where C is the set of all categories in Figure 3(b) upon which C_2 is derivable.

Since '$\sum$*Income*' and 'Cardinality' both are additive functions, following the definition of S'-join, Sum-of-Income$((R_1 \otimes R_2) \cap (C_1 \otimes C_2)) = 438$, and Cardinality$((R_1 \otimes R_2) \cap (C_1 \otimes C_2)) = 9$.

The 'Average-income' is a computed statistical function whose component functions are '$\sum$*Income*' and 'Cardinality'. Thus, Average-income$((R_1 \otimes R_2) \cap (C_1 \otimes C_2)) = 48.67$. □

3.7 Database Update

After a database is updated, the stored *statistics data* might be inconsistent with the database instance. The following proposition defines a property whereby *statistics data* of additive functions can be updated directly without recalculation from the raw database. The summary of a particular category after updating is equal to the summary before the update, plus the amount attributed to the data inserted, and minus the amount associated with data removed. In this model, since every tuple is only in a category instance, every update to the database needs to modify summaries of at most two category instances.

Proposition 7 Given R and R', which are relation instances before and after update, an additive function S, and the summary $S(R \cap C_i)$, the summary of the instance of category C_i after update, can be calculated as:

$$S(R' \cap C_i) = S(R \cap C_i) + S((R'-R) \cap C_i) + (S((R-R') \cap C_i))^{-1}$$

where the inverse is with respect to addition.

Proof From the definitions of set operations, $R' = (R \cup (R'-R)) - (R-R')$.
Taking the intersection of the both side by C_i, we obtain

$$(R' \cap C_i) = ((R \cup (R'-R)) - (R-R')) \cap C_i = ((R \cup (R'-R)) \cap C_i) - ((R-R') \cap C_i).$$

Since S is additive, $S(R' \cap C_i) = S((R \cup (R'-R)) \cap C_i) + (S((R-R') \cap C_i)^{-1}$.
Thus

$$((R \cup (R'-R)) \cap C_i) = (R \cap C_i) \cup ((R'-R) \cap C_i), \text{ or}$$

$$S(R' \cap C_i) = S(R \cap C_i) + S((R'-R) \cap C_i) + S((R-R') \cap C_i)^{-1}. \blacksquare$$

4. STATISTICS DATA SECURITY

The statistical database security problem has been studied in [5] and other papers. The proposed model also offers a methodology which could prevent unexpected information being revealed to unauthorized sources. This characteristic provides, in advance, full control over the information which may be accessed by users without tracing the queries by users or enforcing access control mechanisms. Basically, knowledge about the database can be categorized as *instance independent* or *instance dependent*. For example, for a company, the knowledge that "the salary of a manager is higher than his/her employees" is an instance independent knowledge since it is applicable to all possible relation instances, while the knowledge, "the total salary of the EE Department is $500K", is instance dependent, since it is only true for certain instances. Since it is difficult, if not impossible, to detect whether users have some particular instance dependent knowledge, instance dependent knowledge cases will be ignored in this paper. Proposition 8, below, indicates that if a category is not derivable from a category set, then the summary of a category cannot be obtained with full confidence from the existing summaries, unless some particular instance independent knowledge is available.

Proposition 8 Let $C = \{C_i\}$ be a generating category set, S be an additive statistical function, I be a set of indexes of C, and C_1 be a category. It follows that
$\forall I$, if ($\exists t$, $t \in ((C_1 \cup (\bigcup_{i \in I} C_i)) - (C_1 \cap \bigcup_{i \in I} C_i))$ and $S(\{t\}) = 0$), OR $C_1 = \bigcup_{i \in I} C_i$, then

$$C_1 = \bigcup_{i \in I} C_i \text{ if and } \textit{only} \text{ if } \forall R,\ S(C_1 \cap R) = \sum_{i \in I} S(C_i \cap R)$$

Proof Given an I, assume ($\exists$t, t$\in$ ($(C_1 \cup \bigcup_{i \in I} C_i) - (C_1 \cap \bigcup_{i \in I} C_i)$) and S({t}) = 0), OR $C_1 = \bigcup_{i \in I} C_i$ is true.

[*if*] Since $C_1 = \bigcup_{i \in I} C_i$, it is true that $\forall$R, $C_1 \cap$R $= \bigcup_{i \in I}(C_i \cap R)$.

By the additive property of S, $\forall$R, S($C_1 \cap$R) = S($\bigcup_{i \in I}(C_i \cap R)$)

$= \sum_{i \in I} S(C_i \cap R)$.

[*only if*] the original proposition can be rewritten as

($C_1 = \bigcup_{i \in I} C_i \Rightarrow \exists$R, S($C_1 \cap$R) = $\sum_{i \in I} S(C_i \cap R)$).

Assume $\exists R_1$, such that S($C_1 \cap R_1$) = $\sum_{i \in I} S(C_i \cap R_1)$.

Let $R_2 = R_1 \cup ((C_1 \cup \bigcup_{i \in I} C_i) - (C_1 \cap \bigcup_{i \in I} C_i))$.

If S($C_1 \cap R_2$) = $\sum_{i \in I} S(C_i \cap R_2)$, then the proof is complete.

Otherwise, S($C_1 \cap R_2$) = $\sum_{i \in I} S(C_i \cap R_2)$.

From the given, $\exists$t, t$\in$ ($(C_1 \cup \bigcup_{i \in I} C_i) - (C_1 \cap \bigcup_{i \in I} C_i)$) and S({t}) = 0, and noting that $R_3 = R_2 - \{t\}$,

it follows that S($C_1 \cap R_2$) = S($C_1 \cap R_3$) or $\sum_{i \in I} S(C_i \cap R_3) = \sum_{i \in I} S(C_i \cap R_3)$ exclusively.

Finally, S($C_1 \cap R_3$) = $\sum_{i \in I} S(C_i \cap R_3)$) can be concluded. ■

In the above proposition, if the instance independent knowledge ($\exists$t, t$\in$(($C_1 \cup \bigcup_{i \in I} C_i$) − ($C_1 \cap \bigcup_{i \in I} C_i$)) and S({t}) = 0) is not true, i.e. for S, all the elements of C_1 and $\bigcup_{i \in I} C_i$ except their joints have values zero, then a summary might be obtained even if its associated category is not derivable.

5. CONCLUSION

While the relational model provides a harmoniously integrated approach to data organization and list-type queries, statistical queries still have to be handled in an ad-hoc fashion. The proposed approach allows *statistics data* and statistical queries to be integrated within the relational model through the use of categories. The new class of queries can also be serviced in a reasonably performing fashion.

We are currently extending this work to include additional functional capabilities, and are planning to implement the proposed system through a commercially available relational DBMS. This will allow us to evaluate the utility and performance of the proposed system. We are also working on estimating the outcome of additive function in the case that the category is not derivable.

6. ACKNOWLEDGMENT

This research was funded by the IBM Almaden Research Center. The authors would also like to thank Dr. Kwan Wong and Dr. Sakti Ghosh of IBM, for their support and valuable discussions.

References

[1] Bates, D., Boral, H., and Dewitt, D., "A Framework for Research in Database Management for Statistical Analysis," in *Proceedings ACM SIGMOD* (1982).

[2] Chen, M., "NP-hardness of Derivability Problem," Internal Report, CS Department, UCLA, (1987).

[3] Chen, P., "The Entity-Relationship Model--Toward a Unified View of Data," *ACM Trans. on Database Systems* (March 1976).

[4] Codd, E., "A Relational Model of Data for Large Shared Data Banks," *Communications of the ACM* (June 1970).

[5] Denning, D. and Schlorer, J., "Inference Controls for Statistical Databases," *IEEE Computer* (July 1983).

[6] Denning, D., Nicholson, W., Sande, G., and Shoshani, A., "Research Topics in Statistical Database Management," pp. 46-51 in *Proceedings Second International Workshop on Statistical Database* (1983).

[7] Fortunato, E., Rafanelli, M., Ricci, F., and Sebastio, A., "An Algebra for Statistical Data," in *Proceedings Third International Workshop on Statistical Database* (1986).

[8] Fredman, M., "The Complexity of Maintaining an Array and Computing Its Partial Sums," *JACM* (January 1981).

[9] Garey, M. and Johnson, D., *Computers and Intractability,* Freeman (1979).

[10] Ghosh, S., "SIAM: Statistics Information Access Method," Tech. Rep. RJ4865, IBM, (1985).

[11] Ghosh, S., *Data Base Organization for Data Management, 2nd edition,* Academic Press (1986). Chapter 9.

[12] Ghosh, S., "Statistical Relational Tables for Statistical Database Management," *IEEE Trans. on Software Engineering* (December 1986). Also published as IBM RJ4394, 1984.

[13] Ghosh, S., "Statistical Metadata: Linear Regression Analysis," in *Foundation of Data Organization*, ed. S. Ghosh Y. Kambayashi K. Tanaka, Plenum Press (1987). Also published as IBM RJ4444, 1985.

[14] Ghosh, S., "Category Numerical Relational Operations for Statistical Database Management," Tech. Rep. RJ5780, IBM, (1987).

[15] Hebrail, G., "A Model of Summaries for Very Large Database," in *Proceedings Third International Workshop on Statistical Databases* (1986).

[16] Ikeda, H. and Kobayashi, Y., "Additional Facilities of a Conventional DBMS to Support Interactive Statistical Analysis," in *Proceedings First International Workshop on Statistical Database* (1981).

[17] Klug, A., "Equivalence of Relational Algebra and Relational calculus Query Languages Having Aggregate Functions," *ACM JACM* (July 1982).

[18] Koening, S. and Page, R., "A Transformational Framework for the Automatic Control of Derived Data," in *Proceedings VLDB* (1981).

[19] Nwokogba, I. and Rowan, W., "A Model for an Integrated Statistical and Commercial Database," in *Proceedings COMPSAC* (1984).

[20] Nwokogba, I. and Rowan, W., "A Statistical Parameterization Model for an Integrated Statistical and Commercial Database," in *Proceedings Computer Science and Statistics: the Interface* (1986).

[21] Sato, H., "Handling Summary Information in a Database: Derivability," in *Proceedings ACM SIGMOD* (1981).

[22] Shoshani, A., "Statistical Databases: Characteristics, Problems, and Some Solutions," in *Proceedings VLDB* (1982).

[23] Shoshani, A. and Wong, H., "Statistical and Scientific Databases Issues," *IEEE Trans. on Software Engineering* (October 1985).

[24] Smith, J. and Smith, D., "Database Abstractions: Aggregation and Generalization," *ACM Trans. on Database Systems* (June 1977).

[25] Srivastava, J. and Lum, V., "A Tree Based Statistics Access Method(TBSAM)," in *Proceedings International Data Engineering* (1988). Also published as IBM RJ5399, 1986.

[26] Thorey, T., Yang, D., and Fry, J., "A Logical Design Methodology for Relational Databases Using Extended ER Model," *ACM Computing Surveys* (June 1986).

[27] Walker, A., "On Retrieval from a Small Version of a Large Data Base," in *Proceedings VLDB* (1980).

[28] Yao, A., "On the Complexity of Maintaining Partial Sums," *SIAM J. Computer* (May 1985).

PANEL DISCUSSION

Question: Whether there is a point for including statistical data in the physical operations of database management system to improve performance ? We found out that for very common operations such as sampling, if we put into the database management, such things as indices and B-trees, then we can reduce the sampling selection problem, instead of having to execute a whole query. Can we gain the same performance with such an approach as we would have gained by inserting operations into a system, such as, INGRES ?

Answer: I agree, if you have such functions that are mainly dependent on data and can be integrated into the database management system. It's really complicated. It is not simple to decide which functions should be included. I don't think it's necessary to have all the statistical computations reinvented in statistical database management. You have to be aware that everything that is redone could be done better, or it is not meaningful to have the same thing done a second time, unless we can say "well it is absolutely necessary, and those procedures might be more easily implemented in database management systems".

Ghosh: I would like to back up what has been said. In statistical databases, certain types of statistical set operations, such as, sampling, is really not a mathematical operation, it is an art. Such operations are dependent on the category aspect of the data through sampling frame. My feeling is that, these type category operations in statistics, would serve better if we look at the category part separately from the numeric part. In that case, operations like, indexing, even repeating group operations, would provide better performance. We have to find, what aspects of the operations can be done better at the database management and logical processing level, and what aspect of the operations can be done efficiently by inventing some sort of basic numeric processing, which are efficient at the operating system level.

Comment: It has historical reasons that we have both statistical database management systems and the concept of the database management in statistics. These are different kinds of entities and the only thing that I am saying is that, I think your analysis is very nice. Everything that was said I agree with, except that this should be so in the future also. It is irrational not to have so. I think that in the future, allowing sufficient time to elapse, these two system categories have to merge together.

Podehl: He made my point. I hope the discussion is only temporary thought. In the early 70s it was the database management problems, so the database management problem was solved by the database management systems people. Then there was the statistical processing, which was solved by statistical packages, like. SAS, etc. All the time many of those vendors found to their advantage to join their packages to DBMS, for example, SAS now offers database links to many of their statistical packages. However on the other side, if you look at the commercial success of SAS and you would take a survey of its application, I would say that the majority of the applications (the bread and butter of the statistical applications) like tabulation and graphics, as opposed to the more advanced statistical functions in DB2, you have to know what you are doing and these were not intended for ordinary statisticians. I hope, I assume that all the time, this type of numerical operations which have precise definitions are entered into various statements and language. After all SEQUEL is a language in which things have to be done by enhancing codes to establish a formal language. I think that's the way things will go on. We have discussed a lot of temporary phenomenon.

Ghosh: What you are saying is an experience analogous to what we had in our institution. A matrix based statistical language was developed in the early sixties. This language was adapted by all engineers, who did statistical processing. It was also evident that this language could not solve all their problems, but the engineers were afraid to change to SAS. Same is the story with GRAFSTAT, which has nice graphic features attached to statistical data analysis. None of these systems can talk directly to DBMS. I have a feeling that, once statistical database people can provide the tools for doing real time statistical data analysis, which makes life easy, the engineers will accept it.

Klensin: Since there is unanimity of this point, I would like to argue the other position. It's very easy to talk about embedding statistical operations in a DBMS, particularly in a relational DBMS, as long as you're talking about the relatively simple statistical operation that we've been talking about so far, what amounts to a simple univariate summary. Once you want to go off and take half a dozen of them and compute multivariate correlations it may be alright. But when you consider a little bit more serious problems and you move towards more complex dimension transforming, variable transformation statistical procedures, you are into an environment where you may be able to make these computations in the DBMSs. You're going to do significant violence to the nature of the DBMS environment and impose significant difficulties on yourself in terms of getting back into the DB environment, growing out of the procedure or want to utilize them in some other way. Even the matter of doing things that are statistically simple and common place today, as bootstrap procedures, is an incredibly difficult problem to think about in a continuously updated summary statistics, because bootstrap procedure involves computing a lot of estimates of some particular statistic and then looking at the distribution of those estimates. It's not that you can't figure out a way to do this in the database system but rather, in the process of doing that, you may distort your database management environment to a sufficient degree that, you may ask the question whether this is ultimately worthwhile? Due to data representation, the need for semantics becomes, and semantics themselves become, much more complex when we are dealing with relatively raw data, either at the level for simple categories or a larger number of categories. You are in a different domain, a different kind of extracted metadata, and you may pose very difficult problems on DBMSs which may not be worth doing. I'm not saying it couldn't but you have to concentrate on data managers.

Ghosh: I remember in 1968, Charlie Buchmann and Ted Codd were debating about network data models and relation models. At that time people had no idea of integrity constraints, distributed databases, or semantic constraints and so forth. Within the next 20 years (i.e. today) at least, category processing of data can be done very well with our present DBMS, so I agree with John that the problem that he is talking about, i.e. integrating statistical operations with DBMS, will have all these problems. I agree with him 100%. I am confident that many of the members here as well as those who couldn't make it, will be working on these problems, because if we don't do it, we have a very very bad future for statistical databases, not only professionally but literally in the real world. Today microprocessors are used to collect data in various environment, not only in the field of manufacturing but also in the field of medicine and so forth. The amount of data is enormous. One of the presidents of a division of IBM has called this the data jail, i.e. "after data is generated, it just goes one way (in), and there is no way of extracting information out of the storage". We should start looking at these problems. How can we solve statistical problems at a DBMS level, where users do not have to write programs to do basic statistical operations?

Malvestuto: I feel that this is a basic data theory issue that should be incorporated into the DBMS. I think that if Codd had developed the relational theory independent of the evolution of DB systems then, we could ask ourselves "Should we introduce the relation of data theory into the management systems ?" I proposed in a recent paper, a universal table model for for statistical databases in order to provide a universal table interface. I was obliged to use a sophisticated statistical model known as log linear interaction model. This is the natural price that we must pay if we want comprehensive knowledge of data.

McNamee: We are very much aware of this problem of the operational phase of the statistical relational DB and in fact, we just look at the work we've done so far as preliminary. Until we go in and instruct the system and evaluate the computed results that we get, we can not make definite conclusions. Hopefully in the next conference we will be able to give some results.

Wittkowski: I completely agree with Sakti Ghosh that the user shouldn't need to reinvent statistical procedures again using some primitives that are available in statistical analysis systems. So there are two possible solutions. First: Is to integrate all the power you have in the analysis systems into the DBMSs or use instead of integration, interaction between DBMSs and statistical analysis systems. And in my view the interaction concept using well defined interfaces is more promising than integration especially, if you are thinking of rank tests, computing residuals, in multivariate models and boot strapping techniques that are extremely computation or intensive in computational terms. I'm not sure that DBMSs will be able to handle those problems with the same efficiency as statistical analysis systems.

Shoshani: It sounds like we are repeating things that I heard four to six years ago. It's whether statistical data should be expanded to include more data base functions or should relational systems or DBSs to be expanded to include statistics, or should we have an interface? Of course, the argument for having a combined system was that you gain more efficiency if you do both together. That wasn't really much of an argument, as I remember, but what I am hearing now is that certain things you will be doing better with DBMSs, and for certain things you need a different model. So to speak, like Malvestuto was saying, a different data model that relational systems can not handle, for example, might be needed or you might as well leave them out. Well, probably statistical DB management research is: to figure out how lot of things have to be done together. It's not a secret that, if you take some particular application and you try to shove it into the relational system, you will run into problems; so everybody knows that. So you modify a little bit , extend a little bit, but that's not necessarily the only way. It could be that: the way for statistical DBMS is to actually identify a model that is not of a relational nature and define it in such a way that, it does both DBM well and statistics part well. It is not the question of reinventing the wheel but it is a question of reinventing statistics. It's a question of taking what we know about the statistical packages, and putting them together into a single system for efficiency purposes and that's the question. I don't hear an answer about that. It is really inherent that, you have to do it in two different parts. In the one part where you handle the data or certain functions like selections, may be sampling, and then there is another part: where you do a lot of difficult statistical functions, for that you have to have another model. So you really want to have data for two kinds of models: one for this purpose and then get out, and then do other functions. Or really, Is it possible conceptually to have a single model, regardless of whether it's relational or not, whereby you can do all the functions, statistically and DB management, efficiently for system? I don't hear an answer for that. Any opinions?

Klensin: Assuming if it's possible to construct a single model and then, that model will be neither as efficient with the DB applications as, a more DB orientated model would be, nor will it be as efficient for the statistical applications as a class of set of statistical models would be.

Shoshani: Is this application, the database application for the statistics ?

Klensin: Yes.

Shoshani: Why don't we set a restriction to distinguish between database application and statistical applications ? You needn't do both.

Klensin: Well, you need to do both but once the data analysis is done then it's not a problem with applications but the way that many of these things get done is that: we go to the database environment and can carry out some set of operations which involves selections and subsetting and arranging; and those kind of things we've always thought of as traditional DB operations. Then we take some of those data out in the traditional statistical model, and for a large number of cases, for a small number of variables, we can take advantage of some data structures which traditionally don't go easily with the kind of data bases, traditionally we talk about, such as: high order voltage flow rate.

Shoshani: You can not separate database of high order flow rate.

Klensin: I'm not suggesting that you can not build a database system with high order multidimension slow rate. I'm suggesting that if you are countering the DB with high order multidimension flow rate, that you're either going to need a whole separate class of operator to deal with the domain down association inherited with multidimensional race problem, which may be unnecessary complication, or you going to run the risk of making that DB system less attractive, less efficient using the sort of conventional relational way, we have been thinking about. What we do those high order multidimensional slow rate arrays, is data bashing of a very severe sort. We will be going into some class of variables and out into another class of variables. We think of concocting joining operations on those things rather differently than, concocting joining operations which results in defining different observations. We have been doing that for years. After we get through bashing, we get out some things, not for surrogate variance and those things would make variable one come back into the DB system , but there isn't in that pattern of use, which is quite independent of the art effect of how we build the system. In that pattern of use, the need for high level integration between the DB management functions and the inferential statistical function is not demonstrated and do we want to do it? Yes, but the claim would be to cut performance penalties and no sooner you begin to say something like, "we have this DB layer which is somehow the heart of the hardware and how far we put the statistical applications into it?", then you can ask the question in a way which implies an answer, and that answer is: we can possibly get the statistical stuff down there. But still it is not clear that, that answer is seen as a less bias question.

Wittkowski: I'd like to make a small remark so that I am not misunderstood. I agree the user should have all the power. It has in a very integrated environment, that is user friendly interface. No problem. The problem is: Should it be a monolithic system or an integrated system? In Germany, we could say that: "we have an integrated cow, which is an animal that produces eggs, fuel, milk and beef." Or, should it be an interactive system with several components communicating with well-defined interfaces, so that the user can connect those components that provide him with the most efficient environment? I think for the second approach. They have several components and you can choose from those components that fits your purposes best. This is the better solution than that I have in the example.

Svensson: Well, I think that maybe the debate has been locked into a corner due to historical semantics or whatever. I would like to see in the future data analysis systems, not DB systems, not statistical systems, but one kind of statistical system, which is a system for data analysis. Then there may be statistical distribution systems and whatever but what I'm interested in is data analysis system in the future and these I would envisage as being homogeneous as seen from the user. The user sees one universe with a homogeneous set of concepts. Then I certainly agree with Dr. Wittkowski about desirability of modularity in such an environment and that it may be absolutely important that you are able to add the user groups seamlessly. The single users are able to add to this environment in a way that is, if you like, there are no ugly limits between what I add and what the producer of a system originally included. That's very important but the problem is that, with a constant DB, there goes a lot of sacred axioms that have to be fulfilled. I don't like that - such as being able to lock multi user updates and whatever. Therefore, I think that the terms are meaningless when we're talking about these things.

Lurkoff: I would also agree with the proposal of Dr. Wittkowski that, this line is similar to our development, i.e. to have a strong modularity and I think it will also be in the future. So that nobody will be able to develop such very complex homogain system from the software point of view. We need relational operations or space operations from statistical point of view. Now we are mainly transferring data from one package to another package and we have to extend this not only data also metadata and I think later transfer knowledge. If I use a select or join, then it's not only to transfer data but also maybe some kind of knowledge about this new relation. If I continue, then it's another package, and this knowledge could be used. If now, we are in a sense to standardize data format, then I think, we need to standardize alternate concepts, how to standardize metadata and how to standardize knowledge in the future, to get the software developed first and the possibility to connect to such standards? Then I think, it would be easier to connect them; and from the user's point of view, he should not or he should have to be less aware about internal structure of the package, he is using. Rather functions should be developed giving homogeneous user friendly interface to the user package.

Ghosh: I would like to make a proposal. Here we are assembled, as the world's leading people in statistical DB. In the 1960s the CODYSAL DB task force people played a very important role in the development of DBMSs as practical tool. I would like to present: Whether this group thinks, it would be appropriate to form such a task force to understand the basic fundamental technical problems and system architecture problems in the of statistical DBMSs? May be before we adjourn on Thursday, we should make a decision; work towards that.

Cubitt: Can I cloud the issue with a couple of practical things - I'm afraid I have a habit of doing this. The first is that, the reality of the situation as I see it is that, the statistical analysis that is being done, as the demand for data that already exists in certain administrative contexts, is to be used for statistical analysis and there are very few organizations whose prime role in life is to collect data for statistical purposes. There's more and more need to use data that already exists than to perform some sort of needs to use it for statistical needs, such as: trends, predictions and things like that. So, one of the problems is that, this data exists in DBMS whose role in life is to support administrative or financial operations within a particular context. I'm thinking of management information systems, DBs on vehicle licensing, where you want to do some statistical analysis but it is something which is in addition to the role of the operations system, which exists and I think that in that way I would tend to come into line with Knut's idea: that one of the things that's also happening is that, because people want to do statistical evaluation, one of the problems they have: is finding the data and discovering the semantics of the data they are dealing with and to that end, they are justified. I think that one of the things is, that there is already a growing need to extend factual semantics with data where as really, you don't need it, you are doing this to control the system. There's an awful lot of information you don't bother to capture because to run your floor control system, you don't need it. But when management, in a management information system context says, " By the way we'd like to have some information about predictions, about modelling, about econometric and so on"; suddenly, there's beginning to be a great pressure. So I think that in the medium term, the objective must be to start, and there is a demand there for more and more metadata semantic information to be captured at source so that at a later date, it can be analyzed and we can pass more complete pictures of the data over to whatever analysis we want to do. Maybe at a later stage, we may then be in a situation where because of even more pressure, there'll be more and more need to look at a more global system and perhaps to include some of the functionally within the DBMS, but at the moment the reality of the situation is that companies and countries with who, I have more contact with, are using more and more administrative data which is collected for purely other reasons for statistical needs because they can't afford to do it any other way. The immediate pressure is to connect more semantic information with that, so that when they do get down to doing the statistics, they have enough information to do something sensible; and I think this is going to be a medium to long term type of view that: there is the immediate need to connect the semantics, to find the data primitives, perhaps after that, there may be more pressure to build more into the model, in terms of functionaries.

Ghosh: We had the same problem when we talked about data jail. The collection of data becomes so large and because it takes so long to analyze the data, that the engineers are turned off, and they don't do it. So performance is an important issue. So the huge volume of data that have been collected today, if we could put it in a state that can answer statistical queries fast, then they will, likely, use it; and I don't think that's a long term problem. I agree with you that the data, that will come in the future would need much more sophisticated analysis and ways of handling and processing and at least we can do something which can make it much more faster.

McNamee: I would like to again refer to this chart (see McNamee's figure on system architecture). What I've heard people say, is basically what we're sort of talking about here. We have a knowledge base system, a rule base, some management system, statistical processor. It represents what we are talking about.

Cubitt: Well, this is essentially my approach and your model tells you about the database and this is a primary concern. Statistical functions is of secondary importance,

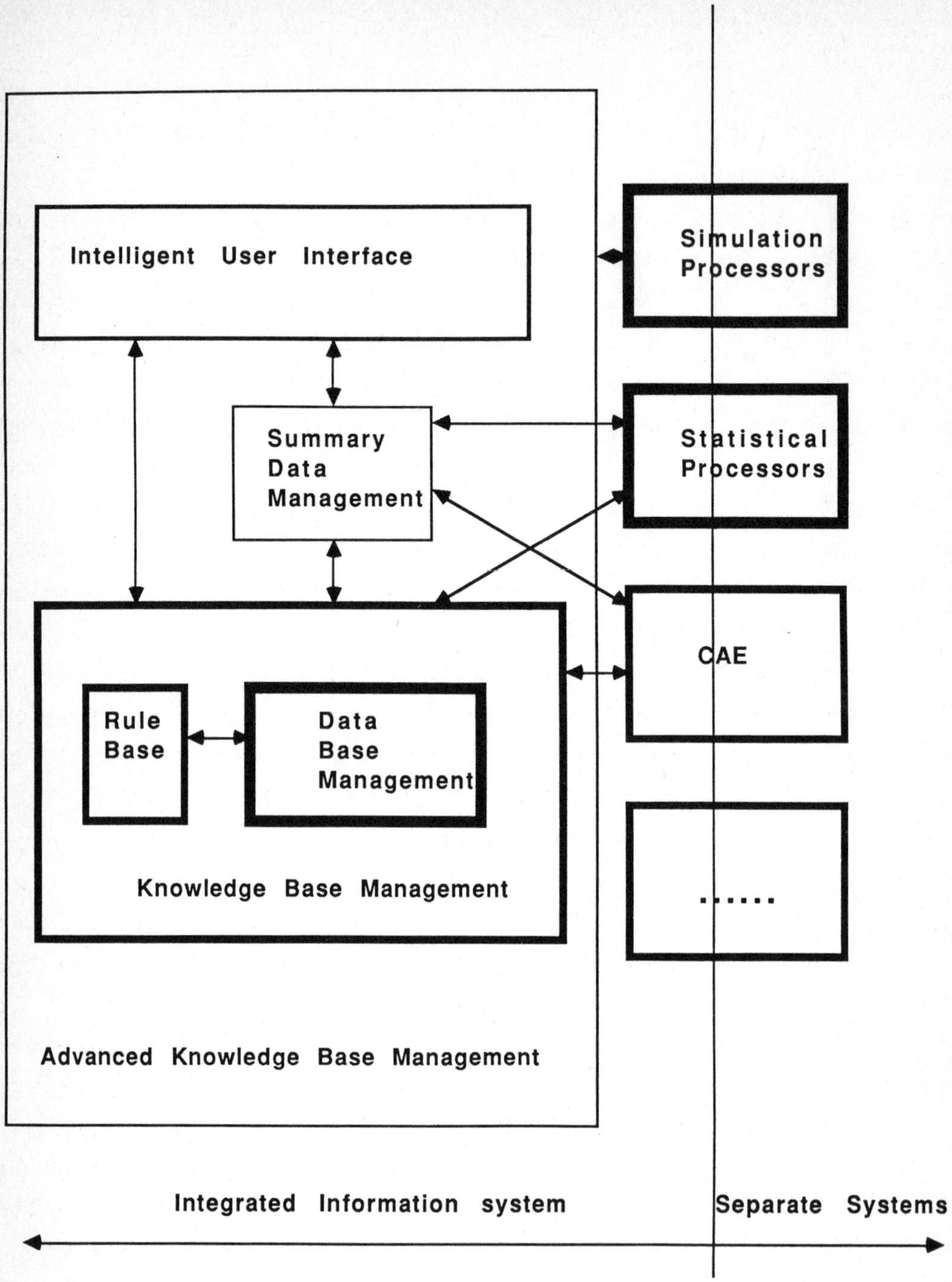

Prof. Lary McNamee's chart on System Architecture.

the primary is the metadata, the knowledge, and the rules associated with database management techniques.

Shoshani: Essentially, I do not have a problem with modularity. Certainly modularity has been around for a long time. The problem, I have is; "what I think I hear, which is a claim that if you deal with the data, you have one view of it, a conceptual view of it, and if you deal with the statistics, you have another conceptual view". For example, if you deal with the data, you may want to look at the statistical tables in a relational sense or something; and then you get it out and you want to think about it as two dimensional structure. That means that, once you think about a multidimensional structure and you manipulate it, or generate other multidimensional structures, they are not any more relational, which means, that a user has to learn two kinds of systems- one is this kind of data structure with it's operators, another is the other kind of data structure with it's operators and you do this first then you do the other thing. Conceptually, I have problems with that. I think modularity within a single system is fine, I have no problems there. But I think there is nothing sacred about a multi-dimensional table, unless I don't understand anything the statisticians are saying and I'm not a statistician. But I think there's nothing sacred about a multidimensional table that can not be expressed as a data structure and if you can express the data structure through the multidimensional table, and that's what you need, then that's what you should deal with. Nobody forces you to look at it as a relational DBMS. If multidimensional data is what you need, then that's what you should have. So, if I was a user of daily physical analysis, I'd like to deal with with the most appropriate single type of structure available to me. I don't want to be forced into a data management structure and to have a statistical analysis structure. I want to try to have a single homogeneous view regardless of what you do with this modularity and how you deal with it. It is a single view, that I can now apply to both data management functions and statistical functions and when I apply one I get the same type and when I apply the other I get the same type and it should do this, in as much as homogeneous language as possible. My problem with this approach is that: it's not the modularity problem, it's the separation, as if you have to count the functions and they cannot be merged. They cannot be viewed as the same kind of structure and function and I just don't quite follow the logic that says that, "this is true". I don't think you have shown me, one example that convinces me of that.

Westlake: Can I come back on that, I think there are two different views. I think one is that, we have the conceptual view of the data and there's a great need for enrichment for that, and the work you're doing. Temporal data shows that there is a lot of progress that can be made and those improvements to the conceptual view of the data improve the way a statistician can work, but there is also the conceptual model of statistical process and I think that's completely different from the data. In the conceptual model of the data, and I'm not optimistic that, it is possible to construct a conceptual model of the statistical process. It's essentially something, which is very varied and very subjective to change. You look at things in different ways and from different points of view that may require that, you use different models for the same data but the statistical process itself is something that is essentially different from the data. I share your optimism that it is possible to find conceptual models of the data but I don't really believe that it is possible to model the statistical process in the same sort of way. I think the statistical process can be very much enriched by some of the expert system ideas that are being put forward. It can be enhanced by having better data structure. But I don't think that, I know enough, to assume that. That's why I don't think we'll ever get to the stage where a complete data analysis system can be built into the DB. There are lots of statistical processes which include very little analysis that are basically straightforward summary application operations and there's no reason, why those shouldn't be stored in a database system and there are great advantage in doing it that way, but that's one end of the statistical spectrum. At the other end are the esoteric statistical investigations, which I think are probably beyond this sort of conceptual model.

Klensin: Complete agreement.

Shoshani: Well, if you look at this figure here (McNamee's figure on system architecture) the summary data management, whatever that is right now, it will be some kind of the conceptual model that the statistical processor can relate to (Westlake: Yes), which is different from the conceptual model that the DB management has.

Westlake: Why ?

Shoshani: That's what I heard you say. Could you speak in those terms.

Westlake: That summary data management does not encapsulate all statistical knowledge.

Shoshani: So that won't do.

Westlake: No, but it will do a lot. It would be very useful to have that structure and in fact the summary data management could be much more closely linked to the database, but summary data is not all statistical knowledge.

Klensin: Again, I completely agree, but to fully answer your question, and I tend to agree with your outline of the difference. We have to take some major decisions, which is the part of the problem. It becomes beyond the summaries and summary covers a lot of what everybody thinks most of the time and a lot of work is involved. It is one of those sectors of outlets, and there will always be summaries, such as the frustrated one put in. It's a different way of thinking of data, that conceptual model is very very difficult and we can't talk about a single distributional model because there is the data, that is becoming extremely complex and this is a field, which will be struggling for several more hundreds of years than DB management. It's getting worse.

Shoshani: What you're saying is that basically, all we can hope to do in this field of statistical database management is really only summary DB management.

Klensin: No, you can do the summary and you can provide very good support for those other operations and some of the things that Malvestuto has been talking about and I have been talking about. Some of those semantic issues have to do with very good and very important support for those other operations, but it's supporting activity and who's the application and who's the set of service functions depends about how you look at that world and it's not very interesting work to subsume into the DB management context. All this statistical context and that way of thinking about things is, to address a terribly difficult problem, which is not very interesting.

Shoshani: Is it straightforward in your mind, what is the interface between the statistical conceptual model and, I don't know what you call it now, the DB conceptual model? What is the interface ? Is it really summary data?

Klensin: It is not straight forward in my mind. I can give you heuristic about where some of the controversy might lie and heuristic, that we have been talking about, might explore, in the last half an hour or so. My heuristics would be different from several other people.

Svensson: I would like to refer to the result of the discussion we had two years ago in Luxembourg where we had designed a knowledge base statistical analysis system of several layers with the conventional DB and programming system ideas. Statistical functions could be handled on a level to facilitate access to the data. That's a level for summary data and then we have designed methods where the knowledge of what we have actually done and then, we have models where we have the knowledge of how we want to interpret it, and then we have the domain and what we want to do with the data, so I think we will have several levels and the resulting system will be of considerable complexity. Nevertheless the user interface has to be unified, and we have to think a lot about how to present the data and the results, to the user and interact with the user but I think this is the approach we could follow to structure the analysis system in a very unified way.

Klensin: I partially agree with this entire very difficult problem. This picture comes out of one statistical paradigm and and it's a statistical paradigm with which I agree and sympathize, but one could easily now try another picture of purely orthogonal statistics, representing a somewhat different statistical paradigm and somewhat different data orthogonality, and those differences start from the degree of control as of the date, when they are collected, but there are other portions of the difference also. As soon as their orthogonality goes in, we think of four dimensional viewgraph problem as distinct from merely the four dimensional system problem. Statistics is not even a required means of SAS, which has been branded around today, represents a rather good realization of what a view of statistics is about and a perfectly lousy representation of several other views of what statistics is about, which is not a criticism of SAS.

Ghosh: I would like to take one more question because I see that the crowd is thinning out and we have run almost an hour. The last question ?

Mulvestuto: I think that it's easier to reveal the lower levels of this picture for the software design for these functions, which are quite stable and you know what you want to do in the design. I think in the user interface, I don't believe that, all statisticians should have to say: what they do on the data. So I think this should be very open and this umbrella function should make different images of the lower levels, different type of results, and I think that should be the most. I don't think we can agree about one unified interface that all statisticians or other users have to go to this interface. We have to have different interfaces.

Shoshani: I agree, that's why we have the user knowledge base.

Ghosh: I would give each member of the panel a last chance for concluding comments.

Malvestuto: No special comments.

Svensson: I think we have to agree to integrate more semantics into relational DBMSs and relational models. I think we can't do without more semantics in these systems.

Ghosh: As concluding remarks of this of this session, I would like to say definitely that for most of us, this has been a very interesting, stimulating technical discussion for the last two and a half hours and I hope it won't be only talking. I hope, we will get down to doing something about solving these problems. It's not very simple, as you all agree, you have been debating for the two and a half hours to get an insight into the problems. So please see what we can do to solve these problems in a realistic manner. Thank you very much for all the discussion. This session is closed.

Database Management Systems for Statistical and Scientific Applications: Are Commercially Available DBMS Good Enough?

Members of the Panel were:

Per Svensson, Swedish Defence Research Establishment (chairman)
Martin Podehl, Statistics Canada
Geoffrey Stephenson, Logica S.A., Luxemburg
Michael Cawson, Steward Observatory, University of Arizona

1. POSITION PAPER BY PER SVENSSON, SWEDISH DEFENCE RESEARCH ESTABLISHMENT

Of the four panelists, I might well be the most biased since I have during the last decade or so promoted the design and development of a special-purpose SSDBM system, the relational data analysis system Cantor. In spite of this bias, I have accepted to serve as chairman of this panel. By selecting user representatives rather than system designers for the other three panelists, I hope that my impudence can be forgiven.

In fact, this series of conferences can be said to exist because some people, most notably Arie Shoshani and his group at LBL, have vigorously and successfully fought the standpoint that statistical and scientific database management problems share a set of characteristics which distinguishes them from the mainstream, "commercial" dbms applications.

While this was certainly true in 1981, when the "First LBL Workshop on Statistical Database Management" was held, one can ask to what extent this standpoint is still valid, and more specifically, whether modern commercial DBMS (CDBMS) have not in fact evolved to a level of sophistication where they can easily cope with most SSDBM problems.

Clearly, some existing systems are good enough for many statistical and scientific applications. It makes little sense to point out to their happy users that specially designed SSDBMS's could be made better in one respect or another, such as query evaluation performance, user language power and the like.

There is however a set of important concepts and facilities that are lacking in commercially available dbms and statistical systems.

Among these are:

- an algebraic user language adequate for scientific and statistical applications (since statistics is the primary toolbox of observational science, such a combination is obviously needed); such a language must include the concept of vector spaces with its relevant operators as well as sets and relations and provisions for recursion and deduction.

- a "seamless" extensibility in the database environment, that would allow application-dependent modelling facilities to be added through user-defined object type mechanisms. Such facilities should allow the user to describe in a uniform way not only external data but all kinds of facts and relations that have to be considered in the application. In the case of exploratory data analysis, this includes base data descriptions, analysis history descriptions, dependency structure of derived data, descriptions of how successive analysis steps affect statistical significance, etc.

- the ability to handle large-scale special-purpose databases, such as databases over particle physics experiments or molecular biology databases of great complexity.

In statistical and scientific database management technology, there exists now a broad spectrum of basic methods developed for various purposes. There are also end-user oriented commercial products, such as Oracle or SAS, which in combination can be effectively used to solve many SSDBM problems.

But we are still far from a situation where scientists can work with data analysis problems within a homogeneous non-procedural language environment that includes data management and basic mathematical and statistical structures and procedures.

To further discuss this question, we need to look at some of the main "SSDBMS arguments" [1]:

1. Usage characteristics

Neither SSDBs nor CDBs have homogeneous usage characteristics, a fact that has perhaps not been pointed out sufficiently clearly in the SSDBM literature. Originally, CDBs were characterized by centralization, the archetype being the "enterprise model" through which an organization keeps track of all its business transactions. This archetype is no longer generally valid - if it ever was - although there are certainly many organizations in which a centralized database plays a very important role. Today, database systems have entered the world of end-user computing. Small groups or even individuals can now afford to manage their records with the aid of a workstation DBMS.

The relevance of this phenomenon to the SSDBM debate is that one of the arguments for developing specialized SSDBMS has been that their user groups are often small, with little or no support staff available, and therefore with the development and maintenance responsibility residing with the end users, or possibly with a part-time consultant or programmer. This situation has now become commonplace and can no longer be said to separate SSDBMS and CDBMS usage.

To confuse the issue, there are important subclasses of statistical and scientific applications which are centralized. Systems for data dissemination are commonly maintained by statistical offices, and they are of considerable importance also in many scientific fields, where they have replaced printed tables and catalogues. Such systems have more in common with information retrieval systems than with end-user database management. New low-cost storage technologies like optical disks and high-density cassettes, together with the proliferation of powerful PC workstation hardware and software, are providing increasing competition for these centralized distribution systems.

Most statistics dissemination systems have poor analysis facilities and, indeed, do more to protect their data from unauthorized access than encourage their analysis and productive use. This is part of the reason why I think they should gradually be replaced by straight-forward media distribution - in a way a return to the pre-computer era, although the media are now machine-readable. Of course, in some situations the most cost-effective distribution medium should be telecommunication, but the role of the distribution systems will mainly be to support the finding and ordering of appropriate information, billing the customer, and sending out the media. Since users can be expected to have access to data retrieval and analysis software within their own workstation, and since data media are cheap, it should in general make more sense for a data deliverer to offer a limited set of prepackaged data products than to encourage customers' browsing through structurally complex meta data bases before placing orders.

Whatever the source of statistical and scientific data may be, and whatever method one uses for their dissemination, it is an obvious fact that their scientific use, in distinction to their use in journalism or education, say, in general requires combination with other data and subsequent analysis according to the needs of the scientific problem at hand. Thus, data analysis (or statistics production, as some would say) is the central activity common to professional SSDBMS users. I do not intend to say that professional users do not need good metadata facilities, only that their importance is secondary to powerful manipulation and querying facilities.

Professional SSDBMS usage will increasingly take the form of end-user computing in powerful personal data analysis systems. The most important issues for the SSDBMS designers are those of user communication and control in a dynamic database environment, where the rate of change for individual data records is moderate, where sequences of data transformations are frequently produced and have to be kept current, and where the course of an analysis should be automatically recorded in a manner that allows for easy modification and repetition.

In this last respect, professional SSDBMS usage is quite similar to CAD or CASE systems usage, where dependencies between procedures have to be kept track of, to allow automatic recompilation or reevaluation of intermediate results whenever an ancestor in the dependency structure has been changed. In the realm of programming environments, such a mechanism is called a configuration control or version handling subsystem. The analogy goes further. A data analysis system needs a query development environment, complete with query text and interdependence database, version manager, language-sensitive editor, parser, query optimizer, code generator, and evaluator.

2. User interfaces and logical modeling

Designing user communication and control subsystems for future data analysis systems will include finding answers to the following questions:

- what are the appropriate data models?

- what are relevant user language design objectives?

- what degree of integration between database management and other analysis tools, in particular statistical or geometric analysis procedures, is appropriate?

- what is the proper role of expert systems technology in a data analysis system?

I am convinced that scientists and statisticians need quite different database management tools than clerks or managers in a company. Their data handling and analysis tasks are in general much more complex, they can afford to spend the time needed - but no more - to penetrate an important problem, and their educational tradition encourages, indeed requires, thinking in terms of abstract models and concepts. The scientific user should therefore be given much more advanced tools, including abstract user languages, whenever such tools would improve their problem-solving ability.

Peter Buneman once pointed out that the task of solving the general cubic equation, carried out by the Italian mathematician Tartaglia in the 16th century, must have been a very difficult one, because algebra as we know it was unknown to him. Written down in natural language by Tartaglia's contemporary Cardano, the proof is utterly difficult to follow. Although most users of computers do not care about such problems, those who do certainly need appropriate formal languages. Current trends towards "friendly" graphical or natural language interfaces can clearly benefit the professional data analyst, but such interfaces are far from sufficient when difficult problems must be solved.

Much critique of the relational model as a basis for a powerful, formal language for scientific data analysis seems to me to be based on a too narrow interpretation of its concepts, or on conceptions about the relational model as a static, even petrified paradigm. I do not see why this should be true. However, although the relational model, properly interpreted, provides a sound framework for elementary set operations, current relational languages lack the ability to handle tasks that require "higher" mathematical concepts. Two important examples are recursion and euclidean vector spaces. Recursion is syntactically trivial but semantically difficult to incorporate efficiently in a system. Still, its importance has long been recognized and considerable research has been carried out. Practical applications include the ability to express in a declarative language such problems as the shortest path through a network.

There have been few if any attempts, however, to add more mathematical structure to the relational basis to support elementary geometry and linear operators, so that transformations and queries in the space-time domain can be formulated and evaluated. This would require a considerable research effort, since concepts to be introduced in this way must be efficiently implementable to be useful.

For geographical database applications, which are rapidly becoming of major practical importance in many sectors of society, and many of which are of a scientific or statistical nature, such concepts are needed. Present systems rely on procedural, more or less ad hoc, solutions to solve geometric and topological problems.

Our group has tried to apply Cantor to several geographical database problems. Our conclusion from these attempts is that many more geographic problems than one first tends to believe can be solved within the conceptual framework of relational algebra, sometimes quite easily, sometimes only with considerable intellectual effort. The solutions are however frequently unavoidably inefficient. One example is restriction with respect to distance, which must be expressed by a general algebraic formula. There is no way for the query optimizer to recognize such expressions as optimizable.

Another example involving vector spaces is linear interpolation in a table of function values, a procedure which can be expressed in SAL, Cantor's relational query language, but whose solution is so complex that hardly anyone would consider it usable in practice.

3. Statistical operators

Any serious SSDBMS has facilities for expressing descriptive statistics, such as average, coefficient of variation, standard deviation, etc. However, to integrate the great number of statistical methods that are incorporated in a system like SAS requires a major effort which no SSDBMS designer has yet embarked upon. Still, the integration of selected statistical analysis methods into a modern SSDBMS is in my opinion a very worthwhile project. Several authors have complained about the relational model's lack of support of the concept of matrices. But a matrix can easily be mapped onto a relational table. Semantically, the explicit labelling that the relational model can provide is much to be preferred to the integer indexes of the matrix concept, unless arithmetic on the indexes is an issue. Integration of computational procedures into an SSDBMS may lead both to an improved user environment through better uniformity of the user model, and to improved computational performance, as pointed out by Khoshafian et al. in [2].

4. Physical organization and access methods

A number of papers have been written about physical database organization and search in SSDBM systems. I would like to contribute here some empirical data to give an indication of the level of performance improvement that can be obtained in practice by using database organization techniques specially designed for such systems. A few specific examples can not prove any general statement about system performance, but with background knowledge about what has been measured and how it was done, I interpret the figures as showing that there are SSDB storage structures that provide improved performance for many, but not all, operations of importance for an SSDBMS. But the improvement is seldom more than an order of magnitude and may not be sufficiently large to motivate a user to change system. Of greater importance for data analysis applications is the fact that the studied SSDB structures are also entirely self-organizing and quickly created.

For my conclusion, I have to assume that the comparisons were made against systems whose performance can be considered representative for state-of-the-art CDBMS. I believe that this is true, but further tests would be needed to prove the case.

Our group at FOA has made two empirical performance studies where special purpose SSDBMS were compared with commercial systems. The first study was carried out in 1978 and involved our prototype storage and access system called Prelat, which was compared with a commercially available flat-file dbms, called 1022, and developed for the Dec System 10 and 20 series of computers. The second study was carried out about a year ago and involved our relational SSDBMS Cantor and the Swedish commercial DBMS Mimer.

Data for only a few basic operations are included. Attempts to compare performance for higher-level operations, such as entire analyses, were largely unsuccessful due to functional limitations in the comparison system.

In both cases, the comparison systems were built on the "flat file with secondary indexes" concept. In addition, Mimer uses a B-tree structure with respect to the primary key of each stored relation.

Prelat was the direct ancestor of Cantor, and they both employ a blocked CFTOF storage structure, although the compression algorithm of Prelat was more advanced but also more than five times slower than that of Cantor.

The two experiments produced generally consistent results with respect to the cost relations between our CFTOF-based systems and the commercial systems. Among these four systems, three are portable (written mainly in Pascal and Fortran, respectively), whereas 1022 was written in assembler language.

Prelat vs. 1022 experiment (1978)

The comparison figures in b) - e) below are (cpu time for 1022) / (cpu time for Prelat). They refer to operations on a table with 8568 tuples and 10 attributes. Attribute cardinalities varied between 3 and 1000. Comparison of 1022 with indexes vs. Prelat without is motivated by the fact that usually, indexes provide no improvement in Prelat.

a) Storage space ratio (space in 1022 / space in Prelat)

	restricted range*)	full range
without indexes	1.5	7.1
with indexes	1.6	3.7
Prelat without, 1022 with index	4.3	10.0

*) 1022 used maximum-byte-length packing such that attribute values can not be updated outside their predetermined value range

b) Load time ratio

without indexes	with indexes	Prelat without, 1022 with index
0.4	0.9	3.2

c) Columnwise sequential access read performance

	attribute 1 columnwise	attribute 10 columnwise	all data rowwise
cost ratio	250	19	2.7

(1022: restricted range)

d) Columnwise direct access read performance (block of 64 randomly selected items in one column)

	attribute 1 columnwise	attribute 10 columnwise
cost ratio	13	2.8

(1022: restricted range)

e) Range search performance. Cost relative to Prelat without indexes

(% hits)	query 1 (0.6 %)	query 2 (0.7 %)	query 3 (0.18 %)	query 4 (0.18 %)	query 5 (2 %)
Prelat with indexes	1.3	0.7	0.8	1.0	0.8
1022 with indexes	2.5	1.4	1.0	0.9	0.6
1022 without indexes	88	15	18	14	29

Cantor vs. Mimer experiment (1987)

The comparison figures represent the ratio (cpu time for Mimer)/(cpu time for Cantor).

a) Range search over a single key attribute (before the interpolation search algorithm, presented at this conference, was implemented).

(% hits)	query 1 (0.1 %)	query 2 (1.0 %)	query 3 (10 %)	query 4 (50 %)
file size (ktuples)				
12	0.10	0.51	3.2	7.9
20	0.16	0.82	3.6	8.8

b) Projection over a varying number of attributes. 5 ktuples, 10 attributes in table. Attribute 1 is primary key, attribute 2 is candidate key. Either one is always selected, i.e., no dupli- cates exist.

% of the attributes:	10	50	90
proj. over candidate key	6	4.8	4.5
proj. over primary key	13	6.8	5.1

c) Randomly selected updates in a table with 20 ktuples and 5 attributes.

% updates:	0.05	0.5	5	50
	0.4	1.1	2.4	7.1

d) One tuple randomly selected with respect to a key.

file size (ktuples)	primary key	secondary key
12	0.1	1.4
20	0.1	1.9

Conclusions

Commercial DBMS are in general not good enough for statistical and scientific applications, but neither are the special-purpose systems that have currently been developed.

Commercially available so-called geographical information systems (GIS) enjoy a commercial attention that other kinds of SSDBMS have never experienced and provide an interesting example. These systems are frequently built on top of a CDBMS, relational or otherwise, but their user language constructs are invariably as ad hoc with respect to the geometric concepts which underlie most operations on geographical data as were the first commercial database management systems with respect to set operations.

Present systems for statistical and scientific database management are thus in their conceptual infancy and much research has to be carried out to improve the situation. Unfortunately, few scientists seem to think that this is a worthwhile research area. Instead, what development there is, seems to be driven mostly by market forces. These forces will rarely favour investments in research and development for science's own benefit.

Storage structures and associated access algorithms, specially developed for SSDBM systems, can provide significant performance improvement for problems occurring frequently in data analysis applications, but only the user can properly assess the value of such improvements in relation to other advantages that the use of a CDBMS may entail.

References

1. Shoshani, A. and Wong, H. K. T.: Statistical and scientific database issues.
IEEE Trans. on Software Eng., Vol. SE-11, No. 10, Oct. 1985, pp. 1040-1047

2. Khoshafian, S. N., Bates, D. M., and de Witt, D. J.: Efficient support of statistical operations.
IEEE Trans. on Software Eng., Vol. SE-11, No. 10, Oct. 1985, pp. 1058-1070.

2. SUMMARY OF PANELISTS' STATEMENTS

2.1 Martin Podehl, Statistics Canada

Are commercially available statistical database systems good enough?

Of course, the answer depends on circumstances.

Whether one use a commercially available software package or whether one develops ones own software is a "make or buy" decision.

Such a make or buy analysis will take into account various factors: the requirements for the software; what the market offers and what compromises one has to make in order to use a commercial package rather than an in-house developed one; the amount and source of funding available for either of the make or buy alternatives; an evaluation of operational costs and of performance of software alternatives on existing or newly to be acquired hardware; the life expectancy of the project; the number of users using the system and thus the requirements of robustness of software and of software documentations; and so on.

The distinction between statistical databases and non-statistical databases is difficult to make.

Considering the former, we basically distinguish between

a) applications of DBMS's in the survey processing environment (that is to convert respondent data received on questionnaires or captured through CATI to clean survey data, which then can be tabulated and disseminated),

b) the application of DBMS in storing and disseminating aggregated statistics which are offered to the public through some sort of online information service or (in the future) on diskettes and CD-ROMs.

Considering the latter, an indication that the distinction is not a precise one is the fact that SAS, which is one of the most popular stattistical analysis packages, has been interfaced to many commercial database management systems. Even if the original purpose of the database is for other than statistical applications, once the operational concerns have been solved the potential for statistical information from such a database can be exploited easily through SAS.

Specific Cases in Statistics Canada

In the following I will outline several cases in Statistics Canada which are relevant to the question to be addressed by the panel.

Case One: CANSIM

CANSIM is a Database Management System to store aggregate statistical data in time series format and to allow remote access from users through terminals and online retrieval software. The development of CANSIM started in the late 60's as a batch system which then was converted to an online interactive system in the early 70's. At that time no DBMS system was available which could do the job; consequently all code was developed in-house. Even today, if we were to redevelop CANSIM, it is not clear what commercial DBMS system would be suitable for the job as the requirements of time series are rather unique and online retrieval systems have to be fast. Looking at other time series databases which are offered to the public, to my knowledge, they are all more or less based on own developments rather than commercial packages.

Case Two: RAPID

For the 1971 Canadian Census of Housing and Population an innovative approach was pioneered by storing the complete census data in micro form (that is, not aggregated) and to provide powerful flexible tabulation tools linked to this data. This was the origin of the so called string file concept which in 1976 was converted and packaged into a software product called RAPID. Since 1976 RAPID has been used on every Canadian Census of Housing and Population (1976, 1981, 1986, and planned for 1991). RAPID's retrieval flexibility, power, and economy of machine utilization is unsurpassed by any commercially available package. However, it cannot be said that RAPID is a complete DBMS system, as it lacks much of the end-user functionality which today's DBMS's have. In addition to the Census, several other applications were development based on RAPID in the late 70's (and are still operating), but none recently. We have rather shifted our focus to commercial packages which may not quite have the speed and elegance of internal access to date as is available in RAPID, but provide much more functionality and flexibility in terms of entering data, maintaining data, and retrieving data.

It would be too expensive to develop this functionality in-house and add it to RAPID, although some steps in this direction were taken by the Statistical Computing Project of the ECE. This was the basic reason for halting further enhancements to RAPID, although we will maintain the current version of RAPID indefinitely.

Case Three: ADABAS

In the late 70's, when it became apparent that RAPID would not offer all the functionality desired, Statistics Canada shopped around for a commercial DBMS and chose ADABAS. Since then about fifteen applications for processing or in support of processing survey data have been implemented with ADABAS.

The most recent development, in which ADABAS is being used, is the redesign of our Business Register, referred to as the Central Frame Database. However, while the database itself is based on ADABAS, complex software had to be written to allow a hierarchical and network view of structures of companies and to support temporal views. In temporal databases you carry sort of historical information about every entity with all information labelled with two types of dates: the date at which the facts changed (or were supposed to have changed) and the date at which the change was applied to the database. To our knowledge no current database management system commercially available in the market is supporting temporal views and thus we had to develop extensively this facility ourself. The result is a very complex package which is not yet completed and which, no doubt, will cause maintenance concerns. However, the benefit of that system will be to increase productivity in the application and save many person years in the ongoing operations.

Case Four: ORACLE

In 1986 we reviewed the situation of DBMS's in Statistics Canada and in the market. We found that relational database theory (which was, sort of, implemented in RAPID and which ADABAS can support) and SQL as a language to interface to relational databases had become de-facto standards.

We also discovered that some DBMS offer portability across different computing environments which had become one of the key criteria of our software strategy. We chose ORACLE as the relational database with the best fit for our needs and we have not regretted the decision yet. Although no major ORACLE applications is running yet on the mainframe, many microcomputer based applications using ORACLE have been developed. Some of these have been tried experimentally on the mainframe and these experiments have shown that an application developed under ORACLE on a microcomputer can be easily moved to the mainframe and other computing environments. Particularly, we have chosen ORACLE as one of the corner stones in our new attempt to generalize survey processing software for Statistics Canada.

Question from Arie Shoshani

Why have Statistics Canada dropped the development of the Rapid system?

Podehl: The system was oriented towards efficient storage and retrieval. Its query facilities were not sufficiently powerful by current standards. To add such facilities was considered too costly.

2.2 Geoffrey Stephenson, Logica S.A., Luxemburg

Use of commercial database management systems for statistical databases

One should distinguish between different aspects of this question. First in relation to size of database, second in relation to online versus batch operation, third in relation to data handling per se versus operations on the data. These aspects interact in addition to their individual characteristics. One should also remember that the distinguishing mark of "statistical" databases is that they are used, not in a transaction processing environment, but in environments where random sub-sets of the data are involved in individual operations.

The number of applications where commercial databases simply cannot handle the volume of data has fallen with the growth in the density of disk media, so that this is rarely in itself a problem. However the transfer of large data volumes and the cost of loading them into a DBMS can be a problem. For example the transfer of data between national statistical offices is still largely by posting magnetic tapes, and where the volumes are large (e.g. of the order of 100 MB per months), significant delays may occur in the use of the data if it has to be loaded into a DBMS before use. One solution to this problem lies in extractions from the raw data for high priority analyses, before general processing is allowed, The main problems with data volumes arise because of the interaction with response time requirements in online applications. The problems of size are also important where the data is intended for distribution, for example in archives. DBMS typically structure the data internally, so that loading and unloading has a cost penalty. In these circumstances it is difficult to see the advantage of bearing this penalty when the data is only to be stored for selective retrieval.

The principal difficulty with the use of DBMS for statistical processing, lies in the paucity of processing capabilities available. The commands directly supported in, for example, SQL are not suitable for much apart from data retrieval and simple reporting. The obvious solution is to provide interfaces to transfer data from the DBMS environment to the statistical environment whether it is a major package supporting a variety of facilities or a purpose written program to carry out some specific algorithm. The drawback is that the process cannot easily be carried out online, as it is usually not possible to integrate the two environments sufficiently. Thus the advantage of the DBMS facilities is lost and there seems little left to justify the use of storage programmes other than those provided by the main statistical packages.

A main concern for those working with large online data sets is response time. The problems are created by the difficulty of combining a general retrieval capability in a multi-dimensional data set, with fast responses when several thousand data items have to be accessed for the production of one table or summary statistic. Experience with this situation suggests that the overhead imposed by the inflexibility of commercial DBMS (in storage strategies, lack of compression facilities and choice of access path) is sufficiently high to make it worth considering special purpose data handling software. Some work with ADABAS and ORACLE in comparison with a purpose written FORTRAN system, indicate an improvement may be achieved in favourable circumstances of two orders of magnitude or more (this was with two projects involving Gigabyte data volumes with a multidimensional data set).

It is probably not reasonable to assume a change in this situation in the near future. What would be required on the interface side would be a statistical analysis package of the complexity of SAS or SPSS fully integrated with a database system like INGRES. A more hopeful possibility is that the advent of object-oriented database systems will make it much easier to add new functions to database packages.

2.3 Michael Cawson, Steward Observatory, University of Arizona

Selection of a Scientific Database Management System: a case study

1. The Application

The main requirement is for the storage, retrieval and interrogation of numeric data pertaining to the variability and colours of hundreds of thousands of stars and galaxies within a fixed survey region of the sky.

Several hundreds of megabytes of raw image data are collected each clear night, processed and analysed to yield brightness and shape parameters (and confidence levels) for all detected objects.

Data from different nights are merged together using measured x-y positions (i.e. noisy compound keys), to produce summary databases from which statistics and identifications can be made using ad hoc selection criteria depending on the scientific questions to be addressed (e.g. faint extended sources, objects with no near neighbours, highly variable point sources, etc).

2. SSDBM selection criteria

Versatility

The ability to use a common database system for managing the nightly data files in both raw and calibrated forms, controlling and logging the processing of the nightly data itself, and managing the summary databases, in order to unify the engineering, software and scientific aspects of the project, was seen as being very important.

Embedded data access - binary internal representation

The ease of assessing data in the system directly from within special-purpose application programs (i.e. the programs which create, update and verify the data) is essential. The interface to the SSDBM could not therefore be constrained to an ASCII string format, for instance, but an embedded query language or subroutine library would be required, preferably with a FORTRAN interface.

Data compression

Due to the very large data volumes, data compression techniques would be useful but not of overriding importance. However, using data-types matched to the information content of attributes is viewed as essential (i.e. integers and reals of different word-lengths).

Data-types

It is not only essential to provide a variety of atomic data-types but also to allow these to be put together into repeating groups and sequences where the order of entries is important. Ideally, complex data-types consisting of n-tuples of compound data-types should also be allowed.

Tertiary storage management - generic database descriptions

Even with data compression techniques, the total database size is still far too large for all the data to reside on primary and secondary storage simultaneously, so it would be necessary either to have a system which could manage tertiary storage media (such as magnetic tape or optical disks), or to break down the system into a number of separate databases, one for each night, and others for the summary information, and allow the multiple nightly databases to share common descriptions (i.e. generic database descriptions). In this way, only the databases pertaining to the observations currently being processed (and the summary databases) would need to reside on fast random-access read/write media at any given time.

Metadata

The extreme homogeneity of the data make them well suited to having many instantiations of generic descriptions of the databases, combined with small amounts of metadata to capture the important differences (e.g. the date of observation, etc). However, the management of the metadata and its availability within the query language are essential.

Interactive definition of derived data

The ability to support complex iterative queries (interrogations) on the data where each iteration may depend conceptually or even analytically on the results of previous queries is seen as essential to the scientific method (e.g. select entries in k-space which are more than 5-sigma above a least-squares fit to a model of the data entered interactively and refined experimentally at query time). A subset of this requirement is the ability to define derived (virtual) attributes on-the-fly using mathematical and statistical operators (such as LOG, SIN, MEAN, etc) without the need for the database system to append the additional attributes to the permanent copy of the databases (since these derived attributes will vary in definition greatly from application to application and are likely to be needed only during an interrogation session).

Physical data organization

Support for 'range queries', tolerances on matches and inexact searches are extremely important when the keys are measured properties and therefore subject to measurement uncertainties and noise. Range queries are also often given on several attributes simultaneously (multidimensional range queries). Ideally physical data organizations which support such queries efficiently should be available. In any event, exact match queries, which most commercial database systems are set up to evaluate efficiently using inverted files, hashing and tree-indexes, are relatively rare.

Graphics output and input

The integration of interactive graphics into the data interrogation language is important in order to allow the visualisation of complex information and the specification of queries graphically in terms of regions-of-interest on plots.

Statistical summaries

The integration of statistical summaries (such as HIGHEST, LOWEST, MEAN, VARIANCE, CORRELATION, etc) into the query language to help guide interrogations and see the overall patterns in the data is important.

Sampling operators

The integration of sampling operators (such av EVERY n, FIRST x% or RANDOM y%) into the query language would be extremely useful so that approximate results can be obtained quickly without the need to scan an entire database.

Summary

In short, a system is required which will support, and actively encourage, the scientific method of investigating inferences which can be drawn from large volumes of numeric data, with as many different types of tool available as possible (i.e. graphics, statistical analyses, summaries, regressions, modelling, etc).

These requirements may be viewed as applications which might sit on top of a more traditional database system, and therefore be beyond the realm of the database system itself. However, the user interface must be fully integrated to avoid the necessity of a context switch when data manipulation tasks and data analysis and presentation commands are intermixed as is so common in a scientific research environment. Furthermore, the retrieval and analysis aspects of data processing are so closely linked that the database system and the users' view of the data (both conceptually and graphically) should be streamlined as much as possible for greatest efficiency.

3. The decision

To build a very simple data I/O package (subroutine library) supporting a variety of atomic data-types and also linear fixed-length arrays. Descriptions of file types were to be entirely generic and linked to the file-name suffix to encourage system-wide file-naming conventions. Each generic file description should include a definition of the record structure of the files and a description of a header block which would contain metadata to distinguish different instantiations of the file type.

To build a number of utility programs for performing general-purpose operations on data files such as browsing, printing, sorting, searching, comparing and merging, as and when the need for each arose.

To build a single, general-purpose, powerful data interrogation program for performing the required interactive, statistical and graphical queries and analyses on the data from any file.

4. Conclusion

The high-level data interrogation capabilities provided by our in-house system far outweigh the cost or inconvenience of creating the system from scratch, and they have led to a completely new methodology for performing scientific investigation on computer-based data, namely the in-depth interactive data interrogation session, which replaces the writing, debugging, running and maintenance of countless special-purpose data verification programs.

Commercial systems abound, yet they provide very limited operators for performing statistical analyses and summaries. SSDBM systems abound, but only in the form of special-purpose code with very limited facilities for database management, high-level query language and relational database operators.

In general, scientific applications are never easily or accurately specified in advance, particularly in a research environment. For this simple reason the most flexible and versatile system should be chosen.

In some respects this is a vote in favour of commercial systems, since they have to address the needs of a large user-base in order to amortise their development cost, and so they may have features which, although they do not initially appear to be particularly useful or relevant, may later prove to be of great value.

However, I believe that most commercial systems are inadequate for scientific experimental and laboratory purposes, and offer extremely constrained query languages. Until this changes, the advantages of database technology for providing data independence and relieving the scientist of the laborious overhead of data management, will not be felt in any significant way except in projects which are sufficiently large to write and maintain their own systems.

Question from Arie Shoshani

It seems that astronomers want to use graphical information when specifying queries. Can you give an example?

Cawson: Yes, we would like to take graphical output and use parts of it as a component of a query predicate. For example, a subset of a picture may constitute a star cluster. One might want to mark the clusters graphically and then formulate a statistical query comparing the cluster members with the other stars in the picture with respect to some function of some attributes whose values are retrieved from the database.

2.4 Hideto Sato, Institute of Social and Economic Research, Osaka University

Statistical Database Management Systems in Japan at present and in the future

1. Present situation

There are many statistical DBMSs working in Japan. The configurations of some of them are summarized in Table 1, separating DBMSs, DD/DSs, query language processors, and linked statistical packages. In this figure, [C] means a commercially available software, and [S] a specially developed one. As seen, none of them do without a CDBMS, and none do only with a CDBMS; i.e. all of them are combinations of [C] and [S]. This is due to the following reasons.

A mere CDBMS is insufficient for an SSDB

(1) data model

CDBMSs do not support data models characteristic of SSDBs. Herein, the term "data model" is used for representing data structures in a DB and data manipulation facilities applicable to them. When a DBMS does not provide a statistical data model, a user of a statistical DB is confronted with troubles such that it is difficult to find data they need, misuse of data may be incurred, and necessary manipulation of data may not be executed.

For these reasons, Case A and D in Table 1 provide query language processors specially ordered for time series DBs. Case B and C provide DD/DSs specially developed and query language processors supported by the DD/DSs.

(2) linkage with statistical packages

For SSDBs, linkages with statistical packages are important since statistical data are used mainly in research and planning actitivites. However, as most packages for cross sectional data analyses have rich

Table 1 Major Statistical DBMSs working in Japan

case	data type	year of development	instance data management (DBMS)	meta data management (DD/DS)	query language processor	linked statistcial package	
						type	transmission of data
A	time series	about 1980	[S]+VSAM	-	[S]	[S] TSP like	direct
A	cross section (table)	about 1980	[C]+SQL like	[C] OODB + [S] under development -	[C] SQL + [S]	[C] Business [C] SAS like	direct file to file
B	time series & cross section	under development	[C] Model 204	[C] Model 204 +[S]	[S]	[C] SAS + [S] TSP like	file to file
C	cross section (table)	1986	[C]? SAS	[C] Model 204 +[S]	[S] + SAS	[C] SAS	direct
D	time series	about 1975	[C] ADABAS like	-	[S]	[S] TSP lika	direct
D	cross section (table)	about 1975	[C] ADABAS like	-	[C] + [S]	[S] Business [C] SAS like	direct file to file

N.B. Case A: National Land Agency B: a bank C: Hiroshima University D: a government branch
[S]: a special purpose program peculiarly developed
[C]: a commercially available software
Business (in "package type"): tabulation, drawing graphs and maps, etc.

functions and provide original data management facilities, time consuming file to file transmission is inevitable for a linkage of a DB and a package.

For this reason, Case C adopts a statistical package SAS as a tool for managing instance data and makes a CDBMS manage metadata.

A CDBMS is used as a part of an SSDBMS

(1) An SSDBMS and a CDBMS need similar facilities

As for physical data management such as authorization (security control), concurrency control, backup, and error recovery, facilities of a CDBMS can be used for an SSDB.

(2) usage of a CDBMS makes system maintenance easy

Accompanied with technological progress in hardware, update of a DBMS is always important; e.g. introduction of cache memories and RAM disk. As a CDBMS is updated by its developer, usage of a CDBMS reduces system maintenance load on the information management section.

(3) an information management section does not want to increase the number of DBMSs to be managed

An ordinary organization needs not only an SSDB but also operational business DBs. It becomes a heavy burden for an information management section to manage two or more DBMSs.

Conclusion

In the present situation, the following two solutions seem to be realistic.

Solution 1: Like Case B and Case A, a CDBMS is employed as a tool for physical data management and a DD/DS is developed in use of special ordered programs on top of a CDBMS, supporting a data model peculiar to an SSDB.

Solution 2: Like Case C, an application program (a statistical package) is employed for managing instance data and a DD/DS is developed on top of a CDBMS for the sake of sharing data.

2. Future prospect

Presumption

(1) Workstations will be more popular and more powerful
(2) OOBMSs (object oriented DBMSs) will be established and commercially available.

Influence of workstations

Most applications including statistical packages will be located at workstations. An SSDBMS will reside in a central system whose function is specialized as a data server. Downloading of data from the central DB to the workstation DBs will be unavoidable.

Accordingly, time consuming file to file transmission will be left even in the future, and Solution 2 above will lose its generality.

Influence of OODBMS

(Since the importance of OODBMSs for SSDBs in now under investigation, the following remarks must be considered as tentative.)

OODBs have many interesting features, among which the concept of abstract data-types is especially important for SSDB applications. Under this concept, many data types having different sets of applicable operations can effectively coexist in a single system. This means that a single OODBMS can support many different data models.

In case of statistical DBs, many types of data manipulation are needed for different purposes. An OODBMS, which supports the above mentioned abstract data type and inheritance mechanism, may provide an effective solution.

Tentative conclusion

Commercially available object-oriented DBMSs supporting enhanced and extensible data models will appear in the near future. Then, a special purpose DBMS such as an SSDBMS will be rather easy to develop by adding special classes and methods. The resulting SSDBMS will look like an intermediate between a ready-made DBMS and a custom-made one.

However, since it is inefficient for each organization to define such special classes and methods, standardization of classes and methods for SSDBs, or standardization of SSDB data models, will become important.

3. OPEN FLOOR DISCUSSION

Svensson: There was also a contribution from Hideto Sato which is perhaps one of the most interesting. I asked Dr. Sato if he wished to be on the Panel, but he declined because he didn't have enough time to prepare (a slightly edited version of this paper is included above).

We're now open to comments from the audience.

Sakti Ghosh, IBM Research: We started this conference on statistical and scientific database systems about 8 years ago. Even today, we do not have any commercial SSDBMS's. I don't know of any research institute that has even built a prototype. I would like the Panel's views on why this area of statistical database systems has been neglected for so long, but first let me give you some background as to what I think are the reasons.

To develop a useful "Statistical Database Management System", we have to cover the needs of two professions: statisticians and database managers. Both are highly paid professionals. Each does not wish to learn the other's subject and that is a reason why it becomes an expensive proposal to do research in this area. Apart from the people of this audience, there are very few who do research in this area. Is this research neglected because we do not wish to put resources into it, or for some other reason? I would like the comments of each of the Panel members.

Podehl: Very good point. This problem occurs not only in research, but also in the daily work of implementing another survey or another statistical application system. I represent the systems methodology side of Statistics Canada where we have 100 to 200 programmers, analysts and highly paid specialists. In addition we have another group of 150 survey methodologists who know all those formulas and understand the probability issues and it's a continuing problem to bridge the gap between them. We have one group that designs surveys and so on, and we have another group which designs databases. They're trying to talk to each other but their culture and background is such that it is very difficult. I don't have the answer, but that's a reason why we need this conference.

I would like to add also that we are trying to hit a moving target. Did we abandon methods that occupied us in the 70's? No, we didn't abandon them, we adjusted them and developed them further, so it's a moving target.

Ghosh: In 1977 we had more papers on database management than we have on SSDBM today and that is because we have not been able to attract people to do research in this dual subject. I would like through this body to suggest a joint effort to bridge the gap, otherwise we will just have to carry on complaining.

Svensson: Twelve years ago, in my Institute, hardly anybody knew what a database management system was when I proposed the design of a completely new system which would be of use for the scientific data analysis work that I saw around me. After a while we managed to convince various financing organisations to embark on a development project which up to now has absorbed 25-30 man years of effort. And yet it's a commercial failure. I think the reason for this is not technical, it's a mixture of project management failure, and an education or marketing problem. Nevertheless the system is now used successfully in important projects, and is still being developed.

Ghosh: I'm not in a position to give figures, but in the development of the relational model, research institutes spent more than the manpower you (Svensson) have already spent. But when it was put into the commercial arena, more than 5 times the research effort was spent before a production system was developed. It's not cheap to develop a commercial system, especially not one which is built on new concepts.

John Klensin, MIT: Ghosh's question is based on a pair of incorrect premises and the implication from that pair of premises is even more disquieting. One of the realities we have to face is that on the statistical side of the problem, and to a lesser degree on the scientific side, there have been data management facilities going back to the mid-1960s, for example the data management facilities in the BMD and BMDP series of packages and in SAS and SPSS. The data management facilities in these packages are lousy in our evaluation. But they nevertheless seem quite satisfactory to a very large number of users.

As a consequence, one of the reasons why we have not been successful, is that people who have found themselves able to perform the data management and data manipulation which they need in, say, SAS, have no reason to go off and do something else which is more expensive both in cost of acquiring and in learning.

The second incorrect premise is that there are no commercial statistical or scientific database management systems in the field. For example, there is a package called SIR which is widely distributed and commercially available and heavily used. If SIR is not a statistical and scientific database management system, then I don't know what it is. Again, one can debate the quality of these things. SIR is not relational and it's not a lot of other things.

The Consistent system which came out of the second generation work at MIT in the 70's is in production use for the analysis of statistical data on a daily basis. Those of you who come from the Netherlands have your social security checks processed through a statistical database management system. We know that General Motors uses a statistical database management system to make decisions about automobile production among other things. These are without question commercial applications of production quality SSDBMS's. Why there aren't more is an interesting question and why these things are not more widely disseminated in the market is another, but declaring that there are no statistical database management systems is just not true.

Ghosh: Everybody knows that a database management is a sub-system of the operating system. (Interjection from Klensin, "I disagree"). (An argument then developed between Ghosh and Klensin as to whether DBMS's are part of the operating system or not).

Svensson: I think we have a problem here with terminology. Perhaps we should use the term "data analysis system" or something like that, instead of "statistical database management system".

The term DBMS has certain implied meanings and maybe we are not willing to accept all of these when we are talking about SSDBMS.

Ghosh: My definition of a statistical database management system is that part of the operating system which can perform statistical, semantic processing. You talk about SAS, this is an application program, and one of a large class. But application programs will not run as efficiently and an SSDBMS will not reduce costs if run at the application level.

Lars Rauch, Statistical Office, German Democratic Republic: I come from the Statistical Office of the GDR and I am responsible for system development. We have our own database system which we started developing in about 1981 but we also look for commercial systems. We took a decision to put many years of effort into our own software development. Our loosely coupled software system allows us to put commercial software packages into places where they are suitable and where we cannot justify spending our own effort.

We have RAPID from Statistics Canada, which we have used not as a DBMS, but for its metadata capability. What is missing in the RAPID system are the relational capabilities but we have implemented such capabilities ourselves with international co-operation.

I think that today more or less all institutions have DBMS's, either commercial systems or in-house systems. We are in a position now where it would be very difficult to agree on a commercial system. When we finish our own system next year, then I'm sure that we won't be able to say 'Now we'll start using a new system'. We shall have to use it for several years.

Let me come back to this claim that the database system is very close to the operating system. The DBMS manages data access. To this we have to add new layers for handling metadata. Later on we shall add a knowledge-based system to the existing software system. Here I see future possibilities for community co-operation in developing new concepts and standards.

Knut Wittkowski, University of Tubingen, Federal Republic of Germany: I think we're faced with several problems. First a definition problem, what is a statistical database management system? There are several different opinions. One view is that the principal statistical operators need to be integrated into statistical database management systems. I have a different view. To me a statistical database management system is a database management system which satisfies certain requirements arising from statistical data analysis.

Next, what is statistical data analysis? And again, we have two different areas. First we have census data, several gigabytes of data used primarily for counting against different criteria. In my view this is not exactly statistical analysis. Statistical analysis is modelling and drawing conclusions and I think we need different systems, one for handling very large databases for tabulations and one for modelling where we have, say, three megabyte data sets, but where we want to use complicated methods.

Some years ago, we thought we needed relational database systems, but they were too complicated and too slow and could not be efficiently implemented. Nowadays we have relational database systems, so there has been some progress. We are now faced with a need for more integrated semantics and more metadata. There are some prototype systems, but there is a need to agree on the requirements otherwise the authors of commercial database systems will not know what to supply. They sell what people want to buy and so we have to formulate requirements.

We also have to educate people who use statistical database management systems because if people using them don't understand statistics, then the system can produce nonsense. So I think we have to do two things. First we have to specify requirements and then we have to educate users.

Ghosh: First of all, I don't agree that there has been no progress. There are things happening, but you have to look very carefully to see the movement. It's a bit like astronomy.

There is a slight case of the 'Not Invented Here' syndrome. You can see systems like Tau, which was done on the budget of OPCS in the UK. You see that the UN has the statistical computing project SCP. A lot of work was done in that. It picked up RAPID and it picked up a lot of other work. At the end of the day, the project stopped and nobody is using any of it; maybe, one or two people are, but it is not universal.

Martin Podehl said that RAPID and CANSIM have stopped being developed. That happened to "Cronos" and it died. If you don't develop a software system to meet the changing needs of the users, and the changing technology, the system will die. For example, some years ago we could count the number of our PCs on the fingers of one hand. We now have roughly one PC for two users. With the way we are working now and expectations that users have, the old architecture with a central engine and a star of terminals doesn't stick anymore. The users want to have their data cleaned and put somewhere so they can pull it back down to their workstation and work on it.

Also, we are a non-homogeneous set of users. Some want systems for finding data and some for analysing data. In our environment we want systems for producing clean data, and then systems for getting it to other people who go on to use it. In our organization we're actually splitting these two activities. In one of those areas we have found products. We intend to put our simulations and our time series into the Aremos system and use it for dissemination. It's a commercial product and we get the support. We have also discovered another product from the IMS world called ACUE which is for multi-dimensional data. We are also going to take that. But it doesn't handle the problem of taking raw micro data, cleaning it up, aggregating it and getting it ready to put into a system to disseminate it. These two worlds at the moment don't seem to get any closer together. Packages like SAS and SIR are moving towards SSDBMS systems but they are not moving fast enough.

Roger Cubitt, Statistical Office, European Community: Podehl mentioned systems developed by Statistics Canada in the 60's and 70's. At SOEC we have developed Cronos. There is a lot of organisational pressure to use commercial products to get away from the terrible burden of in-house system maintenance. The problem is that in order to use commercial database management systems, you have to write extra code on top of the DBMS to do your actual application. You may then find that you had to spend so much time making the system do the job you want it to do, that in the end you didn't gain very much. We have been down that path and now we're rethinking it.

Cawson: If I was a potential supplier of SSDBMS's coming to this conference and listening carefully to what everybody has said, I wouldn't see a consensus of opinion on requirements. And yet I would see all these government organisations using massive amounts of data. Clearly there must be potential for selling lots of systems and I would ask myself what it is I have to put in to make it usable. I think we are finding it difficult to agree what should be in such systems, but whatever it is it's something that has to be packaged and supported in the commercial sense. As a potential producer I would also be worried as to whether there was a market for my product. That's my answer as to why there isn't an SSDBMS available.

Stephenson: A number of points have come up. On the question of marketing costs, I think the estimates made earlier were under-estimates, if anything. The cost of developing and writing a piece of software to your own specification is probably only 2 to 3 percent of the cost of moving the product into the marketplace. Developers look for a small number of facilities that are common to many applications to make the market as large as possible. For example, the data handling concepts of DBMS products are very general. When you narrow the application area, for example to statistical database management, the number of requirements grows. At the same time, the market gets smaller. These two effects combine to lower the profitability and thus deter product investment.

Ghosh: Every government in the world collects statistics and there are millions of reels of statistical data. The reason the money is not spent is that producers are unable to supply a system which can tell them within 5 minutes what is, say, the industrial production of Italy. If we could do that, then many social statistical problems could be solved. There is a huge market, but we are not able to satisfy it.

Svensson: It is not unknown in history that special purpose systems for advanced scientific usage have been developed and successfully marketed. There are many successful companies, such as Hewlett Packard, which started with the esoteric idea of marketing very expensive and sophisticated equipment. Even if the market is small there is evidence that it may be possible to succeed commercially.

Ghosh: I agree there is a special market. I don't agree that there is no market for statistical database management systems. Statistics has been around since the birth of Christ. (Interjection from Stephenson, "So why hasn't IBM produced a product yet?" Laughter.)

Geographical Database Systems and Statistical Information

Edited by Andrew Westlake*

This paper is a report on a Panel Session of the same name held during the IVth SSDBM meeting. The material has been reorganised to have more structure but is all based on discussions and verbal and written presentations from the session. The Panel members were David Rhind (Birkbeck College, London), Georgio Gambosi (IASI, Rome) and Andrew Westlake, (LSHTM, London), and major contributions to the discussion were made by Robert Laurini (INSA, Lyon), John Klensin (INFDS, Cambridge MA), Arie Shoshani (Lawrence Berkeley Labs, Berkeley CA) and Roger Cubitt (Eurostat, Luxembourg)

1. Introduction

Any data with a geographical component is a legitimate subject for geographical study. Since statisticians and computer scientists are intimately involved with geographical data one would expect that links between disciplines would be profitable. However, few seem to exist, and a list of some 5000 references on Geographical Information Systems maintained at Birkbeck College, London, has almost **no** overlap with the references included in the papers presented at the SSDBM meeting. The purpose of the panel session was to explore the additional input that can come to Geography from computer scientists and statisticians, and also to look at what Geography can add to statistical and scientific applications. The report is organised in three sections, picking up points which seemed of major concern to the three disciplines.

In the 1960's geographers tried to create a spatial science which exhibited regularity. This effort failed. They developed lots of theory, but had no data, no computing facilities and no statistical methods which make use of the spatial component: all standard statistical methods are **a**-spatial.

In the 1980's many things have changed. Almost everyone is involved in geographical data, and the topics are seen as conceptually simple. The number of applications has grown enormously, with a variety of professionals bringing together data and technical developments from different disciplines to create new systems. Geographical Information Systems (GIS) are an important area for software development and are seen by some commercial firms as a major market. In the past the systems have concentrated on retrieval and display problems, and are only just beginning to build in analytical and modelling tools. However, there are still no statistical methods for spatial data, and the explosion of geographical data has revealed major database problems.

---

* Small Area Health Statistics Unit, London School of Hygiene and Tropical Medicine, Keppel Street, London WC1E 7HT, UK

2. The Geographer's Perspective

2.1 Geographical Structures

Geographical data can relate to a number of different structures. Points can exist as entities in their own right, but are more often some sort of abstraction (such as the centroid of some larger structure) or as samples from a continuum (such as the point at which height is measured). Lines are important and are often not continuous either in themselves or their derivatives. Closed areas are referred to as regions or polygons, and the two names reveal a terminological problem which also has a conceptual component: should we think of areas as bounded by (arbitrary) lines, or as part of a regular tessellation of all space.

All of these structures can have a variety of attributes, and relationships obviously exist between them. We have a problem about the conceptual representation of geographical concepts, and several alternative proposals have been made. Even if you can come to terms with the concepts in one set of data you often find that you need to use (and link) several sets to pursue particular analytical objectives. The physical representation of continuous location measurements is always approximate, and (for example) arbitrary choices can be made about the set of straight lines used to produce a polygon boundary for a region. Thus the linking or integration of data presents great difficulties. Current methods are all based on polygon overlays, usually using different area bases (tessellation systems), and involve fuzzy matching. This is an area where geographers need considerable input from other disciplines.

There are many proposals for the physical implementation of the different representations, none of which fit together well. Extensive discussion of this topic is presented by Prof. Rhind et al. in **[RhOG88]** which appears elsewhere in this volume and so is not discussed further here.

2.2 Theory-less data exploration

There is great demand from geographers for methods to automatically explore large data sets. At present there is no statistical theory available and so geographers are inventing pragmatic and heuristic methods. Everyone recognises the need for good statistical input to develop methods with a sound theoretical underpining, and for input from computer scientists to provide the efficiency for these large scale tasks.

The Geographical Analysis Machine (GAM) described by Openshaw in **[OCWC87]** and summarised in **[RhOG88]** is a reaction to the lack of statistical theory for exploratory geographical analysis. It omits the statistical model fitting stage and instead detects anomalies directly by looking for clustering of extreme attributes. The detailed methodology is still under development but the GAM approach has generated much general interest, and considerable hostility from some statisticians.

2.3 The Current Situation

Geographers currently **have**:

Working **GIS software** available off the shelf. At least 15 different packages are available from major vendors.

Data warehouses stuffed full of geographical data, from census, commercial, satellite and other sources, much of it currently unused.

Many **real applications**, both pragmatic and theoretical.

Geographers still **need**:

Methods for describing the quality of their end results.

Encoding of rules and good practice.

Data access, for example the census data available on-line with tables produced on demand (subject to confidentiality restraints).

Data driven algorithms for integrating diverse data sets.

Modelling superstructures integrated with retrieval systems.

Some theory to underlie the current pragmatism.

3. The Computer Scientist's Perspective

3.1 Models

The Computer Scientist's first reaction is to ask whether there is anything still to be done in the field of Geography. The geographers seem to have spent a long time producing various systems, and if those systems only need a little improvement or refinement then there is little interest for real computer scientists to do. An alternative view is that geographers have been re-inventing some of the computer scientist's wheels, and what they really need are some well-designed tools added to existing DBMS models to support their particular applications. Or does Geography really present a **new** problems and **new** data types to be modelled?

The thoughtful response seems to be that the problems in Geography are very real, in some ways like those in Statistics. The geographical location is a new (non-atomic) data type, and the structures mentioned above represent new entity types, requiring new operators, and all can also carry statistical attributes of all types.

The problems of physical organisation of geographical data is central in geographical database systems because of its great impact on the physical size of the database and on the efficiency of query execution. A list of many of the different physical organisations appears in **[RhOG88]** together with a Figure indicating their properties in different contexts.

Unfortunately, the importance of physical considerations seems to have been the dominant factor in the development of the conceptual level. Extension of the relational model is not simple because the closed domains of relational attributes cannot handle the potentially infinite resolution of the location of points and structures. Some authors have distinguished between the extensional and intentional approach to this problem, the former explicitly storing (say) all the points which make up a line, whereas the latter stores only the end points of a straight line, together with a rule from which you can implicitly compute all the points on the line. This is similar to the problem of representing **time** (see **[SeSh88]**), but more complex because of the two-dimensionality (or sometimes three) of location.

Topological relationships between geographical entities (intersection, inclusion, adjacency) can sometimes be handled explicitly within the relational model (at the expense of storage efficiency). However, conceptual problems arise from the fact that the intersection between two regions, as well as being a relationship between the regions, is a region in its own right. It is again possible to extend the database to include all such implied structures, but it seems likely that the intentional approach of computing the relationships (and such implied entities) when they are needed will be more useful.

What is not clear is whether it is necessary to have a fundamental model which recognises the difference (and the different types of entity) or whether this is just an implementation issue. Several authors promote the separation of different types of entity. For example, Gambosi et al **[GaNT88]** propose a conceptual model with classes for attribute data separate from the geographical components, and Laurini and Milleret **[LaMi88]** recommend separating the surface representation (using Peano curves) from the structures defined by lines.

Quite a lot of papers have been written about conceptual models, but there is a view amongst geographers that these have been too remote from the implementation issues. This view may arise because geographers have identified so many implementation issues as difficult, or it may be that the conceptual models really are not good enough. But it is also fair to say that many geographers have been data or application driven, and have not tried to break down the gap between the conceptual and practical issues.

3.2 Resolution

This is an example of a problem with many aspects, and one which is perhaps not found in most application areas. At the logical level a change in resolution can cause a change in structure. An entity which appears as a point in a broad view (a village on a national map) becomes a region when we increase the resolution and refine our view. We understand implicitly that different questions about Geography imply different levels of resolution, but we must design our databases to handle a variety of questions: one of the advantages of computer models is that they force us to say explicitly things which previously were merely generally (but perhaps not universally) understood.

Recognising the problem does not always simplify it, and the resolution problem has significant volume implications. Many structures have a limiting level of resolution. For example, travelling routes (eg roads) usually exhibit smoothness once one reaches the resolution level of the objects (eg cars) which use the route. But at coarser resolution smoothness often decreases as resolution increases: roads look straighter on 1:625,000 scale national maps than they do on the local 1:50,000 maps, and these maps reveal additional minor roads. And at the limiting resolution for smoothness of a road you begin to be able to observe other structures such as faults in the surface and services running along and beneath the road. It is perhaps a reasonable working assumption that the amount of information available about a linear structure is very approximately proportional to the resolution of the structure. This implies that if you increase the resolution of observation by a factor of ten the information content will increase by ten for linear structures and up to one hundred for areal ones.

The resolution problem has important practical implications and serious errors can occur if your information is not available at the correct level. For example when looking for a new ski slope one could scan a database looking for areas at least a kilometre in length with an average slope of 5%. But that average does not say anything about the smoothness of the slope at the resolution of a ski-length, and this also affects its suitability. As in Statistics it does not make sense to report an average without some measure of variability, but in Geography you also need to report the resolution at which that variability was observed.

4. The Statistician's Perspective

4.1 Introduction

Geographical data represents a great challenge for the statistician, but it is a challenge which can often be sidestepped. Much summary data about populations is organised using hierarchically grouped geographical areas (administrative tessellations). It is possible (and may be sufficient) to ignore the topological concepts such as adjacency and just use the hierarchical structure in a relational system.

In one respect Geography is like Time: it provides an attribute which has uniform meaning across a variety of differing entities. This allows us to relate the attributes of one entity to another one at or close to the same place. Knowing that two pieces of information relate to the same place thus provides a link between them which might otherwise be difficult to establish.

Knowing **where** places are allows the concept of nearness between pairs of places, which can lead to *natural* aggregation definitions, such as *all people who live within 10 Km of a particular point.* Nearness can also be taken as implying similarity between places, which leads on to statistical models which include spatial correlations. Note that different applications may require quite different metrics for nearness.

So Geography provides the Statistician with an additional tool for deriving attributes for statistical entities, either through linking or as a method for defining aggregations, which is clearly very useful. But is there anything else, does geography affect the way in which the data is used?

4.2 The statistical process

The basic data units (eg people or registration districts) are unlikely to be of statistical interest in their own right, though there will be situations in which units are selected for unusual attributes and reported directly. More often the units will be subject to mathematical and geometrical operations to **derive** further attributes. (This corresponds to the window relation in the model of D'Atri and Ricci in **[DARi88]**.)

Attributes from the data units will be aggregated to produce **summaries** for the larger aggregate areas, usually more than one. Most attributes of aggregate areas are counts (or other sums) or rates (or other ratios) and these are themselves aggregated by further addition. The reporting of such summaries is the end of many statistical tasks.

Otherwise we proceed to generalise the summary results by fitting a (statistical) **model** to the data (and the model fitting may not need the summary step). The fitted model may be reported as an end in its own right, or we may close the circle by returning to the original data units looking for outliers or anomalous units.

How, then, is this statistical process altered by the geographical context?

The reporting of most statistical results can be enhanced by **geographical presentation**, particularly mapping. This is not strictly a statistical issue, but it is a point where GIS are strong and can certainly contribute to the statistical process. There is, however, the danger that the eye is quick to see spurious patterns in a visual image.

Geography can be used to define aggregation areas for data, or to classify and distinguish groups of data, or to link units in different data sets, ie as an extension of the derivation stage. After such use the geography is usually ignored and the statistical analysis proceeds as for any other set of data.

We may use location as a surrogate for some unobserved attributes. If we are convinced that an anomaly represents a real and significant effect we may try to find some geographical explanation to add to the model. For example with medical data we may look for a geographical structure which could be a source of hazard. More often location is ignored in the analysis. Alternatively we can use location as part of the statistical analysis process through the concept of nearness between data units. In the first approach Geography affects only the data, whereas in the second it also affects the analysis.

Usually when we fit a statistical model we end up by partitioning each original observation into a **fitted** part explained (or predicted) by the model and a **residual** part attributed to a combination of individual and unexplained variability. In the geographical context we can go a step further and hypothesize that some of the unexplained variability is due to unobserved geographical factors which are smoothly distributed. This means that these factors will have a similar effect on near areas so near residuals will be similar, and from these the geographical component for each area can be estimated. The work by Clayton and Kaldor **[ClKa87]** using Bayes posterior means gives a lead for this. These geographical components (with the observable effects already removed) should vary much more smoothly than the raw residuals (when there is a real geographical effect) and should be more precise since they contain contributions from a larger area.

The use of all pairs of distances is potentially an n^2 problem, but is probably not so in practice. Any reasonable model for the nature of a nearness effect will be inversely related to some function of distance, so that even with a weak decline in effect it will be possible to delimit an area of influence around each unit outside which any contributions can be ignored.

4.3 Statistical distributions

In most statistical situations we look on our data as a (more or less complex) sample from some hypothetical super-population, and this justifies the use of theoretical probability distributions for statistical tests. In a geographical context we often have information available for all areas within the region of interest.

With a complete population we cannot call upon the usual statistical theory to justify the use of theoretical distributions. Instead we can observe the complete sampling distribution of a test statistic by computing it for the whole population. This is the standard distribution-free procedure of computing a randomisation distribution, and corresponds to the hypothesis that the test units are randomly chosen from the whole population. We may discover that a statistic does indeed follow a theoretical form, which can provide insights about the underlying mechanism generating the data.

5. Summary

Geographical data reveals a great need for cooperation between disciplines.

Computer Scientists and Geographers need to cooperate to develop conceptual models which recognise the special characteristics of geographical data and draw on the extensive work already undertaken on physical implementation. This is similar to the situation with statistical data, where progress is being made slowly. Statisticians have much to gain from using the geographical components of their data to greater effect, whether for better linking or for truly spatial analysis. Geographers will be particularly pleased if the latter happens, and they are also looking to Statisticians for better methods for linking data represented using fuzzy area concepts.

References

[ClKa87] Clayton, D. and Kaldor, J., 1987. Empirical Bayes estimates of age-standardised relative risks for use in disease mapping. Biometrics 43, 671.

[DARi88] D'Atri, A. and Ricci, F.L., 1988. Interpretation of Statistical Queries to Relational Databases. IV SSDBM.

[GaNT88] Gambosi, G., Nardelli, E. and Talamo, M., 1988. A conceptual model for the representation of statistical information in a geographical data model. IV SSDBM.

[LaMi88] Laurini, R. and Milleret, F., 1988. Spatial Data Base Queries: relational algebra versus computational geometry. IV SSDBM.

[OCWC87] Openshaw, S., Charlton, M., Wymer, C. and Craft, A.W., 1987. A Mark 1 Geographical Analysis Machine for the automated analysis of point data sets. Int. J. Geographical Information Systems, 1, 335.

[RhOG88] Rhind, D., Openshaw, S. and Green, N., 1988. The analysis of Geographical Data: Data rich, Technology adequate, Theory poor. IV SSDBM

[SeSh88] Segev, A, and Shoshani, A, 1988. The Representation of a Temporal Data Model in the Relational Environment. IV SSDBM.

THE ANALYSIS OF GEOGRAPHICAL DATA:
DATA RICH, TECHNOLOGY ADEQUATE, THEORY POOR

David Rhind*, Stan Openshaw~ and Nick Green*

ABSTRACT

This paper describes the current and near-future situation in geographical data processing from several viewpoints. It commences with a summary of - so far as is known - user needs in relation to functionality and other characteristics of the necessary software (Geographical Information Systems or GIS). The burgeoning availability of certain types of spatially referenced data is described; the data structures used or developed to date are summarised, together with their advantages and disadvantages. The most powerful uses of GIS - modelling or spatial analysis - are presently rare but one example, based upon Openshaw's Geographical Analysis Machine, is presented; based on experience with this tool and much else, the shortcomings of existing software tools are described and future collaborative research to remedy these shortcomings is outlined.

INTRODUCTION

Geographical data are those referenced in space, whether by postal address, zip-code, cartesian or non- cartesian co-ordinates or by any other means which may be mapped into co-ordinate space - even if some degree of approximation is involved. The first examples of handling these by computer were published in the late 1950s but, as late as 1980 (Rhind 1981), most of the work was either in replicating traditional mapping procedures using the computer or 'one-off' statistical analysis (which, in many cases, did not explore the spatial nature of the geographical data but simply involved multivariate analysis of the attributes of geographical entities).

Since the early 1980s - and based upon research in such organisations as the Canadian Geographical Information System (Tomlinson, Calkins and Marble 1976), the Harvard Computer Graphics Laboratory (Chrisman, forthcoming) and the Experimental Cartography Unit (Rhind, forthcoming) - integrated software tools for handling many different types of geographical data, often in combination, have emerged. These are generally known as Geographical Information Systems or GIS. Since 1983, in particular, commercial vendors of GIS have proliferated. The factors driving the expansion of interest and use of such systems are highly diverse, with different factors being of paramount importance in different countries (see Chen Shupeng 1987, Kubo 1987, Rhind 1987, Tomlinson 1987). What is striking is the diversity of applications to which the same tools can - and already are - being put: these

---

* Birkbeck College, University of London, 7 - 15 Gresse Street, London W1P 1PA

~ Department of Geography, The University, Newcastle upon Tyne NE1 7RU

range from global monitoring of the environment (e.g. Mooneyhan 1988) to micro-scale planning (e.g. involving individual houses) in urban areas.

The only known, substantial attempt to study these developments and to anticipate what will happen next is that of the UK government's Committee of Enquiry into the Handling of Geographical Information (HMSO 1988); this followed earlier enquiries into remote sensing and digital mapping (SCST 1984) and into the national mapping organisation - Ordnance Survey (HMSO 1979). Certain of the results of the Committee of Enquiry, resulting in what is now known as the Chorley Report, are highly relevant to this paper. In particular, that report stressed the need for additional research and development in GIS - but R and D which would be focussed on meeting the needs of users, rather than driven solely by curiosity of the researchers. We now consider briefly some aspects of meeting users' needs through the creation of a GIS with all the necessary functionality and performance. On that basis, we then review data structures in relation to functionality and itemise the shortcomings of existing GIS.

HOW DO WE KNOW WHAT USERS NEED?

General issues in user studies

Notwithstanding attempts to develop a structured approach to system design (e.g. Tomlinson 1972, Calkins 1983, Marble 1983), the formal design of Geographical Information Systems is presently more of an art than a science (Chrisman 1987, Smith et al 1987). Part of this is because we know relatively little in quantitative terms about the real needs of the user community, as opposed to those facilities which are provided by commercial vendors. This is true both in terms of the total functions required and the frequency of use of individual functions - unless we assume that the relatively small number of studies adequately documented to date (e.g. Tomlinson et al 1976) is a reasonable sample of the multiplicity of systems now being operated or planned.

Furthermore, the scientific community in particular is a particularly difficult customer base so far as planning a GIS is concerned. Though often highly educated and motivated individuals, scientists are by their very nature committed to exploring topics which may necessitate totally new tools; routine and repeated applications are not the stuff of work at the research frontier in research councils or in universities. To that extent, the curiosity-driven scientist may be at the other end of the spectrum from the individual in a production (e.g. forestry) agency where the tasks are repetitive and even cyclic, linked to specific (often monitoring) tasks which are integrated with the commercial or government business itself. Even in such organisations, however, uncertainty exists in terms of the tools and data bases required for high level tasks for senior management, as opposed to low-level monitoring tasks. In summary, then, planning a GIS for an organisation is usually much more of a creative activity than might be imagined.

Two factors complicate most 'user need' studies so far as computer- related facilities are concerned: the rapidity of technical change and the growth (from a low base in some cases) of user appreciation of what can be achieved. In any one organisation, however, a combination of other, individual factors may have a significant bearing on the final solution, i.e. though many of these factors are encountered elsewhere, their combination is idiosyncratic. In many cases, the first of these factors is the difficulty of defining the bounds of the likely users. Another such problem may arise from the spectacular internal diversity of an organisation whose component parts may differ in their thematic interest, geographical areas of concern, scientific methodology and historical evolution. One consequence of this is often the existence of opposing 'centralist' and 'highly distributed' concepts of how data sets are to be held, maintained and exploited.

Perhaps the most significant of all characteristics of many scientific organisations like universities and research councils (at least in the UK) is the much heavier reliance now put upon contract research for the outside world: for instance, no less than 40% of NERC income was generated in this way in 1985/86 (NERC 1987), the remainder being obtained from the traditional science vote. Many of these contract research projects are quite short term and typically therefore the building and assembly of data bases has to anticipate the projects or else be compiled on a 'quick and (relatively) dirty' basis. This manifestly suggests the need for robust and tolerant software and a 'tool-kit' from which users can re-assemble software modules very rapidly to suit their own immediate needs.

Finally, though the concerns of this paper are with spatial or geographically referenced data, it is impossible to ignore the other data held by any organisation; these are usually intimately associated with those which are spatially referenced. In general, geography often forms only one of the retrieval keys used. The logical consequence of this is that the GIS has to fit within a more general database strategy in any organisation.

Functionality and user needs

The list of functions which a 'fully fledged' GIS needs to be able to carry out is a matter of some dispute in detail but there is general agreement over most of the functions. Based upon work by Rhind (1981), Dangermond (1983), Guptill (1985) and Rhind and Green (1987), Table 1 sets out the key functions involved. It is self-evident that different organisations and individuals will attach importance to one function rather than another in relation to specific tasks. In general, however, it is possible to illustrate the nature of the needs of users of geographical data with a series of simple, generic questions:

- what is at location .....?
- what is adjacent to .....?
- where is ..... true/ to be found?
- what has changed since ..... and where has this occurred?
- what spatial pattern(s) exist(s) and where are the anomalies?
- what if .....?

Driving the system: the user interface

Functionality is not the only consideration of importance to the user. Performance, reliability and ease of use are often also of critical importance. In particular, one under-represented area of GIS research has been the interface between the user and the computer system. Commercial developments such as the Apple Mackintosh and the GEMS software have provided popular alternatives to the traditional menu or command-driven means of 'driving' large software systems. In addition, the UK government's Alvey programme specifically targeted Man/Machine Interfaces (MMI) as a key area of research and development; some 27 different projects were funded under this heading, many with some 'expert system' component, and Green (1986) has drawn upon early results from these studies in an investigation of user needs.

Table 1. A classification of GIS functions (after Rhind and Green 1987)

Data Input and Encoding

- Data capture (e.g. manual or automatic digitising)
- Data validation and editing (e.g. quality checking, detection of digitising errors such as overshoots, etc.)
- Data storage and structuring (e.g. construction of link/node topology, chain coding, etc.)

Data Manipulation

- Structure conversion (e.g. vector-to-raster conversion, quad-trees to vector, etc.)
- Geometric conversion (e.g. map registration, 'rubber-sheet' transformation, scale change, map projection change or image warping)
- Generalisation and classification (e.g. co-ordinate thinning, re-classification, aggregation of attribute data)
- Enhancement (e.g. image edge enhancement and texturing, line fractalisation)
- Abstraction (e.g. calculation of area centroids, proximal features, Thiessen polygons, etc.)

Data Retrieval

- Selective retrieval of information based on spatial or thematic criteria, including 'browse' facilities.

Data Analysis

- Spatial analysis (e.g. polygon overlay, route allocation, inter-visibility, slope and aspect calculation)
- Statistical analysis (e.g. histograms, frequency analysis, measures of dispersion, multi-variate analysis)
- Measurement (e.g. line length, area and volume calculation, distance and direction measurement)

Data Display

- Graphical display of maps, graphs, etc. on both graphical display and on hard copy devices.
- Report writing (e.g. automatic text reporting on database contents in standard form, production of summary tables)

Database Management

- Integrated database management facilities include: support and monitoring of multi-user access to the databases; provision of 'roll-back' facilities for use in the event of system failure; organisation of the database for efficient storage and retrieval without data redundancy; automatic maintenance of database security and integrity; providing the user with a 'data-independent' view of the database.

Note: This classification is based partly on the work of Knapp and Rider (1979), Rhind (1981), Dangermond (1983), Guptill (1985) and Smith et al (1987). In no sense is it intended that all possible variants on all functions need be present at any one site (e.g. automated digitising facilities are probably best concentrated).

The key to user satisfaction is not only whether the system performs what it is claimed to do but whether it does it in a fashion agreeable to the user. Green reviewed the obvious elements of user psychology-information overload, task complexity, delays in response times and inadequate user control - and defined the response characteristics of different classes of users, based upon other studies in the literature (Martin 1975) and observations of the use of GIS. In addition, he considered the need for feedback to the user, for help facilities, for system interface consistency, for response time control and for good error handling. Based on all of this, he examined the various types of user interfaces. User-initiated dialogue was taken to include both command-driven and forms-oriented interfaces; of particular importance in the former is the prospect of a Natural Language Interface (see

Frank 1982, Robinson et al 1985, 1986). In contrast to such user-led interfaces are program-initiated dialogues, such as menu-driven and program-prompted approaches. His conclusion was that no one user interface was suitable for all users and tasks: thus multiple access paths to the same information and tools are necessary.

DATA AVAILABILITY

It is common experience that both the absolute amount and the proportion of data being made available in machine-readable form is increasing rapidly in most fields of research. This is certainly true of geographical data handling. Socio-economic data sets (such as those from the Censuses of Population) are now often either most cheaply or only available in such form. Data sets compiled in the course of routine administration in a business are increasingly being spatially aggregated and sold as 'spin-offs' to new customers as a means of recouping revenue; we expect to see much more of this occurring in the UK in the next few years, notably by utility organisations, credit and retail agencies and even by central government itself. The most serious problems encountered in creating such spin-offs are the spatial referencing (if none already exists) and the maintenance of confidentiality; the first problem is often minimised by use of postal address or postcode/zip-code as a locational reference and a look-up table between the addresses and co-ordinate space.

If we can look forward to a situation where the great bulk of such data is in machine-readable form and readily obtained through the sign-posting services urged by Chorley (HMSO 1987), the computational problems arising from these data are likely to be relatively small. The UK Population Census Small Area Statistics data for 1981, for instance, approximates to 2Gb - of the same order of magnitude of the raw household questionnaire returns but necessarily disguised in table form to prevent disclosure of information about individuals (but see Rhind and Higgins (1988) for a description of an on-line rules based system to permit any aggregation from the 'raw' data).

Many other geographical data sets, however, present much greater problems either from their bulk or from the inference which is necessary to convert the data into a form suitable for many routine purposes. Thus digital map data can be highly voluminous: one single sheet of 1/ 50,000 scale topographic map data made available by Ordnance Survey in topologically structured form amounted to about 35 Mb when stored in a proprietary GIS using a relational database; this implies a total for the UK of about 7Gb and - though the figure is little more than an informed guess - about 70Gb for the European Community as a whole for this scale and type of map alone. Such data needs to be updated periodically: some nations re-publish their map sheets on a fixed (e.g. 10 year) cycle. Since smaller scale maps often cannot be used to answer the same questions as larger ones (their content differs and they contain many deliberate simplifications or generalisation - see Rhind and Clark 1988), this implies that browsing and querying such volumes of data will be necessary. Moreover, recent experiments using proprietary data bases and carried out by Sanderson at the UK Natural Environment Research Council have demonstrated times of 24 hours+ to load and build data bases from individual map sheets.

If the data volumes involved in dealing with maps (and those cited above are by no means the most extreme - the UK has total map coverage at 1/10,000 scale and 70% coverage at 1/1250 or 1/2500 scale) are considerable, other factors also complicate computer processing. Of particular note is the difficulty of ensuring that the data are 'clean' in the topological, geometric and attribute senses. Whilst considerable progress has been made in devising and using software for checking topological completeness - first manifested in ARITHMICON and its predecessors in the US Bureau of Census - this alone is rarely sufficient to prevent major problems for subsequent processing software.

All such data base problems, however, pall into insignificance in comparison with those arising from remotely sensed data. At present, it is commonplace to work (at best) with one satellite image. If this is produced by the latest Landsat satellite, it occupies about 300Mb; some 11,000 of these cover the land areas of the globe as one near-synchronous data set. The nature of satellite imaging, however, is that the data collection is repetitive: even so, existing cycles of observation often miss important events (Chernobyl, for instance, was observed by good fortune by civilian satellites). NASA's plans for the next generation of remote sensing - as exemplified in the EOS program - anticipates a daily collection of up to 1 Terrabyte per day. When it is appreciated that much of this data is of limited value until the multi-spectral data is classified and that increasingly this will be achieved using contextual classifier's rather than the computationally simpler spectral signature-based, 'one pixel at a time', traditional classifier's, the data processing requirements are obvious. NASA intends to meet this by providing a new, integrated processing system for all of its data collection streams (see e.g. Estes and Bredekemp 1988).

Finally, it should be noted that geographical data are inherently fuzzy in several respects; most obviously, several commonly used areal descriptions are not formally defined (the Alps cf Italy). A knowledge of the likely extent of such 'fuzziness' or at least of the source of the data (and hence of its likely reliability - see Rhind 1988) should be central to assessments of the results from any analysis, especially if this was preceded by the integration of data from different sources.

SPATIAL DATA MODELS AND DATA STRUCTURES

Since the data volumes and processing tasks normally involved are non-trivial, it is important to consider the extent and form of data organisation optimal for each of the functional requirements already identified. This is best done by considering Peuquet's (1984) model of spatial data representation, extended to include five levels of abstraction, and arranged below in terms of increasing specificity:

(a) reality;

(b) an analogue abstraction or model of reality;

(c) a data model, being the human conceptualisation of reality without any implementation conventions or restrictions and comprising defined sets of entities and relationships between them;

(d) a data structure, being a representation of the data model designed to reflect the recording of the data in the computer;

(e) a file structure, the particular representation of the data in computer form.

In practice, the data structure is the most important of these for our present purposes and the properties of individual data structures are now examined. Green and Rhind (1986a) give many more details and examples of existing systems than can be included in this brief overview. Our classification of data structures is given in Table 2.

Table 2: A classification of data structures for GIS
Simple vector structures
Non-topological structures (i.e. 'spaghetti')
Simple topological structures (e.g. in GIMMS)
Directed topological structures (e.g. DIME)
Hierarchically indexed topological structures (POLYVRT)
Hybrid vector structures
Network indexed structures (e.g pre-TIGRIS Intergraph)
Geo-relational structures (e.g ARC/INFO, System 9)
Regular tessellations
Grid or raster structures
Hexagonal structures
Triangular structures
Irregular tessellations
Thiessen (Dirichlet) polygons
Delaunay triangulation
Nested regular tessellations
Regular quad-trees
Linear quad-trees
Edge and line trees
Nested hexagon structures
Nested irregular tessellations
Point quad-trees
K-d trees
Strip trees

Vector data structures

Simple Vector Data Structures

The vector data model describes points as a single location defined by an explicit x,y co-ordinate; lines as a string of x,y co-ordinates; and areas as a closed loop of x,y co-ordinates, a feature often referred to as a 'polygon'. Traditionally, vector data structures have been used to represent spatial data for five main reasons:

(i) It is relatively easy to capture data in simple vector form. Most conventional manual digitising systems are designed to obtain data in this manner.

(ii) Vector data are easy to store and relatively efficient in terms of data volume, particularly where the entities to be encoded vary greatly in size.

(iii) A large software base for handling vector data has been established. Algorithms for storing, accessing and manipulating vector data are well-understood and widely implemented.

(iv) Vector data are ideal for many types of spatial analysis and manipulation. Modification and alteration of data is usually simple.

(v) The representation of vector data provides realistic, high quality and traditional-type cartographic output.

Non-Topological Vector Structures

The most simple vector data structure is termed an unstructured or 'spaghetti' structure. The data are spatially defined but no spatial relationships are explicitly retained. Hence, points are represented by a single x,y co-ordinate, a line by a string of x,y co-ordinates and a polygon (area) by a closed loop of x,y co-ordinates. Where polygons are adjacent, the common boundary is stored twice, once for each polygon. The lack of any stored spatial relationships means that data validation other than area closure requires visual checking.

A slightly different case of 'spaghetti' is where boundaries are only held once but are not coded as boundaries - though they might appear as such on visual examination. Such 'spaghetti' does not include the concept of an area. Despite these limitations, the use of unstructured vector data is still surprisingly common where the quality of the graphic output is more important than the utility of the data for other forms of spatial analysis.

The Ordnance Survey DMC digital map data format (Ordnance Survey, 1983) provided until recently for customers is a good example of an unstructured vector model. Individual map or cartographic entities are stored as features, each of which has a single feature code used primarily to allocate the symbolism for graphic display or to exclude the totality of the feature on substantial reduction in map scale. Without additional processing, the data are of limited utility for other than mapping purposes: features can be selected by their feature code and the map scale can be changed within limits but the great bulk of the more complex operations listed by Green et al. (1985) cannot be carried out.

Topological Vector Structures.

The most useful form of spatial relationship to store with a vector data model is the spatial adjacency or topology. One of the main benefits of storing the topology is that, unlike unstructured vector data, shared boundaries of adjacent polygons need be stored only once and questions such as 'find all the areas of woodland adjacent to lakes' can be readily answered without exhaustive calculation. The topology of vector data stored in this manner has been incorporated in many cartographic and geographic systems. GIMMS (Waugh and Taylor 1976, Waugh 1980) is an example of an early system designed to handle data in this manner. One restriction commonly encountered in operational systems (such as GIMMS) is that, in order to simplify the construction of a table of the topology, all line segments have to be recorded with explicit identification of area to the left and right of each line segment.

Directional Topology.

An extension of the topological model involves the explicit recording of directional information in addition to polygon adjacency. By identifying line segment end and line intersection points (commonly referred to as nodes), the direction of a line segment can also be defined. The best known model incorporating this data structure is the GBF/DIME (Geographic Base File/Dual Independent Map Encoding) model developed by the U.S. Bureau of Census which has been modified and extended in the TIGER (Topologically Integrated Geographic Encoding Referencing) model for the 1990 census (Broome 1985, Dewdney and Rhind 1986). Developed as a means of expediting the gathering and tabulating of census data (U.S. Bureau of Census, 1970), each boundary in the original GBF/DIME file was represented as a series of straight line segments with each line segment containing a census tract and block identifiers for the blocks (polygons) on either side (see figure 4). By explicitly assigning direction to line segments by encoding 'from' and 'to' nodes, the data could be automatically scanned for missing segments and other errors. Indeed, in the earliest DIME files, no geometry whatever was included. However, one of the major drawbacks of the GBF/DIME system was that segments were not stored in any particular order; thus, to retrieve any particular line segment, an exhaustive sequential search had to be performed on the file.

Hierarchically Structured Topology.

To overcome the retrieval problems associated with simple topological models such as GBF/DIME, Peucker and Chrisman (1975) developed and implemented a topological data structure entitled POLYVRT (POLYgon conVERTer) which separately stored each type of data entity in a hierarchical structure. In their file containing polygon names, there is a pointer to a 'chain list', that is a file containing the names of the 'chains' forming the boundaries of each polygon (a 'chain' being their term for a line segment). The chain file stores the left/right topology of adjacent polygons and also the to- and from-node identifiers. The points forming a chain are held separately from the topology, as are the co-ordinates of node points. The separation of line co-ordinates from the topology has some important implications: in particular, selective retrieval of specified classes of data is possible and the co-ordinate definitions of lines need only be retrieved when they are required for particular operations such as plotting or calculation of area.

The POLYVRT data model can also be used to represent more complex data structures. For example, by some modification of the basic data structure new levels can be added to the hierarchy without violating the basic data model; thus new lists could be added to link additional attributes with features. Within a single level in the data structure, different levels can also be defined by modification of polygon or chain codes. For example, by adding a prefix to chain codes, a hierarchy of different type of polygon can be defined in the same file. As an example, if a six figure numeric code is used as the identifier for individual chains, the first two digits might be used to identify polygons falling in the highest level of a hierarchy, the next two digits used to represent polygons falling at an intermediate level, and the last two digits representing polygons at the lowest level. Such coding is used, for instance, in describing line segments comprising the 1981 wards, districts and counties of Britain created for the Department for the Environment.

Hybrid Vector Data Structures

One of the major problems arising from use of a hierarchically structured vector data model such a POLYVRT is that there is no provision for the linking of physical relationships between items. To incorporate such linkages requires the utilisation of a data base in association with the topology. In addition to the hierarchical approach, relational and network data base schemes have been used to extend the power of the topological vector data model.

Network Vector Data Structures.

The separation of attribute and geometry data allows larger amounts of vector and attribute data to be handled in a more efficient manner, given that individual user queries may only require access to one of these data sets. Network data models are a natural extension of hierarchical models and were the first to be incorporated in GIS. The main advantages in using a net work structure are that physical relationships are established between entities, allowing relationships to be quickly navigated and hence high performance for specified queries and rapid interactive response times can be achieved. The main disadvantages are that the data structure is relatively inflexible, some degree of redundancy has to be built into the data base and that the effort to specify all the relationships on setting up the system can be considerable.

Relational Vector Data Structures.

Perhaps the most advanced form of vector data structure currently employed in GIS is the hybrid vector/relational or geo-relational model which incorporates relational database concepts in addition to a hierarchical file structure (van Roessel and Fosnight 1985, Deucker 1985). The main advantages of a relational structure over a hierarchical or network structure are that logical relationships can readily be established between entities which allows both processing flexibility and minimal redundancy of data. Furthermore, due to the flexibility of the data structure, well defined non-procedural query languages can be applied.

Tessellated data structures

Regular Tessellated Data Structures

The term 'tessellation' is used to describe the sub-division or 'tiling' of a plane into discrete sections. Two basic types of tessellation are recognised, regular tessellations and irregular tessellations. A tessellation is said to be regular if all tiles are equivalent under the symmetry group of tiling and each tile is a regular polygon (Bell et al., 1983). There exist 81 isohedral types, but only 11 with different adjacency structures. The three best known tessellations from this group are the square, triangle and hexagon. All three of these basic types have been used as the basis of spatial data models.

Grid or Raster Tessellations.

The regular square or grid mesh has been the most widely used of all the regular tessellations. The reasons for the dominance of this model are numerous, but some of the more important are:

(i) The simplicity with which grid-based structures can be handled by high-level programming languages such as FORTRAN since the data can be simply stored in arrays and processed sequentially.

(ii) The simple relationship of a grid structure to co-ordinate geometry.

(iii) The ease of data capture and display. Data capture devices such as mass digitising raster-scanners and remote sensing devices such as the Landsat MSS capture data in gridded form. Numerous output devices, even of high resolution, now work thus.

Hexagonal Structures.

Very few attempts have been made to develop operational systems employing regular hexagonal tessellation data models (Gibson and Lucas, 1982). The primary advantage of a hexagonal mesh is that the centre points of all neighbouring cells are equidistant. In a grid mesh, diagonal distances are 1.414 greater than distances between cells measured in the cardinal directions, this is claimed to have important implications for spatial search and retrieval. Hexagonal addressing mechanisms (Lucas and Gibson, 1981) have been used in the handling of geographic information, the processing of raster data into vector form (e.g. in the SAGE system), and the integration of data from multiple sensor systems

Triangular Structures.

Very few attempts have been made to utilize regular triangular tessellations in a data model. One unique feature of both regular and irregular triangular tessellations is that the orientation of the triangles along bounding edges are reversed. By assigning a height value to each vertex, the facets to the triangle represent the slope and aspect of regular portions of the land surface. However, although calculation of contours is more consistent on a regular mesh, most of the triangular-based systems developed thus far have used irregular rather than regular triangular structures. This preference is probably due to the fact that most data collection for terrain modelling is on an irregular sampling basis and some form of interpolation would be required to generate a regular mesh, thereby possibly biasing the surface characteristics (see Grassie 1982).

Irregular Tessellated Data Structures.

The essential difference between regular and irregular tessellations is that an irregular mesh can be built from the data and eliminates the need for data redundancy or interpolation. Hence, regular tessellations arise directly from point sample data collected by most scanning systems whilst irregular tessellations arise most commonly from manual (e.g. field) data collection or those procedures where data density reflects the data collection process (e.g. contour sampling). The spatial resolution of the irregular tessellation model is variable and reflects the density of data, with data cells becoming larger where data are sparse. The size, shape and orientation of data cells is a reflection of the distribution of the data points themselves and, as such, is useful for visual interpretation of the data. The main strength of irregular tessellated data structures is also their greatest weakness; because the data structure reflects the spatial organisation of the data, the data model is said to be 'data dependant'; any alteration in the location of an entity can change the data structure - a situation which may make editing a difficult process. There are two basic forms of irregular tessellation; simple irregular tessellation such as triangulated irregular networks (TIN's) or variable area (Thiessen or Dirichlet) tessellations and nested irregular tessellations such as point quad-trees and k-d trees (see below).

Thiessen (Dirichlet) Structures.

Thiessen polygons, also called Voronoi diagrams or Dirichlet tessellations, are formed by calculating areas of equal adjacency between points distributed on a plane. Rhysburger (1973) described Thiessen polygon formation as the propagation of a circle at a constant rate from a number of points scattered across a plane. Circle growth continues until the boundary of a circle intersects another circle or the boundary of the plane. Thiessen polygons are most useful for the calculation of adjacency and proximal analysis. The analytical derivation of Thiessen polygons has been studied by a number of people (e.g. Kopec 1963; Rhynsburger 1973) but undoubtedly the best current method was described by Green and Sibson (1978) and was implemented in the MAPICS system (MAPICS, 1985). MAPICS is one of the few examples of a commercially available general purpose system employing Thiessen polygons as the basis of a data structure. The Thiessen model is used as the basis for the interpolation of a grid matrix which is subsequently employed in surface modelling applications. The weighting of points such as to influence their surrounding area and produce a weighted Voronoi diagram was suggested by Boots (1979) and generalisation of the Voronoi diagram to handle disjointed line segments, intersecting line segments and other features was put forward by Drysdale and Lee (1978).

Delaunay Triangulation.

Probably the most widely used form of irregular tessellation is the Delaunay Triangulation, which is also known as the Triangulated Irregular Network, or TIN model. The TIN model is the standard means of representing terrain data for land-form analysis and display. The model is generally built directly from data points irregularly distributed over space and, as with the regular triangular tessellation model, the facets of the triangle reflect the landscape slope and aspect. The TIN model, however, also represents a direct dual to the Thiessen structure as the perpendicular bisectors of the edges of the triangulation also give the boundaries of Thiessen tiles. One major draw-back with the TIN model, and many other irregular tessellations, is that these data structures tend to be more difficult to generate than do regular tessellations. The highly data dependant nature of these models also makes up-date or editing difficult.

Nested Regular Tessellated Data Structures

Regular square and triangular meshes can each be sub-divided into smaller cells of the same shape. Only the square, however, retains its shape and orientation on sub-division. Recursive sub-division of a triangular structure produces triangles with alternating orientations and single cell comparisons become more complex. Hexagons can not be sub-divided into other hexagons, but the basic shape can be approximated and Ahuja (1983) described how hexagonal 'rosettes' can be constructed to form a hierarchy. In general, however, spatial data models based on the recursive sub-division of the plane are receiving increasing attention. One of the most studied of these models is the quad-tree, which is based on the recursive sub-division of the grid and has proved useful in a number of applications including Knowledge Based Geographic Information Systems (KBGIS). We now consider different types of quad-tree.

Region Quad-trees.

Peuquet (1984a) outlined some of the principal advantages of the quad-tree and Samet (1984a) gave a comprehensive discussion of the various types of quad-tree and associated forms. The three principal advantages of the quad-tree data structure as claimed by Peuquet are as follows:

(i) The recursive sub-division performed in the generation of a quad-tree results in a regular, balanced tree structure with each node possessing four sons. Tree structures have been well researched - the implementation, storage, compaction and addressing schemes for tree structures have all received considerable attention and hence implementation is a predictable task and efficiency can be predicted with some certainty.

(ii) The changes in scale between different levels in a quad-tree hierarchy are based on the power 2. Scale changes between these built-in scales can be simply achieved by retrieving data stored at higher or lower levels in the tree. Storing the same data at different degrees of resolution at different levels in the hierarchy can also be used as a simple means of automatic map generalisation. Thus, coarse data can be retrieved from higher levels in the tree to produce small scale maps, while for larger scale maps, more detailed data are retrieved from lower levels. The storage of the same data at different levels of resolution, however, increases the over-all storage requirements.

(iii) The physical distribution of data in a recursively sub-divided data structure is such that browsing and retrieval of data can be very rapid. Windowing operations are also very efficient, especially if the window is coincident with areas represented by the quad-tree cells.

In addition, quad-trees can be stored in a compacted form and, as with grid structures, spatial relationships are implicitly contained within the data model. One advantage of quad-trees over grid formats is that the processing time to perform operations using the structure depend on the number of leaves in the tree rather than the number of pixels in the image. Holdroyd (1988), however, has shown that routine processing of substantial areas is often faster with row ordering than through use of a quad-tree structure.

In practice, the term 'quad-tree' has been applied to the entire class of recursively generated hierarchical data structures. A quad-tree results in a decomposition of space into equal-sized parts. This is in contrast to the point tree (Finkel and Bentley, 1974) and k-dimensional binary search (k-d) tree (Bentley, 1975) where the decomposition is governed by the input (data dependant) and can thus be classified as nested irregular tessellations. Pyramid structure-based exponential reduction of an image by successive quartering without explicit inter-level connections (Tanimoto and Pavlidis, 1975; Shapiro, 1979) also constitutes a different form of recursive decomposition. Most quad-tree data models are based on traditional tree storage techniques which rely on pointers. An alternative approach using direct addressing mechanisms has been examined by a number of scientists (Abel and Smith, 1983; Gargantini, 1982) which allows data to be organised in a linear fashion. For this reason these structures have been termed linear quad-trees (see below).

Quad-tree concepts have also been applied to the representation of multi-dimensional objects. The best known of these approaches is the oct-tree or three dimensional quad-tree which has a branch ing factor of eight (Jackins and Tanimoto, 1980) which is of particular interest to scientists working in computer vision and pattern recognition (Woodwark, 1982). Different classes of data can also be registered by individual quad-trees and spatially registered to form multiple layers. This process is known as generating a 'forest' of quad-trees and has been applied to the handling of attribute data in Knowledge Based Geographic Information Systems (see Green and Rhind 1986a).

Linear Quad-trees.

Most early work on quad-trees used explicit pointers to represent the quad-tree data structure (Dyer, 1982; Tanimoto and Pavilidis, 1975; Rosenfeld and Samet, 1979; Samet, 1980). More recently, an alternative addressing mechanism termed linear quad-trees has been adopted (Gargantini, 1982; Able, 1984; Samet et al., 1984b; Mark and Lauzon, 1984; Mark and Able, 1985). In a linear quad-tree, each leaf node is given a unique key number (which may also express the level of the node), based on an ordered list of the node's ancestors. Thus the quad-tree is simply represented by a list of all the leaves sorted by that key. The finding of neighbours and spatial search by traversing the tree are achieved by modular arithmetic or by bit addressing of individual bits in the keys.

Samet et al. (1984a) described the implementation of a GIS using a quad-tree structure. Area, line and point features were encoded using a variant of the linear quad-tree and a memory management system based on B-trees (Comer, 1979). Database functions for the editing, searching and integration of data layers were also implemented. A region quad-tree is constructed by repeatedly subdividing an array of pixels into quadrants, sub-quadrants, and so on until blocks (or even single pixels) of the same value are obtained. This process is represented by a tree in which a root node corresponds to the whole array, the four sons correspond to the quadrants and terminator nodes to the pixels.

Space-filling curves or "Peano scans" have been shown to have some useful properties for the searching of tessellated data structures (Stevens et al, 1983). A Peano curve fills space in such a manner that every element in the data set is passed through and elements which are close together in the curve are close to each other in space. One of the first systems to use Peano curves as a means of geographic indexing was the Canadian Geographic Information System (CGIS) (Tomlinson, 1973). The CGIS data base was divided into frames indexed using a number scheme termed the Morton matrix. This scheme has the additional advantage that Morton matrix address can be directly computed from the interleaving of the binary representation of x and y co-ordinates. In a linear quad-tree constructed using Morton numbers, consecutive records having the same attribute number can be effectively combined without losing any of the inherent quad-tree topological or spatial relations. The addressing schemes for linear quad-trees devised by Abel and Smith (1983) and Mark and Lauzon (1984) differ from this scheme in that the address also indicates a level in the quad-tree structure.

Gargantini (1982) and Able (1984) reduced storage requirements for linear quad-trees by making the level of a leaf node implicit in the linear key and by storing only the leaves of one value. More recently, two-dimensional run-length encoding (2DRE) techniques have been adopted as a mechanism for the compaction of linear quad-trees (Lauzon et al. 1985; Cebrian et al., 1985; Mark and Lauzon, 1985). Whenever the linear quad-tree has consecutive leaves of the same colour (value), only the key of the last leaf in the 'run' needs to be stored. Mark and Able (1985) have now developed an algorithm to produce a 2DRE file directly from a vector representation of a polygon. The space saving in using 2DRE Morton file is typically 50% or better than that for a standard quad-tree representation and the computational complexity of the algorithms is not very different from those reported for other pointer-free file structures (Mark and Lauzon, 1985). Research into linear quad-tree structures and the development of algorithms to store and access them is one of the fastest growing fields in GIS (Stewart, 1986).

Edge and Line Quad-trees.

Edge trees (Shneier, 1981) represent an attempt to store linear information from an image in a manner analogous to storing region data. A region containing a linear feature is recursively sub-divided into quadrants until a section of curve that can be approximated by a straight line is detected. Each leaf node contains information on the intensity, direction, intercept and directional error (i.e. the error induced by approximating a curve with a straight line). The points at which an edge terminates are recorded and a flag set on the appropriate node. Small leaves are required in the vicinity of curves or edge intersections, long edges are represented by large leaves and many leaves contain no information at all. The edge tree representation is preferable to one suggested by Hunter and Steiglitz (1979) who proposed the building of a quad-tree directly from a polygon, with nodes generated at each line edge intersection. This representation has the disadvantage that the line width is represented and any shift operations result in information loss.

Samet and Webber (1984) approached the problem of encoding polygon data stored in image form by recording blocks of pixels that have no lines passing through them. Each leaf node then records which of its edges is coincident with a boundary. The storage required for this form of boundary representation is similar to that required by a region quad-tree representation of the same area. The resulting quad-tree is referred to by Samet and Webber as a 'line' quad-tree. The main drawback of both edge and line quad-tree representation is that they are only an approximation of vector boundaries and certain properties of a polygon map cannot be represented. For example, it is impossible in the line quad-tree representation for five lines to meet at a vertex. In addition, these structures are highly sensitive to shift and rotation and, as most lines have to be represented at the pixel level, the quad-trees have to be nested fairly deep (Samet, 1984a). For these reasons, the representation of polygon vertices using point quad-trees or k-d trees is the generally preferred approach.

Samet and Webber (1983) represented polygonal maps using a variant of the PR (point-region) quad-tree which exactly represents line edges thus avoiding the edge width problem associated with the Hunter-Steiglitz (1979) method (see above). A polygonal-map or PM quad-tree stores each vertex of a polygon map in a PR quad-tree and records how polygon vertices intersect the quadrant boundaries. Efficient algorithms exist to perform point-in-polygon, line insertion, map overlay clipping and windowing operations. Samet and Webber (1985) describe a modification to the PM quad-tree, termed a PM quad-tree, whereby decomposition continues until there is only one vertex in each quadrant. A closely related structure called the PMR quad-tree, described by Samet et al. (1986), is now used in their quad-tree-based geographic information system, Quilt. The PMR quad-tree uses probabilistic splitting and merging rules to dynamically organise the data. The resulting quad-tree data structure differs from most other quad-tree structures in that the tree is not unique but depends upon the history of manipulations applied to the data. In a comparative test designed to evaluate the performance of various quad-tree structures used in the representation of polygonal data, the PMR quad-tree performed considerably better than all of the others and requires less space (Samet et al, 1986).

Nested Hexagonal Structures.

The principal disadvantage of hexagonal tessellations is that the hexagon cannot be recursively subdivided into smaller hexagons and consequently higher-level aggregations of hexagons can only approximate a hexagon shape (Burt, 1980). That difficulty aside, the radial symmetry of the hexagon has been exploited by Gibson and Lucas (1982) who devised septree (base 7) addressing schemes and algorithms for hexagons which they named Generalised Balanced Ternary, or GBT. Several procedures including distance measurement can be performed directly using GBT addresses without conversion to cartesian co-ordinates. Recently, an addressing mechanism has been devised for a hexagonal or rhombus (HoR) structure based on the observation that an identical lattice is formed by the amalgamation or four hexagons as by the amalgamation of four rhombuses. By using an identical addressing scheme a limiting property of hexagon structures - that they cannot be further subdivided - is avoided, while the adjacency properties of hexagons can be utilised when required (Bell et al, 1986).

Nested Irregular Tessellated Data Structures

The classification of point trees, strip trees and k-dimensional (k-d) binary search trees as nested irregular tessellated data structures is somewhat subjective, but can be justified on the basis that they are essentially data-dependant structures. Although the recursive spatial decomposition performed to generate point and k-d trees is regular for the most part, the resulting data structure depends upon the order in which data are inserted. Similarly, strip trees are generated from a digitised line and essentially represent a vectorised quad-tree, the form of which is entirely dependant of the character of the line. Such data structures can be highly efficient for data storage and retrieval, but the handling of dynamic data insertions and deletions (a capability essential for a GIS) is more complex and may require periodic rebuilding of the data structure.

Point Quad-trees

A point tree is produced by taking a data point as a root and dividing the area around it into quadrants (Finkel and Bentley, 1974). As further data points are added, this process continues, resulting in a fourth degree tree. Although the decomposition is regular, the data structure is highly data-dependent since it varies according to the sequence in which the points are added. As the data structure is a function of the relative location of the points, it is most useful for operations involving spatial search and nearest neighbour analysis. Editing of point quad trees can only take place on the terminating leaves; if any other modifications are required, the point quad-tree has to be rebuilt.

An alternative means of representing point data in a quad-tree is to regard points as non-zero elements in a square matrix. The resulting data structure (termed an MX quad-tree) is similar to a region quad-tree but with leaf nodes corresponding to the presence or absence of points. The MX quad-tree data structure has the advantage that it is independent of the order in which points are inserted. The MX quad-tree, however, cannot be used if the minimum separation between data points is not known. In such circumstances an alternative adaption of the region quad-tree is required which can associate non-discrete point data with quadrants. Samet (1984a) describes a point-region or PR quad-tree which is organised in the same manner as a region quad-tree, but 'black' leaf nodes contain a single data point and its co-ordinates. The PR quad-tree is generated in a similar manner to the point quad-tree. Data points are added by searching for the quadrant in which the point belongs. If the quadrant is already occupied by a point then it is subdivided or split until it occupies its own quadrant. If data points fall very close together this may result in many subdivisions. Like the MX quad-tree the shape of the resulting PR quad-tree is independent of the order in which points are added. The efficiency of the PR quad-tree can be increased by allowing quadrants to fill with data points and only sub-dividing the quadrant 'bucket' when some finite capacity is reached (as described below).

K-Dimensional Binary Search Trees.

The main problem with structures produced by recursive decomposition is that the branching factor increases the storage requirements. The k-dimensional binary search tree, or k-d tree (Bentley, 1975) is one means of reducing storage requirements by simply dividing each area into two parts rather than into four parts at each point, with the direction for divisions rotating at each point. Matsuyama et al. (1984) describe the implementation of a k-d tree and performance comparisons between the k-d tree and the region quad-tree. The k-d tree described was implemented in PL/1 and used as the file system in the MILES geographic information system. To increase efficiency, records comprising the k-d tree are dynamically partitioned into blocks corresponding to a disk page. By storing information concerning neighbouring objects on the same page, responses to queries concerning spatial proximity necessitate the access of fewer pages: hence computer time is reduced. To store lines and areas in addition to points, the centroids of minimum bounding rectangles are stored and a list maintained for each page which records the identification codes of all objects either partially or fully included in the block corresponding to a page. The efficiency of the k-d tree for storing point data was compared by Matsuyama et al. (1984) with that of a region quad-tree. K-d trees were generated from batch input (entering all points at the same time) and by sequential insertion (data points being inserted in sequence to produce an unbalanced tree). Storage efficiency and retrieval times were found to be considerably better than for the region quad-tree representation of the same data.

Strip Trees.

Strip trees (Ballard, 1981) are a method of representing vector data using a hierarchy of bounding rectangles. The data structure is hybrid in that it reflects the cartographic character of lines but the lines themselves are not actually stored. Line curves are represented by a binary tree structure, where lower levels in the tree correspond to finer levels of resolution. Such a use of a hierarchy of strips to represent a line was first used by Peucker (1976) in a model capable of supporting point-in-polygon processing and the calculation of line intersection. A strip tree is obtained by successively approximating a curve using a series of enclosing rectangles. The data structure consists of a binary tree whose root represents the bounding rectangle of the entire curve. Computation of certain search and set operations are possible using strip trees including calculation of line intersection, curve length, the area enclosed by a curve and the intersection of curves with areas (Ballard, 1981).

A structure similar to the strip tree is the Binary Searchable Polygon Representation (BSPR) of Burton (1977) which differs from the strip tree in that all the rectangles have the same orientation and a curve is decomposed into simple sections, each enclosing a single monotonic point segment. The tree is then built by combining pairs of sections, a process which is repeated until the entire curve is represented by one compound section. The BSPR represents a 'bottom-up' approach to curve approximation while the strip tree represents a 'top-down' approach. Both the strip tree and BSPR are independent of a grid system; thus, unlike most quad-tree constructs, they do not have to be spatially registered prior to operations such as geographical overlay. The BSPR has the disadvantage that it is of fixed resolution, similar to hexagonal-based systems.

The main disadvantages of strip tree representations are as follows:

(i) The tree is not tied to a particular co-ordinate system
(ii) Each curve has to be processed independently.
(iii) Any information relating to a line's connectivity or spatial adjacency (topology) is lost.
(iv) Strip trees are not unique for closed or irregular curves.
(v) Strip trees are difficult (if not impossible) to overlay.

The only advantage of strip trees over edge or line trees is that the structure is invariant under shifts and rotations.

Data structures and functionality

By employing a series of evaluation criteria, the suitability of different data structures for different tasks can be explored. Thus Rhind and Green (1987) employed the following criteria:

(a) completeness - the proportion of entities which can be represented;
(b) robustness - the ability of the model or structure to handle special circumstances;
(c) efficiency - the compactness and speed of use;
(d) versatility - the ease with which the model or system can be applied to new situations;
(e) ease of generation - the ease with which raw data can be transformed into a particular structure;
(f) functionality - the range of operations which may be performed on data in this form;
(g) utility - ease of use of the data model or structure.

Figure 1: An interpretation of GIS functionality in relation to different data structures

Data Structures

Primary GIS Functions	Vector			Tessellated				Hybrid	
	Unstructured vector (ie "spaghetti")	Indexed vector (eg network or hierarchical index)	Topological vector (eg POLYVRT)	Regular tessellation (eg grid or raster)	Nested regular tessellation (eg quadtree)	Irregular tessellation (eg TIN)	Nested irregular tessellation	Geo-relational	Hybrid tessellation
Generation	5	4	3	5	4	3	3	3	3
Editing	2	4	3	2	3	2	2	4	3
Structure conv.	2	3	4	3	2	1	2	3	3
Geometric conv.	3	3	4	1	2	2	3	4	3
Generalisation	2	3	3	1	2	2	2	3	2
Retrieval	2	3	3	2	5	2	5	4	4
Integration	1	2	3	5	4	1	3	4	4
DB Management	1	2	3	1	4	2	3	5	5
Analysis	1	2	3	2	3	3	3	4	3
Display	3	4	4	5	4	3	4	4	4

1 = Very Difficult
2 = Difficult/Ineffective
3 = Possible/Routine
4 = Simple/Effective
5 = Ideally Suited

Note: This table reflects a number of qualitative decisions eg it assumes appropriate hardware for display in each case. Some of the cell values may not be self-evident eg retrieval is tabulated as difficult with "spaghetti" data because to retrieve a specified feature would probably require a sequential search of the database.

Clearly, the relative importance of these criteria will differ in different organisations. So far as a general purpose GIS for scientific purposes is concerned, however, those of versatility, robustness and functionality would probably be accorded highest priority. Figure 1 shows how Green and Rhind (1986) believe data structures and functionality to relate. The result is that perhaps the most generally useful data structure (the topological model) is perhaps the least available so far as existing data are concerned and most prone to local implementation variations. In contrast, regular data, based on square grid tessellations are readily available from remotely sensed data and some forms of automatic scanning of maps: in all such cases, inference procedures or human input is then used to add attributes and generation of 'real world' entities or objects can be troublesome.

It is important to note, however, that data may - in most cases at least - be transformed or 'decanted' from one data structure to another though the cost and difficulty of doing this varies greatly (it is, for instance, relatively trivial to convert vector data into regular, square grid (or raster) form for plotting purposes). Such a course of action was advocated by Rhind and Green (1987) in their design of a GIS for NERC which included a 'restructuring box'.

AUTOMATED ANALYSIS AND MODELLING METHODS FOR GIS APPLICATIONS

If low level applications of GIS are now commonplace (involving such tasks as monitoring and inventorying), there has been remarkably little use of them for more sophisticated analysis and modelling. The obvious reason for this - that GIS are currently retrieval engines rather than analytical ones - is only part of the story. The most important reason is that robust methods applicable to 'fuzzy' geographical data and securely founded on theory are either not in existence or are unknown to those developing GIS.

Yet we have already argued that increasing convergence between cartographic and bureaucratic computer systems, made possible by GIS, will create many new opportunities for analysis. In many instances, the purpose will be vague and relate mainly to the fact that the data exist rather than to any more traditional, highly focused, inquiry. The emerging mountain of real and creatable geographically referenced data challenges the conventional manner by which statistical analysis and modelling is performed particularly in a geographical context - but also more generally. The challenge can be viewed as involving the need for an automated and more exploratory modus operandi in a situation which is data-rich but increasingly (by comparison) theory poor. Put another way, the historic emphasis on deductive approaches is becoming less practicable because of the increasing dominance of data-led rather than theory-driven questions that are increasingly being asked. Additionally, most of the geographical data is inherently difficult to handle; for instance, data for points and irregularly shaped and defined zones are far more difficult to analyse in a mathematically rigourous fashion than, for example, lattice data from a remote sensing device when various simplifying assumptions can be made (viz. Besag, 1986). Again, as we have already argued, geographical data are also very seldom "clean", in that fuzziness often affects both resolution and representational accuracy in a manner which is difficult to handle. Possible analogies with sampling errors are not applicable. Furthermore, many GIS operations seem to actually create spatial patterns (by aggregation with the zoning system acting as some kind of filter and pattern detector). Other GIS procedures (viz. overlays) can add additional levels of error and also propagate errors thereby contaminating previous "clean" data sets. There is probably not much that can be done to eliminate all sources and causes of geographic error and uncertainty in the short term, or indeed, ever. Currently, their real magnitude and impact is relatively little understood. What is more obvious is the need to devise methods of analysis which might be able to handle these problems - rather than simply ignore them altogether.

These and other endemic problems characterise so much important geographic data that they have led to a widening gap between what "spatial statistics" (as viewed from a statistical perspective) can offer and what "spatial statistics" (as viewed from a GIS perspective) needs to be able to offer to meet current demands from these systems. We are convinced that geographical data handling will only attain its real potential if formal statistical expertise and understanding of 'real world' problems and 'dirty data' are brought together.

These points are discussed further by reference to two examples: an optimal aggregation technique and an automated point data analysis machine. In both examples it is argued that real progress is only possible by not losing the geography of the problem and by seeking to temper the requirements of statistical or mathematical tractability with geographical reality. It is our contention that much previous work has involved a degree of abstraction which ensured that the results were inapplicable.

An Optimal Aggregation Technique

The existence of the so-called modifiable areal unit problem (MAUP) has been known since the 1920's. Despite this, the problem was until recently dismissed as an academic matter of no great consequence because nothing could be done about it. It arises when data for one level of spatial resolution (either a set of zones or a point data set) are aggregated to another (viz. a larger set of zones or from a point data set to a zoning system). This transformation may change the scale and resolution of the data, amplifying existing patterns, and even create new spatial patterns that are a direct consequence of the interaction between the data being aggregated and the properties of the zoning system used for the aggregation. In short, the data are changed, new relationships are created and old changed, and map patterns altered. The extent of these scale and aggregation effects are basically still unknown but, in limiting cases, can be identified by either simulation (Openshaw and Taylor, 1979) or by empirical study (Openshaw, 1978, 1984, 1987). Some statistical research has also been performed but this tends to concentrate only on the simpler scale problem (viz. modelling relationships as the number of zones change in a nested hierarchical fashion), ignoring the more complex (and, indeed, the principal source of uncertainty which is the) aggregation process itself (viz. which of the many alternative aggregations of N small zones to M large zones do you use); see Arbia (1985).

The availability of GIS now makes this previously academic problem of far greater significance. GIS gives the user the opportunity, within limits, to design their own zoning systems; previously, zones used for statistical reporting tended to reflect administrative boundaries which are neither neutral, consistently defined, or comparable. If the definition of zones influence or even determine the results of spatial analysis of zonal data, then far greater attention needs to be given to the deliberate design and engineering of these zoning systems. The problem is how best to proceed.

There are two basic approaches. The first is to use regionalisation methods to design aggregations that have known properties. The second involves a joint estimation process whereby the aggregation used is estimated as part of the statistical analysis process, in a manner analogous to model

parameter set. The first of these is the traditional regionalisation approach, developed by geographers for descriptive purposes since 1945: the availability of small zone data now breathes new life into this previously unfashionable and previously moribund technology. The classic examples relate to the design of functional regions, daily urban systems, and labour markets; see Coombes et al., 1982; Sforzi et al, 1986; Openshaw et al., 1988). The visualisation offered by GIS is also considered to have some value. In the past, there has been little or no explicit regard given to the possible effects of zone design on the subsequent data analyses, except that there is at least a formal and deterministically applied set of zone design procedures. The zones have known characteristics and there is some justification for the properties they have been given. Clearly, this strategy needs to be developed further and integrated into GIS.

The second approach involves a more general type of optimisation; in this, the zoning system constitutes a set of unknown parameters and these are estimated to optimise some objective function which is used to represent the purpose that the zoning system is meant to meet. The problem can be further complicated by the addition of constraints on either the nature of the zones themselves (for instance, size, shape, heterogeneity) and/or on the quality of the data that are generated (for example, distributional forms, nature of relationships, spatial autocorrelation assumptions etc). Openshaw (1987) reviewed this approach. A simple example should suffice to demonstrate what it is involved.

Consider a large spatial data base resident in a GIS. Suppose the data have been aggregated from one set of small zones to a smaller set of larger zones (for example, municipalities or wards or census tracts). Let us assume that it is desired to build a spatial model of some kind. Now the question arises as how to handle the MAUP with these data. The standard statistical approach has been to take the zonal data as fixed and then to concentrate on the statistical estimation process, perhaps incorporating some geography by trying to handle spatial dependency problems or by using some smoothing procedure. When the results are obtained, it is common for conclusions to be drawn by people who have totally lost sight of the dependency of the results on the zoning systems that were used. A similar weakness affects a wide range of statistics. For example, the use of empirical Bayesian methods to improve the mapping of rates (see, for instance, Clayton and Kaldor, 1987) also suffers from the same fundamental dependency of the results on the scale and levels of aggregation that are used. Since this is assumed away as being "outside the control of the analyst", the validity of the results must also be in some doubt and the utility of the statistical achievement reduced.

A more GIS-relevant approach would seek to incorporate the selection of an appropriate zonal aggregation into the statistical estimation process. One way of doing this is to optimise a common function by simultaneously manipulating both the zonal aggregation and the parameter estimation domains. Openshaw (1978) described how this can be achieved but it is only now with availability of GIS that it is possible to complete the analysis process. For instance, the so-called Automatic Zoning Procedure (AZP) needs a contiguity matrix which can be provided by the topology of a coverage and it needs some way of drawing the regional boundaries that are produced; both are very simple tasks for a GIS employing appropriate data structures. The actual optimisation probably still needs a supercomputer to handle the stochastic optimisation based on simulated annealing with the constraints being handled by penalty functions. Whilst this remains a computationally burdensome approach, the fact is that what was hardly possible ten years ago for small data sets, is now a minor problem with even fairly large data sets.

Towards the automated analysis of point data sets

A second set of problems arise in the analysis of point data. Many of the problems and opportunities being created by GIS are amply illustrated here. Consider an example: a doctor approaches you and asks for assistance in analysing his cancer registry. The data are 'correct' and have been given point references based on postcodes. He wants to know whether his data for leukaemia show any signs of clustering and, if it does, where in particular he should focus his attention. A hypothesis based approach is ruled out because so little is known about the natural history of the disease that there is no basis for specifying hypotheses. Virtually, everything is possible and nothing can be excluded. There is a virtually infinite number of possible but probably specious hypotheses which could be formulated, so where or how do you start? From a geographical perspective, the need is clearly for some means of searching the data for evidence that clustering exists but without knowing where to look or what may be causing it. This type of broadly based, vaguely defined, and generally phrased exploratory objective is likely to be asked with increasing frequency as more and more data sets become available for analysis. The 'see-able' nature of geographic data and the wide availability of mapping systems makes these questions impossible to dodge.

As is usually the case with real-world geographic data relating to human beings, there are two reasonable responses to such problems. The first is to refuse to do anything. In this instance, the locational uncertainty in the data, the inadequacy of small area census-based 'population at risk' estimates, the potentially massive media sensitivity of any results, and the existence of many unanswered yet fairly fundamental methodological questions can all conspire to persuade the cautious that doing nothing is better than risking all by doing something. The lack of any obvious testable hypotheses and the need to develop perhaps innovative forms of spatial analysis in an alien environment, where the risks of errors or mistakes are of possible catastrophic importance, tend to justify conservatism. Such a response, whilst understandable and maybe even justified, is not particularly helpful to the practitioner who sought help. We cannot simply abandon responsibilities so easily.

The solution is simple in principle: it is to use the rapidly increasing computational power of the computer to develop new types of spatial analysis that can cope with the volumes of data, the increasing numbers of data analysis requests, and the methodological problems associated with geographical analysis of noisy and 'real world' data sets, but without making untenable assumptions for purposes of statistical or mathematical tractability. Here one possible avenue of attack is to develop a more efficient, computer automated, exploratory geographical analysis technology which is capable of effective and unbiased searches for evidence of clustering. In the best Holmesian tradition, the data are to be made to confess if they contain unmistakeable evidence of clustering. The particular tool is what Openshaw et al (1987) call a Geographical Analysis Machine (GAM), although it would be equally valid to call it a Geographical Exploration Machine (or GEM).

A Mark 1 GAM has been built to analyse cancer data but it can also cope with any point data set where the search is for evidence of geographic clustering. The basic technology is very simple and fairly obvious. The purpose is to find areas within a study region where a generic hypothesis that the distribution of points is random breaks down. The reason for the breakdown can only be a cause for speculation on the part of an expert in the substantive field. The task here is to provide the locational identification of where to look for clusters. The task of validation and fine tuning of the statistical results is seen as a subsequent stage to the analysis.

The Mark 1 GAM involves the following process. The aim is to examine all possible point locations within a study region for evidence of clustering. This is achieved by covering the study region with a lattice and then drawing overlapping circles, for a wide range of radii, around each lattice point. Data are retrieved for these circular search areas and some assessment made as to whether or not there is firm evidence of an excess number of cases within a circle of a given radius focused on a given point. The circles are designed to overlap by a large degree so as to provide a good discrete approximation to examining all possible point locations and, what is more important, to allow for locational and representational errors in the cancer and census data being retrieved for the circles. With a rare disease, such as leukaemia, it often matters which side of a circular boundary a point lies. In reality, it may be mis-assigned due to geographic errors or refer to an areal unit that actually straddles the boundary. By repeatedly moving the circles along by small amounts these effects can at least be identified and incorporated in the analysis process.

A full description is given in Openshaw et al (1987). The early runs with the mark 1 GAM (now termed GAM/1) provided evidence of a major cancer cluster in an area away from any nuclear installation. The cause might be a non-nuclear form of pollution or even socio-economic factors. The original GAM/1 was developed as a geo-descriptive technique. It offered a comprehensive search over a complete study region for locations where the null hypothesis breaks down. However, it also served to emphasise the importance of hitherto dormant methodological issues. The main contribution was in enhancing the sophistication of a very simple form of spatial analysis. The questions it raised concern the propensity for clusters to appear in purely random data. The overlapping circles made GAM/1 a very sensitive pattern detector; indeed, it is possible that it is too sensitive and merely amplifies random patterns. Common sense suggested otherwise, but the question - once asked - has to be answered. The problem now is the machine time needed to re-run GAM/1 on a reasonable number of randomly generated data sets (say 499) would amount to 5,000 hours or so of CPU time on a large main-frame. Yet, until it is possible to estimate whole map Type 1 errors, then no-one is really going to believe the results!

Because of feedback of this nature, GAM/1 mutated into two variants: GAM/1+ which simulates a 500 observations data set re-run repeatedly but, by careful redesign of the software, run times were reduced to about 45 minutes on a Cray X-MP/48 super-computer. In contrast, GAM/2 uses a non-overlapping lattice, with rotation and repeatedly shifting lattice origins to provide an equivalent degree of sensitivity analysis. Run times for GAM/2 have been reduced from 1.5 hours on the Cray to about 11 minutes. Both methods provide a means of correcting the GAM/1 results for multiple significance tests with non-independent observations (in the case of GAM/1+). The preliminary results seem to validate the original GAM/1 patterns. It can be reported that the approximate Type 1 Error for the observed GAM/1 results being reproduced in a random data set is less than 2 percent. However, this is still not sufficiently convincing. The next step is to re-run the entire process on synthetic point distributions that conform to known statistical processes to investigate power levels and sensitivity.

It is apparent that the GAM style of automated point data analysis may well herald a major new era in the spatial analysis of GIS data. The basic idea - of what amounts to embedding a GIS inside an analysis package - will no doubt become increasingly commonplace. Also the basic mark 1 technology can certainly be developed much further and the statistical aspects can be made far more sophisticated. It would seem to suggest that the 'real world' alternative to a simple analysis science is a far more complex, machine-based science, with large scale, computationally intensive procedures replacing analytical approaches and less reliance placed on the importance of human imagination, statistical theory, and prior knowledge as the basis for inference.

The advocacy of such data exploration approaches, facilitated by the availability of GIS and cheap computing power, by no means excludes the desirability of further developments in traditional statistical analysis tools linked to GIS. We envisage, for instance, that the ability to carry out simple geographical disaggregations of data sets by using other data sets as filters (e.g. into data for urban and rural areas, by parent material, etc) would be likely to raise the explanatory power of traditional tools. Our concern, however, is that the intractable problems of dealing with large volumes of geographical data require either minimum reliance on theory or the development of a much enhanced theoretical under-pinning. If the latter does not come about, gross misuse of existing technologies and new data sets will certainly occur - notably where individuals have no method of assessing the reliability of results from their GIS (see HMSO 1987, Rhind 1988).

CONCLUSIONS

We have tried to demonstrate that geographical data handling is being driven partly by the availability of new technologies, partly by the availability of data in machine-readable form and partly by user needs for better solutions their own problems. It should be evident that the computational problems of dealing with such voluminous and 'fuzzy' data and the statistical problems of assessing the validity of the analyses are non-trivial. We remain convinced that only collaborative research is likely to maximise the value of GIS and we stand ready to work with those from other disciplines to this end.

ACKNOWLEDGEMENTS

Thanks are due to our colleagues for all their efforts in projects in recent years which have helped in providing material for this paper. The UK Natural Environment Research Council and the Economic and Social Research Council kindly sponsored some of the work described in the paper.

REFERENCES

Note: not all of the references in the text can be included in this brief paper. Those interested will find those missing in Green and Rhind (1986a) and Green, Healey and Rhind (1986).

AHUJA, N., 1983, On approaches to polygon decomposition for hierarchical image representation. Computer Vision, Graphics, and Image Processing, 24, 2, 200.

ARBIA, G., (1985), 'The modifiable areal unit problem and spatial autocorrelation problem: towards a joint approach', Metron XL, 325

ARNBERG, W., 1981, Integration of map and remote sensing data. Geografiska Annaler, A63, 319.

ARONSON, P., 1985, Applying software engineering to a general purpose geographic information system. In Proceedings of the Seventh International Symposium on Computer Assisted Cartography: Digital Representations of Spatial Knowledge, Washington D.C. March 11-14, 1985. (Falls Church, VA: ASP and ACSM) p. 23.

BELL, S.B., DIAZ, B.M., HOLROYD, F.C. and JACKSON, M.J., 1983, Spatially referenced methods of processing raster and vector data. Image and Vision Computing, 1, 4, 211.

BESAG J.E., 1986 On the statistical analysis of dirty pictures. J.R. Stat. Soc. B, 192.

BOUILLE, F., 1977, Structuring cartographic data and spatial processes with the hypergraph-based data structure. In Proceedings of the First International Advanced Study Symposium on Topological Data Structures in Geographic Information Systems, Harvard Papers on Geographic Information Systems, G. Dutton (ed.), Laboratory for Computer Graphics and Spatial Analysis, (Cambridge Mass: Harvard University), vol. 1, p. 22.

BURROUGH, P.A., 1986, Principles of Geographical Information Systems for Land Resources Assessment, (Monographs on soils and resources survey no. 12), (Oxford: Oxford University Press)

CALKINS, H.W., 1983, A pragmatic approach to GIS design. In Proceedings US/Australia Workshop on the Design and Implementation of Computer-Based Geographic Information Systems, D. Peuquet and J. O'Callaghan (eds.), (Amherst, NY: IGU Commission on Geographic Data Sensing and Processing) p. 92

CHRISMAN, N.R., 1987, Design of geographic information systems based on social and cultural goals. Photogrammetric Engineering and Remote Sensing, 53, 10, 1367.

CHRISMAN N.R. forthcoming, The GIS work of the Harvard Computer Graphics Laboratory. American Cartographer, summer 1988.

CLAYTON, D., KALDOR, J., (1987),'Empirical Bayes estimates of age-standardised relative risks for use in disease mapping', Biometrics 43, 671

COOMBES, MG., DIXON, JS., GODDARD, JB., OPENSHAW, S., TAYLOR, PJ., (1982), Functional regions, in DT Herbert and RJ Johnston (eds) Geography and the Urban Environment, Wiley, London, 63

CORBETT, J.P., 1975, Topological principles in cartography. In Proceedings of the International Symposium on Computer Assisted Cartography, Auto Carto II, (Reston VA: US Dept. of Commerce).

DANGERMOND, J. 1983, A classification of the software components commonly used in geographic information systems. In Proceedings of the US/Australia Workshop on the Design and Implementation of Computer-based Geographic Information Systems, D. Peuquet and J. O'Callaghan (eds.), IGU Commission on Geographic Data Sensing and Processing, (New York: Amherst) p. 70.

DANGERMOND, J. and MOREHOUSE, S., 1987, Trends in hardware for geographic information systems. In Proceedings of the Eighth International Symposium on Computer-Assisted Cartography, Baltimore, Maryland, March 29 - April 3, Chrisman, N.R. (ed.), (Falls Church, VA: ASPRS and ACSM), p. 380.

DUECKER, K.J., 1985, Geographic information systems: towards a geo-relational structure. In Proceedings of the Seventh International Symposium on Computer Assisted Cartography: Digital Representations of Spatial Knowledge, Washington D.C. March 11-14, 1985. (Falls Church, VA: ASP and ACSM) p. 172.

ESTES J.E. and BREDEKAMP J.H., 1988, Activities associated with global databases in the National Aeronautics and Space Administration. in Mounsey H.M (1988), op cit.

FRANK, A., 1982, MapQuery: data base query language for retrieval of geometric data and their graphical representation. Computer Graphics,16, 199.

GOODCHILD M.F., 1988 The issue of accuracy in global databases. in Mounsey (1988), op cit.

GIBSON, L. and LUCAS, D., 1982, Vectorization of raster images using hierarchical methods. Computer Graphics and Image Processing, 20, 82.

GREEN, N.P.A., 1986, User/System Interfaces for Geographic Information Systems. Report No. 4, NERC Remote Sensing Special Topic: The Conceptual Design of a Geographic Information Systems for the Council. Birkbeck College, London.

GREEN, N.P.A., 1987, An Assessment of some UK-Supported, Commercially Available Geographic Information Systems. Report No. 3, NERC Remote Sensing Special Topic Report: The Conceptual Design of a Geographic Information System for the Council. Birkbeck College, London.

GREEN, N.P.A., HEALEY, R.G. and RHIND, D.W., 1986, A Bibliography of Papers on Geographic Information Systems Published Between 1984 and 1986. NERC Remote Sensing Special Topic: Report No. 5, The Conceptual Design of a Geographic Information System for the Council. Birkbeck College, London.

GREEN, N.P.A. and RHIND, D.W. 1986a, Spatial Data Structures for Geographic Information Systems. Report No. 2, NERC Remote Sensing Special Topic: The Conceptual Design of a Geographic Information Systems for the Council. Birkbeck College, London.

GREEN, N.P.A. and RHIND, D.W. 1986b, Teach yourself geographic information systems: The design creation and use of demonstrators and tutors. In Proceedings of AutoCarto London, M. Blakemore (ed.), London, 14-19 September (1986), (London: ICA), vol. 2, p. 327.

GREEN, N.P.A., RHIND, D.W. and FINCH, S., 1985, User Needs and Design Constraints. Report No. 1, NERC Remote Sensing Special Topic: The Conceptual Design of a Geographic Information System for the Council. Birkbeck College, London.

GREEN, P. and SIBSON, R, 1978, Computing Dirichlet tessellations in the plane. The Computer Journal, 21, 2, 168.

GUPTILL, S.C., 1985, Functional components of a spatial data processor. In Proceedings of the Seventh International Symposium on Computer Assisted Cartography: Digital Representations of Spatial Knowledge, Washington D.C. March 11-14, 1985. (Falls Church, VA: ASP and ACSM), 229.

GUPTILL, S.C., 1988, Feature-based spatial data models - the choice for global databases in the 1990s? in Mounsey H.m. (1988) op cit.

HERRING, J., 1987, TIGRIS: topologically integrated geographic information system. In Proceedings of the Eighth International Symposium on Computer-Assisted Cartography, Chrisman, N.R. (ed.), Baltimore, Maryland, March 29 - April 3, 1987, (Falls Church, VA: ASPRS and ACSM)

HMSO, 1987, Handling Geographic Information. (London: Her Majesty's Stationary Office).

HOLROYD, 1986, Joining tiles in hierarchies: a survey. In Spatial Data Processing using Tesseral Methods (collected papers from tesseral workshops 1 and 2), B.M. Diaz and S.M. Bell (eds.), (Swindon, Natural Environment Research Council), p. 17.

JACKSON, M.J. and MASON, D.C., 1986, The development of integrated geo-information systems. International Journal of Remote Sensing. 7, 6, 723.

KNAPP, E.M. and RIDER, D., 1979, Automated geographic information systems and landsat data: A survey. In Harvard Library of Computer Graphics, 1979 Mapping Collection, Computer mapping in natural resources and the Environment, Harvard Laboratory for Computer Graphics and Spatial Analysis, (Cambridge, Mass.: Harvard University), vol. 4, p. 57.

MARBLE, D.F., 1983, On the application of software engineering methodology to the development of geographic information systems. In Proceedings US/Australia Workshop on the Design and Implementation of Computer-Based Geographic Information Systems, D. Peuquet and J. O'Callaghan (eds.), IGU Commission on Geographic Data Sensing and Processing, Amherst, NY. (Amherst, NY: IGU), p. 102.

MARBLE D.F., 1988 Approaches to the efficient design of spatial databases at the global scale. in Mounsey (1988) op cit.

MARK, D.M., 1979, Phenomenon-based data-structuring and digital terrain modelling. Geo-processing, 1, 27.

MARTIN, J., 1973. Design of Man-Computer Dialogues. (Englewood Cliffs, N.J.: Prentice-Hall)

MOREHOUSE, S., 1985, ARC/INFO: A geo-relational model for spatial information. In Proceedings of the Seventh International Symposium on Computer Assisted Cartography: Digital Representations of Spatial Knowledge, Washington D.C. March 11- 14, 1985. (Falls Church, VA: ASP and ACSM), p. 388.

MOUNSEY H.M., 1988, (ed.), Global databases. Taylor and Francis, London, August 1988

NERC, 1987, The Natural Environment Research Council 1987 Corporate Plan. NERC, Swindon, England.

OPENSHAW, S., CHARLTON, M., WYMER, C., CRAFT, AW., (1987), A Mark 1 Geographical Analysis Machine for the automated analysis of point data sets, Int J Geographical Information Systems 1, 335

OPENSHAW S., WYMER, C., COOMBES, MG., (1988), Making sense of large flow data sets for marketing and other purposes, RRL Report 20, CURDS, Newcastle University

OPENSHAW, S., (1978), An empirical study of some zone design criteria, Environment and Planning A, 9, 169

OPENSHAW, S., (1984), The Modifiable Areal Unit Problem, CATMOG 38, Geo Abstracts, Norwich

OPENSHAW, S., (1987), The aggregation problem in the statistical analysis of spatial data, in CONVEGNO 1987, Informazione ed analisi statistica per aree regionali e subregionali, Perugia, Galeno p 73-83

OPENSHAW S., (1988), 'Building an Automated Modelling System to explore a universe of spatial interaction models', Geographical Analysis 20, 31-46

PEUCKER, T.K. and CHRISMAN, N., 1975, Cartographic data structures. American Cartographer, 2, 1, 55.

PEUQUET, D.J., 1984, A conceptual framework and comparison of spatial data models. Cartographica, 21, 4, 66.

PEUQUET, D.J., 1988, Issues involved in selecting appropriate data models for global databases. in Mounsey H.M. (1988) op cit.

RHIND, D.W., 1981, Geographic Information Systems in Britain, In Quantitative Geography: a British view, Wrigley, N. and R.J. Bennett (eds.) (London: Routledge and Kegan Paul), p. 17.

RHIND, D.W., 1987, Recent developments in Geographic Information Systems. International Journal of Geographic Information Systems, 1, 3.

RHIND, D.W., 1988, A research agenda for GIS. International Journal of Geographic Information Systems, 2, 1

RHIND, D.W. and MOUNSEY, H.M., 1988, The Chorley Committee and Handling Geographic Information. SERRL Working Report No. 5, Birkbeck College, London.

RHIND, D.W., ARMSTRONG, P.A. and OPENSHAW, S., 1988, The Domesday machine: a nationwide GIS. Geographical Journal, 154, 1.

RHIND D.W. and GREEN N.P.A. (1988) The design of a GIS for a heterogeneous scientific community. Intl Jl GIS, 1,4

RHIND D.W. and HIGGINS M.J. 1988, Customer-selected products from the 1991 Census using highly secure technology. BURISA 73

RHIND D.W. and CLARK P., 1988, Cartographic data inputs to global data bases. in Mounsey H.m. (1988), op cit.

ROBINSON, V.B., BLAZE, M. and THONGS, D., 1986, Representation and acquisition of a natural language relation for spatial information retrieval. In Proceedings of the 2nd Inter national Symposium on Spatial Data Handling, Seattle, Washington, July 5-10, 1986. (Seattle: IGU and ICA), 472.

ROBINSON, V.B., THONGS, D. and BLAZE, M., 1985, Natural Language in geographic data processing. In Proceedings of the International Conference of the Remote Sensing Society and the Center for Earth Resources Management, London, Sept. 9-12 (1985), (London: RSS and CERMA), p. 67.

SAMET, H., 1984, The quad-tree and related hierarchical data structures. ACM Computing Surveys, 16, 2, 187.

Sforzi, F., Openshaw, S., Wymer, C., (1986), 'I mercato locali del lavoro in Italia', Seminario su Identificazione di Sistemi Territoriali Analisi della Struttura Socialee e Produttiva in Italia , Roma, Dicembre.

SMITH, T.R., MENON, S., STAR, J.L. and ESTES, J.E., 1987, Requirements and principles for the implementation and construction of large-scale geographic information systems, International Journal of Geographical Information Systems, 1, 1, 13.

SMITH, T.R. and PAZNER, M., 1984, Knowledge-based control of search and learning in a large-scale GIS. In Proceedings of the International Symposium on Spatial Data Handling, Zurich, Switzerland, August 20-24, 1984. (Zurich: Geographisches Institut), vol. 2, p. 498.

TENG, A.T., JOSEPH, S.A. and SHOJAEE, A.R. 1986, Polygon overlay processing: a comparison of pure geometrical manipulation and overlay topological processing. In Proceedings 2nd International Symposium on Spatial Data Handling, July 5- 10, 1986, Seattle, Washington, (Seattle: IGU and ICA),102.

TOMLINSON, R.F. (ed.), 1972, Geographical Data Handling, UNESCO / IGU second symposium on geographic information systems, Ottowa, Canada. (Ottowa:IGU Commission on Geographical Data Handling and Processing).

TOMLINSON, R.F., CALKINS,H.W. and MARBLE, D.F., 1976, Computer Handling of Geographic Data: An Examination of Selected Geographic Information Systems. Natural Resources Research Series XIII, The UNESCO Press, Paris.

VAN ROESSEL, J.W. and FORSNIGHT, E.A., 1985, A relational approach to vector data structure conversion. In Proceedings of the Seventh International Symposium on Computer Assisted Cartography: Digital Representations of Spatial Knowledge, Washington D.C. March 11-14, 1985. (Falls Church, VA: ASP and ACSM), p. 541.

WELLS, M., 1984, The JANET project. University Computing, 6, 56.

WOODWARK, J.R. 1986, Techniques of spatial segmentation in solid modelling. In Spatial Data Processing using Tesseral Methods (collected papers from tesseral workshops 1 and 2), B. Diaz and S. Bell (eds.), (Swindon, Natural Environment Research Council), p. 325.